REVIEWS IN MINE**
AND **GEOCHEMISTRY**

Volume 66   2007

OCT 2 9 2007

PALEOALTIMETRY:
Geochemical and Thermodynamic Approaches

EDITOR

Matthew J. Kohn

Boise State University
Boise, Idaho

COVER FIGURE: Colorized DEM of south-central Asia, home to Earth's highest mountain range and largest plateau. Image provided courtesy of Marin Clark.

Series Editor: **Jodi J. Rosso**

MINERALOGICAL SOCIETY OF AMERICA
GEOCHEMICAL SOCIETY

WITHDRAW\
UTSA Librarie

Library
University of Texas
at San Antonio

SHORT COURSE SERIES DEDICATION

Dr. William C. Luth has had a long and distinguished career in research, education and in the government. He was a leader in experimental petrology and in training graduate students at Stanford University. His efforts at Sandia National Laboratory and at the Department of Energy's headquarters resulted in the initiation and long-term support of many of the cutting edge research projects whose results form the foundations of these short courses. Bill's broad interest in understanding fundamental geochemical processes and their applications to national problems is a continuous thread through both his university and government career. He retired in 1996, but his efforts to foster excellent basic research, and to promote the development of advanced analytical capabilities gave a unique focus to the basic research portfolio in Geosciences at the Department of Energy. He has been, and continues to be, a friend and mentor to many of us. It is appropriate to celebrate his career in education and government service with this series of courses.

Reviews in Mineralogy and Geochemistry, Volume 66

Paleoaltimetry: Geochemical and Thermodynamic Approaches

ISSN 1529-6466
ISBN 978-0-939950-78-2

COPYRIGHT 2007

THE MINERALOGICAL SOCIETY OF AMERICA
3635 CONCORDE PARKWAY, SUITE 500
CHANTILLY, VIRGINIA, 20151-1125, U.S.A.
WWW.MINSOCAM.ORG

The appearance of the code at the bottom of the first page of each chapter in this volume indicates the copyright owner's consent that copies of the article can be made for personal use or internal use or for the personal use or internal use of specific clients, provided the original publication is cited. The consent is given on the condition, however, that the copier pay the stated per-copy fee through the Copyright Clearance Center, Inc. for copying beyond that permitted by Sections 107 or 108 of the U.S. Copyright Law. This consent does not extend to other types of copying for general distribution, for advertising or promotional purposes, for creating new collective works, or for resale. For permission to reprint entire articles in these cases and the like, consult the Administrator of the Mineralogical Society of America as to the royalty due to the Society.

PALEOALTIMETRY:
Geochemical and Thermodynamic Approaches

66   *Reviews in Mineralogy and Geochemistry*   **66**

FROM THE EDITORS – PREFACE

The review chapters in this volume were the basis for a two day short course entitled *Paleoaltimetry: Geochemical and Thermodynamic Approaches* held prior to the Geological Society of American annual meeting in Denver, Colorado (October 26-27, 2007). This meeting and volume were sponsored by the Geochemical Society, Mineralogical Society of America, and the United States Department of Energy.

Any supplemental material and errata (if any) can be found at the MSA website *www. minsocam.org*.

Jodi J. Rosso, Series Editor
West Richland, Washington
August 2007

The idea for this book was conceived in early June, 2005 at a paleoaltimetry workshop held at Lehigh University and organized by Dork Sahagian. The workshop was funded by the tectonics program at NSF, and was designed to bring together researchers in paleoaltimetry to discuss different techniques and focus the community on ways of improving paleoelevation estimates and consequent interpretations of geodynamics and tectonics. At this meeting, some commented that a comprehensive volume describing the different methods could help advance the field. I offered to contact the Mineralogical Society of America and the Geochemical Society about publishing a RiMG volume on paleoaltimetry. Because many of the techniques used to infer paleoelevations are geochemically-based or deal with thermodynamic principles, the GS and MSA agreed to the project. Two years and roughly 1000 e-mails later, our book has arrived.

The chapters

The book is generally organized into 4 sections:

(1) Geodynamic and geomorphologic rationale (Clark). This chapter provides the broad rationale behind paleoaltimetry, i.e., why we study it.

(2) Stable isotope proxies. These 4 chapters cover theory of stable isotopes in precipitation and their response to altitudinal gradients (Rowley), and stable isotopes sytematics in paleosols (Quade, Garzione and Eiler), silicates (Mulch and Chamberlain) and fossils (Kohn and Dettman).

(3) Proxies of atmospheric properties. These 4 chapters cover temperature lapse

1529-6466/07/0066-0000$05.00   DOI: 10.2138/rmg.2007.66.0

rates (Meyer), entropy (Forest), and atmospheric pressure proxies, including total atmospheric pressure from gas bubbles in basalt (Sahagian and Proussevitch), and the partial pressure of CO_2 (Kouwenberg, Kürshner, and McElwain). Note that clumped isotope thermometry (Quade, Garzione and Eiler) also provides direct estimates of temperature.

(4) Radiogenic and cosmogenic nuclides. These 2 chapters cover low-temperature thermochronologic approaches (Reiners) and cosmogenic isotopes (Riihimaki and Libarkin).

Some chapters overlap in general content (e.g., basic principles of stable isotopes in precipitation are covered to different degrees in all stable isotope chapters), but no attempt was made to limit authors' discussion of principles, or somehow attempt to arrive at a "consensus view" on any specific topic. Because science advances by critical discussion of concepts, such restrictions were viewed as counterproductive. This does mean that different chapters may present different views on reliability of paleoelevation estimates, and readers are advised to read other chapters in the book on related topics – they may be more closely linked than they might at first appear!

Last, I hope readers of this book will discover and appreciate the synergy among paleoaltimetry, climate change, and tectonic geomorphology. These interrelationships create a complex, yet rich field of scientific enquiry that in turn offers insights into climate and geodynamics.

Acknowledgments

Many deserve special thanks for their help in this project. First and foremost, the National Science Foundation and Department of Energy provided generous support not only for the volume and accompanying short course, but also for the research of the numerous authors involved in this project.

Alex Speer and Jodi Rosso, at MSA, are both thanked for their tireless and patient efforts and advice through the entire process, from proposal to finished book.

Last, the reviewers' efforts were truly heroic. I still can't figure out how some managed to provide detailed reviews on lengthy manuscripts within a week or two. In addition to several anonymous reviewers and myself, the reviewers included: Ron Amundson, Pierre-Henri Blard, Dan Breeker, Marc Caffee, Chuck Chapin, Andrea Dutton, Todd Ehlers, Dave Foster, Henry Fricke, Carmala Garzione, John Gosse, Matt Huber, Craig Jones, Bruce MacFadden, Larry Mastin, Ian Miller, Andreas Mulch, Joel Pederson, David Schuster, Bob Spicer, and Torsten Vennemann. Their efforts greatly improved each contribution.

Matthew J. Kohn
Department of Geosciences
Boise State University
Boise, ID 83725

August 2007

PALEOALTIMETRY:
Geochemical and Thermodynamic Approaches

66   *Reviews in Mineralogy and Geochemistry*   **66**

TABLE OF CONTENTS

1   The Significance of Paleotopography

Marin K. Clark

2   Stable Isotope-Based Paleoaltimetry: Theory and Validation

David B. Rowley

3   Paleoelevation Reconstruction using Pedogenic Carbonates

Jay Quade, Carmala Garzione, John Eiler

4   Stable Isotope Paleoaltimetry in Orogenic Belts – The Silicate Record in Surface and Crustal Geological Archives

Andreas Mulch, C. Page Chamberlain

5   Paleoaltimetry from Stable Isotope Compositions of Fossils

Matthew J. Kohn, David L. Dettman

6   A Review of Paleotemperature–Lapse Rate Methods for Estimating Paleoelevation from Fossil Floras

Herbert W. Meyer

7   Paleoaltimetry: A Review of Thermodynamic Methods

Chris E. Forest

8   Paleoelevation Measurement on the Basis of Vesicular Basalts

Dork Sahagian, Alex Proussevitch

9   Stomatal Frequency Change Over Altitudinal Gradients: Prospects for Paleoaltimetry

Lenny L.R. Kouwenberg, Wolfram M. Kürschner,
Jennifer C. McElwain

10   Thermochronologic Approaches to Paleotopography

Peter W. Reiners

11 Terrestrial Cosmogenic Nuclides as Paleoaltimetric Proxies

Catherine A. Riihimaki, Julie C. Libarkin

Reviews in Mineralogy & Geochemistry
Vol. 66, pp. 1-21, 2007
Copyright © Mineralogical Society of America

1

The Significance of Paleotopography

Marin K. Clark

Department of Geological Sciences
University of Michigan
Ann Arbor, Michigan, 48109, U.S.A
marinkc@umich.edu

ABSTRACT

Topographic change is one of the most informative measures of continental deformation, yet precise records of past elevation are scarce to non-existent in most orogenic belts. Recent advancement and new development of techniques to measure paleoelevation promise significant progress in understanding continental tectonic processes and how erosion acts on rising topography to shape the Earth's surface. Outstanding issues in continental tectonics include understanding driving forces to deformation, how strain is vertically partitioned in the lithosphere, and how mass is redistributed or conserved in the lithosphere during deformation. Geodynamic and mechanical models that explore these topics make testable predictions of topographic evolution. This review explores key examples of how paleotopography relates to outstanding issues in continental tectonics that motivate development of robust, quantitative measures of paleoelevation.

INTRODUCTION

Topography is one of the most fundamental and tangible aspects of Earth science. Explorers, geographers and scientists alike have long questioned the age and origin of high elevation terrain. Mean topography of the Earth's surface is an important geophysical parameter that most often represents the isostatic balance of the lithospheric column over the asthenosphere. Deformation that alters the thickness of different layers in the continental lithosphere produces elevation change of the Earth's surface through isostasy. In the most simple-minded view, geodynamic models make predictions of mean topography change during orogenesis that can be tested with a quantitative paleoelevation record. As a result, outstanding questions about how the continents deform may be answered. *But…why is this not so simple?*

First, until recently, robust measurements of paleoaltitude have been scarce. Rapid incision of rivers and increases in sedimentation rates have commonly been used as indicators of elevation change, but this viewpoint has been challenged because climate can also influence or drive such phenomenon and increases in relief need not equate to mean elevation rise (England and Molnar 1990; Molnar and England 1990; Zheng et al. 2001). Furthermore, geomorphic approaches to elevation change generally yield qualitative, not quantitative, measures of elevation change unless specific criteria are met (Clark et al. 2006). Second, topography is not a simple product of tectonic deformation, but rather the result of combined tectonic, climatic and erosional processes. Much research in the past 15 years has focused on the complex relationships among climate, erosion, and tectonics that together ultimately shape the Earth's surface (e.g., Koons 1989; Avouac and Burov 1996; Willett 1999; Beaumont et al. 2001; Hodges et al. 2001; Montgomery et al. 2001; Zeitler et al. 2001). For example, tectonically elevated rocks are subject to accelerated erosion rates through slope dependent erosional processes that

DOI: 10.2138/rmg.2007.66.1

may also be subject to focused precipitation and more intense erosion as air masses ascend over mountainous topography (e.g., Reiners et al. 2003; Roe 2005). In turn, the rapid and focused erosion of a mountain belt may affect the thermal structure of the orogen and dictate the mechanics of how the orogen deforms (e.g., Avouac and Burov 1996; Willett 1999; Zeitler et al. 2001; Hodges et. al. 2004). Large elevated regions such as the Tibetan Plateau may also regionally disrupt large-scale atmospheric circulation leading to regional monsoon climates and aridification of continental interiors (e.g., Kutzbach et al. 1989, 1993; Prell et al. 1992; An et al. 2001; Dettman et al. 2001; Kroon et al. 1991). Coupling between tectonics, erosion, and climate fundamentally link the solid earth to its atmosphere, but advancing our understanding of this complex system will ultimately require separate records of climate proxies (such as temperature, precipitation, storminess and seasonality), elevation, relief, deformation and erosional histories.

One of the most exciting recent developments in Earth Science is the development of quantitative geochemical and thermodynamical proxies for paleoaltimetry. This volume covers state-of-the art applications in this rapidly expanding field. The range of paleoelevation techniques covered here measure different aspects either directly or indirectly of topography. This review illustrates the impact that these new methodologies potentially have on outstanding problems in the field of continental tectonics.

DEFINITIONS AND SCALES OF TOPOGRAPHY, ELEVATION, AND RELIEF

In this paper, I define "topography" as the actual shape of the land surface that encompasses two variables: elevation or altitude (the height of a point above sea level) and relief (the elevation difference between to points). Even though topography, elevation and relief are often used in the literature as synonyms, it is useful to distinguish them independently as they represent different ways of measuring the land surface relevant to the techniques described in this volume. I also distinguish between two different reference scales for elevation and relief: (1) "local" and (2) "mean" or "average." A local scale describes measurement at a point or over some small distance, a meter to several kilometers. Relief or elevation measured over this scale is likely supported by the flexural strength of the crust and does not directly represent changes to lithospheric structure that may be caused by tectonic deformation. A mean or averaged measurement of elevation or relief is defined as a scale that is geodynamically significant, i.e., a lengthscale that is greater than the flexural wavelength of average crust in continental orogens (tens to hundreds of kilometers). Thus "mean elevation" is the critical variable that measures the force balance in orogens and reflects the isostatic balance of the lithospheric column.

MEASUREMENTS FROM PALEOALTIMETRY METHODS

The range of techniques discussed in this volume measure local elevation, local relief, erosion and several environmental variables including precipitation, temperature, seasonality, and enthalpy. Relief, erosion and environmental records are related to paleoaltimetry through (1) empirically and theoretically determined climate-elevation relationships, and (2) assumptions about how erosion and relief relate to elevation change. While reading this volume, the reader should consider the specific measurement provided by a particular technique and its sensitivities to other factors. A broad range of approaches provides the opportunity to be both circumspect and comprehensive with tectonic and geomorphic interpretations based on paleoaltimetry data.

Basalt vesicularity, leaf stomatal frequency, and cosmogenic isotope paleoaltimetry can be grouped as techniques that make elevation estimates independent of climate or erosion (although environmental change may possibly influence stomatal density, but not stomatal index) (e.g., Kouwenberg et al. 2007; Riihimaki and Libarkin 2007; Sahagian and Proussevitch 2007). Stable-isotope paleoaltimetry, clumped-isotope thermometry, and foliar physiognomic characteristics from fossil flora measure environmental variables such as temperature, enthalpy, precipitation, and seasonality. These techniques infer paleo-elevation from climate-altitude relationships such as temperature or isotopic lapse rates, moist enthalpy-altitude changes, and species adaptation to climate related changes in altitude (e.g., Forest 2007; Kohn and Dettman 2007; Meyer 2007; Mulch and Chamberlain 2007; Quade et al. 2007; Rowley 2007). Climate-altitude relationships are sensitive to the interaction of topography and climate, especially in the case of extreme relief or directionality of moisture sources.

Some stable-isotopic applications measure compositions derived from regional averages of paleo-water, rather than local sources, and therefore reflect basin hypsometry (distribution of elevation within a catchment) and may include mixing of different climate regimes if the catchment is large (Rowley et al. 2001). A dependence on basin hypsometry suggests greater sensitivity to relief than to mean elevation of the sample location. Application of low-temperature thermochronometry can also constrain the distribution and magnitude of relief in the past and can be used to measure changes in erosion rate through time (e.g., Reiners 2007).

SHAPING THE LAND SURFACE: RELATIONSHIPS AMONG CLIMATE, EROSION AND TECTONICS THAT DICTATE MEAN ELEVATION

The interactions among climate, tectonics and erosion introduce complexity to interpreting paleoaltimetry data in a geodynamic context. Geodynamic models can be tested from measures of mean elevation, but extrapolating mean elevation from local elevation and relief, erosion, and climate conditions requires caution because relief and erosion rates are not related to mean elevation in a simple way (e.g., Molnar and England 1990; England and Molnar 1990; Montgomery and Brandon 2002). Furthermore, tectonic-climate-erosion interactions may alter patterns of elevation change during orogenesis compared to what a tectonic model alone may predict (e.g., Willett 1999; Whipple and Meade 2006). Determination of paleo- mean elevation, and tectonic-climate-erosion effects on elevation change, pose both a challenge and an opportunity to interpreting a paleoaltimetry "result" that affects all paleoaltimetry methods. These challenges are distinct from details of the techniques themselves, which introduce uncertainty in measurements or interpretations, and which are examined in detail by the individual chapters of this volume.

In a simple view, erosional processes are sensitive to slope such that tectonically-driven increases in slope will accelerate erosion rates, and drive an increase in local relief by increasing local river channel slope and hillslope angles. Hillslope angle (local relief) is ultimately limited by drainage density (Hovius 1996) and rock strength (Montgomery 2001). However, climate (precipitation, precipitation seasonality, storminess, glacial/interglacial cycling), basin hydrology (closed vs. open drainage basins, drainage patterns), sedimentary flux, and rock strength also influence relief and erosion rates at all scales. The relationship between climate variables and erosivity is complex, and may vary given the nature of the climate change. For example, greater precipitation may drive higher erosion rates but reduce slope stability (Carson 1976) and generate smoother, lower relief topography by decreasing hillslope steepness or lowering river channel slope (e.g., Whipple et al. 1999; Gabet et al. 2004). The influence of storminess (frequency and magnitude of storms) on incision rates and relief change depends on the degree of change (river discharge) and thresholds required to initiate incision (e.g., Lague et al. 2005; Molnar et al. 2006; Wu et al. 2006). Sedimentary flux can influence fluvial incision rates in mountain streams (Sklar and Dietrich 2004) and is a complicated product of climate, lithology and tectonics (e.g., Langbein and Schumm 1958; Sklar and Dietrich 2004). Drainage spacing and basin integration also control hillslope and drainage basin relief respectively and can be influenced by climate, tectonics or both. For example, tectonically-ponded drainages can keep relief low despite growing elevation (Sobel et al. 2003) and subsequent basin integration can lower baselevel, which will increase drainage basin relief.

High elevation mountain ranges or plateaux can also affect climate and in turn control patterns of erosion and relief. For example, orographic precipitation is caused by the "rainout" of vapor masses that rise over topography and cause increased precipitation focused on windward sides of orogens (e.g., Willett 1999; Roe 2005). The disturbance of atmospheric flow by the uplift of large plateaus, like the Tibetan Plateau, may create monsoon climates and aridification of continental interiors (e.g., Kutzbach et al. 1989, 1993; Kroon et al. 1991; Prell et al. 1992; An et al. 2001; Dettman et al. 2001) causing changes in erosion rates and relief throughout the orogen. Other globally significant phenomenon, such as continental geography, oceanic gateways, and concentration of atmospheric greenhouse gases are proposed to have driven climatic shifts during Cenozoic time (e.g., Zachos et al. 2001 and references therein), which potentially affect erosional processes independent of orogenesis.

Paleorelief vs. paleoelevation

Measurements of paleorelief are typically local and provide a minimum estimate of paleoelevation. However, relating a local paleorelief measurement to a more geodynamically relevant measurement of paleo-mean elevation requires some caution. Local relief is likely supported by the flexural strength of the crust whereas more spatially averaged measures of elevation ("mean elevation") reflect the isostatic balance of the lithospheric column. One must be able to meet two criteria in order to use relief change as a proxy for change in mean elevation: 1) describe how relief change is distributed within the landscape such that the change to the average elevation can be determined, and 2) establish how baselevel has changed or if it has remained the same. Without these criteria, certain data may lead to a false interpretation of elevation gain.

Several hypothetical scenarios illustrate the difficulty of relating relief to mean elevation. For example, climatically driven increases in relief within a fluvial basin may occur with mean elevation remaining unchanged or even decreasing (Molnar and England 1990). When a low-relief, elevated and internally drained region is breached, the ensuing erosional response to that change in baselevel will drive fluvial incision and increase relief without any major change in regional mean elevation (e.g., Pederson et al. 2002a,b). Finally, relief in regions of elevated, but internally drained fluvial basins may significantly underestimate mean elevation. For example, many modern internally drained basins on the Tibetan or Altiplano plateau have low or moderate relief but high mean elevation.

Combining information about distribution of local relief, drainage basin geometries and baselevel from geomorphology may circumvent problems relating relief to mean elevation change. An example from the Sierra Nevada, CA illustrates how a combination of approaches may be used to determine paleo-elevation of a mountain range. The sensitivity of shallow isotherms to surface topography has been exploited to address relief changes through time in modern and ancient orogens (Stuwe et al. 1994; Mancktelow and Grasemann 1997; House et al. 1998; House et al. 2001; Braun 2002; Clark et al. 2005a,b; Reiners et al. 2006; Stock et al. 2006; Stockli 2006). House et al. (1998) interpreted variations in (U-Th)/He thermochronometry ages from the Sierra Nevada, California as bending of shallow isotherms in response to paleo-relief. They interpreted 1-1.5 km relief to have existed across fluvial tributary basins in Late Cretaceous – early Tertiary time. In this case, thermochronometry data alone do not determine range elevation, but provide a minimum estimate from tributary basins. Relief across major tributary streams is less than or equal to the range elevation, which is described by the relief on the trunk rivers (Whipple et al. 1999). Without using additional geomorphic data from the Sierra Nevada, House et al. (2001) extrapolated the tributary relief to a range elevation of more than 4 km using a modern analog from rivers on the western flank of the Andes. This portion of the Andes has a markedly different climate history than the late Cretaceous-early Tertiary Sierra Nevada, and therefore makes a poor comparison because of the effect of climate on local and basin scale relief. Clark et al. (2005b) argued that quantitative stream profile reconstruction of both tributary streams and trunk rivers for the southern Sierra Nevada using geologic evidence of Cretaceous – Miocene baselevel suggested range elevation of only 1500 m and tributary relief of ~ 1 km during late Cretaceous – middle Tertiary time, which was consistent with the thermal data of House et al. (1998, 2001). In this case, the geomorphic data provided robust measurement of the distribution of relief and baselevel, while thermochronometry provided timing and quantitative measure of paleo-relief. Combined geomorphic and thermochronometry data suggests moderate paleo-range elevation for the southern Sierra Nevada (Clark et al. 2005b), which is also consistent with oxygen-isotope paleoaltimetry from the northern Sierra Nevada (Mulch et al. 2006).

The following examples explore how paleoaltimetry data may provide critical information about the evolution of mean elevation, averaged relief, and erosion from different models of continental deformation. However, I consider only changes in elevation, relief and erosion that may be predicted by tectonic models and neglect the influence that climate forcing or erosion related feedbacks could exert on such predictions. As discussed previously, the influence of climate, erosion and related feedbacks on tectonic deformation is important and should not be ignored. However, to consider all the complexities of potential interactions on the elevation record is beyond the scope and focus of this paper. In order to best illustrate the relationships between deformation mechanics and elevation, I review a few example elevation histories predicted by several commonly-cited tectonic models.

TOPOGRAPHY CHANGE PREDICTED BY TECTONIC MODELS

Regions of high elevation are created by both orogenic and anorogenic processes. Most commonly, deformed regions produce mountains that are isostatically compensated by thick crustal roots. Other regionally elevated areas associated with active or recent orogeny have normal thickness crust but are isostatically supported by anomalously buoyant mantle (e.g., Crough and Thompson 1977; Eaton et al. 1978; Saltus and Thompson 1995; Roy et al. 2005). In a few localities, global calculations of mantle flow suggest that elevated regions on the continents, such as the southern African craton, may be supported dynamically by active mantle upwellings that are independent of plate tectonic processes (e.g., Hager et al. 1985; Lithgow-Bertelloni et al. 1998).

In orogenic settings, elevation histories provide constraints on deformation processes in three important ways. (1) Driving forces of continental tectonic deformation can be assessed from the relative contribution of boundary to intrinsic forces set by plate motion and gravitational potential energy (GPE) respectively (e.g., Flesch et al. 2000; Lithgow-Bertelloni and Guynn 2004). Because high elevation terrain can be inherited from a prior phase of orogenesis, the "initial condition" or pre-existing topography to an orogenic period need also be considered. For example, the presence of high elevation and thick crust imply a source of gravitationally derived potential energy that could drive later extensional deformation (e.g., Jones et al. 1996). (2) The degree of rheologic stratification, which dictates crust-mantle coupling, strongly influences the behavior of the lithosphere during deformation by controlling how strain is vertically partitioned (e.g., Bird 1988, 1991; Buck 1991; Royden 1996; Roy and Royden 2000). (3) In a classical view, compressional or extensional deformation observed in the surface geologic record is considered to represent changes in thickness to the entire crust or lithosphere beneath the deforming region. However, if material is added to the crust from mantle melts or from adjacent undeformed regions by lower crustal flow, or if material is lost from the lithosphere by mantle foundering, there may be a deficit or excess of crust or mantle material from what would otherwise be predicted from the deformation recorded in surface rocks. Because thickening and thinning of the deep crust or mantle lithosphere will raise or lower mean elevation through isostasy, elevation histories may provide a way of assessing deep processes that otherwise have no surface structural record (epeirogeny).

I consider three aspects of an orogen's elevation history to be relevant to continental deformation:

(1) What is the initial elevation of the region?

(2) How does the elevation of the deforming region change with time?

(3) How does the elevation of the surrounding (undeforming) region also change?

In the following sections, I review a few, commonly-cited models of widespread compressional and extensional deformation and consider their respective elevation predictions. This review is not meant to be an exhaustive list of all possible mechanisms of elevation change due to continental deformation. Rather, the purpose is to illustrate the role of elevation as a parameter that can be used to constrain mechanisms of continental deformation.

While one may imagine that some tectonic models are mutually exclusive, others may occur in various combinations producing a spectrum of possible elevation histories. In fact, it has been previously argued that several end-member models for the tectonic history of the Basin and Range Province predict geologically unreasonable initial conditions and effectively require hybrid models (e.g., Lachenbruch and Morgan 1990; Bird 1991). For the purpose of discussing how paleoelevation can serve to discriminate between major tectonic processes, I describe the independent consequences of each model and leave creative combinations up to the reader's own conclusions. In all cases, I consider only regionally averaged elevation change, and in some cases, extreme versions of proposed processes that best illustrate distinct elevation histories.

Tibetan Plateau: widespread contraction

Shortly after the birth of plate tectonic theory, it was recognized that the diffuse and heterogeneous deformation of the interior of continents represents a significant departure from the principles of plate tectonics that adequately describes the behavior of oceanic plates (Molnar 1988 for a review). Thirty years later, the relationship between fold and thrust belt deformation developed at a plate boundary and the development of an orogenic plateau away from that boundary is still a first-order question in continental dynamics. Models that describe distributed deformation of the continents contrast "micro plate-like" behavior of the continental lithosphere during convergence (Tapponnier et al. 1982, 2001; Thatcher 2007) with a more distributed, fluid-like viscous response (England and Houseman 1988; Bird 1991; Royden 1996). The surface geologic record alone, which records failure of only the outermost few tens of kilometers of upper crust, may not represent the behavior of the lithosphere as a whole and as such is insufficient to distinguish between competing deformation models. The isostatic response of the surface of the Earth to thinning and thickening of deep crustal and mantle lithospheric layers better estimates the vertically averaged behavior of the lithosphere. But the lack of extensive, precise paleoelevation data hinders the refinement of models describing continental deformation.

The most extensive region of active, intra-continental deformation is the Himalayan-Tibet orogen in central Asia. It is the youngest, most dramatic example of continental convergence on the planet and is an example of diffuse deformation within the continents due to collisional deformation. Since ~50 Ma (Rowley 1996), convergence has produced the Himalayan fold and thrust belt and the Tibetan Plateau, which stands at ~ 5 km average elevation and is largely supported isostatically by 60-70 km thick crust (Fielding 1994; Jin 1994).

In the following section, I review three models that seek to explain the mechanical and dynamic evolution of the 5 km high Tibetan Plateau and compare their respective predictions for elevation change. In all cases, I consider only regional elevation change because long-wavelength topography is considered to be isostatically supported. I neglect consideration of short wavelength topography, such as topography formed over individual structures, as it is likely that this scale of topography is supported by the flexural strength of the crust.

Case 1: Stepwise-growth: a micro-plate model. The first model suggests that strain is localized along vertical faults through the lithosphere at sites of pre-existing weakness, such as paleo-sutures. In its original conception, this model involved convergence of a rigid, undeforming block with a more easily deformable plastic body, assuming plane horizontal strain and the influence of a free lateral boundary (Tapponnier et al. 1982). This model predicted the "extrusion" or "escape" of large, intact blocks toward the free lateral boundary along major strike-slip faults. The assumption of plane horizontal strain in this early model precluded thrust faulting, crustal thickening and the resulting buoyancy forces; however, the model aimed only to explain the origin of major strike-slip faults that are prominent features of the Tibetan orogen. A more recent model proposed that paleo-suture zones that cut laterally across the plateau are loci for concentrated deformation related to both shortening and strike-slip motion (Tapponnier et al. 2001). Subduction of Asian mantle lithosphere occurs at paleo-sutures, and crustal shortening is distributed in the overlying structural block. Tectonic damming of catchments and basin infilling produces a high, flat plateau that expands away from the plate boundary in discrete steps (Tapponnier et al. 1990, 2001; Meyer et al. 1998; Metivier et al. 1998) (Fig. 1A).

A central tenet to the step-wise model is that the continental lithosphere deforms as a series of small plates, and that plate boundary stresses are sufficient to reactivate old suture boundaries as sites of crustal thickening and mantle subduction far from the plate boundary. The continental crust thickens homogeneously above internally undeformed mantle lithosphere that is partially subducted at these boundaries. Localized strain at sites of strike-slip and mantle subduction also requires a mechanically competent lower crust and mantle lithosphere such that stresses are efficiently transmitted from the mantle lithosphere through the crust.

The step-wise model describes the rise of three successive plateaus along discrete vertical boundaries separated in time by tens of millions of years. Attainment of elevation occurs episodically and progresses away from the collision zone at distinct time intervals that are coeval with periods of upper crustal shortening, magmatism related to mantle lithosphere subduction, and high erosion rates that produce sediment and basin infilling to subdue relief (Fig. 2A).

An increase in mean elevation, relief, and erosion rate would be expected on the successive plateau margins and would progress from south to north (Fig. 2A). An increase in relief might be recognized at each paleo- plateau boundary that corresponds with a period of deformation and magmatism. As deformation progresses away from each established plateau boundary and forms a new boundary, relief and erosion rates would decrease in tandem as basin infilling occurs in closed basins and the subdued topography of the plateau surface develops.

Case 2: Continuous deformation: vertically averaged strain. The second and third models take a radically different viewpoint of continental deformation: single faults are not important but rather "add up" to appear continuous over long lengthscales. These models use continuum mechanics to describe the lithosphere as a viscous fluid and consider the ratio of stresses arising from buoyancy forces following crustal thickening and horizontal plate

A) Case 1: Stepwise growth: a micro-plate model

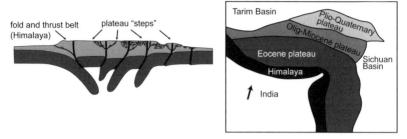

B) Case 2: Continuous deformation: vertically averaged strain

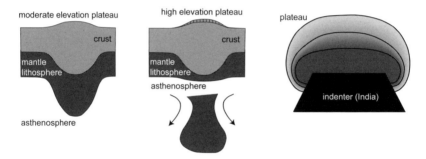

C) Case 3: Continuous deformation: vertically partitioned strain

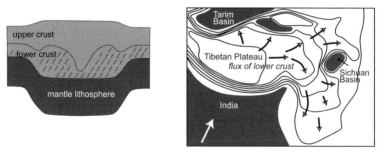

Figure 1. Models of crustal and mantle processes during formation of the Tibetan Plateau (continental compression). A) Strain is localized along vertical boundaries that transcend the lithosphere and form in places of pre-existing weakness. Plateau expands away from the Himalaya (plate boundary) in discrete intervals shown in cross-section (left) and map view (right). Figure modified from Tapponnier et al. (2001). B) Lithosphere deforms as a viscous fluid with vertically uniform strain. Proportional thickening of crust and mantle leads to convective loss of a gravitationally unstable mantle lithospheric root (left and center) (England and Houseman 1988; Molnar et al. 1993). Plateau expands away from the rigid indenter (India) in a gradual fashion shown by representative intervals (black lines) and smooth grayscale shading (right). Figure modified from Molnar et al. (1993) (left and center) and based on model of England and Houseman 1986, 1988 (right). C) Lithosphere deforms as a viscous fluid with vertically partitioned strain. Differential thickening of the upper and lower crust occurs (left). Contour lines represent approximately 1 km elevation intervals and grey shading represents low elevation, relatively undeformed regions (right). Significant crustal material is redistributed within the orogen due to pressure induced ductile flow of low-viscosity lower crust from the area with greatest thickness (Royden 1996; Royden et al. 1997; Clark and Royden 2000). Gravitationally driven flow expands the plateau northeastward and southeastward after the central plateau develops (right). Figure modified from Royden et al. (1997) (left) and Clark and Royden (2000) (right).

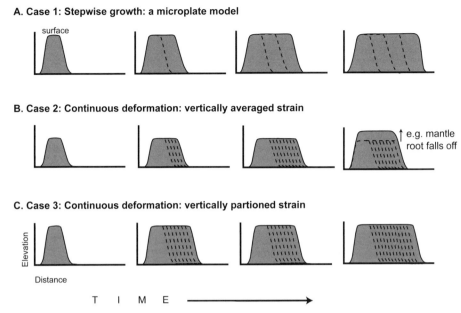

A. Case 1: Stepwise growth: a microplate model

surface

B. Case 2: Continuous deformation: vertically averaged strain

↑ e.g. mantle
root falls off

C. Case 3: Continuous deformation: vertically partioned strain

Elevation

Distance

T I M E

Figure 2. Predicted elevation histories for models in Figure 1. Dashed lines indicate progressive time steps of topographic change. A) High elevations are attained following crustal thickening concentrated on the margins of a lithospheric scale block. Elevation propagates away from the plate boundary in a series of distinct steps bounded by regions of concentrated vertical strain. B) High elevation is attained close to the plate boundary and propagates in a smooth fashion away from the plate boundary through time. Moderately high elevations are attained by wholesale lithospheric thickening. In the late stages of orogenesis, loss of thickened mantle lithosphere results in increased lithospheric buoyancy and plateau-wide elevation increase. C) A narrow region of high elevation near the plate boundary develops first. Areas of thickened crust and high elevation propagate outward, away from the plate boundary in a smooth fashion. No late stage increase in elevation across the entire plateau occurs.

boundary stresses that drive deformation. The difference between the two viscous models discussed here (Case 2 and 3), where deformation is or is not vertically uniform, are not subtle. They predict differences in rheologic stratification of the lithosphere, and mass distribution or recycling into the mantle during continental convergence.

The second model describes deformation of the continental lithosphere as a "thin-viscous" sheet, where lithospheric deformation is vertically averaged thickening of both the crust and mantle lithosphere (England and McKenzie 1982; England and Houseman 1986, 1988) (Fig. 1B). The thin-sheet approximation assumes no shear stresses acting on horizontal planes (i.e., stresses acting on the top and bottom of the lithosphere are negligible, as are the surface slopes); therefore, the behavior of the system depends on the vertically averaged stress acting within the lithosphere. Using a vertical average assumes that rheologic stratification is unimportant and does not allow for large-magnitude differential thickening between layers of the lithosphere. In this model, surface strain is an accurate indicator of deformation at depth.

Thickening at the plate boundary is predicted to occur until gravitational forces balance tectonic stresses and elevation reaches an equilibrium. This balance of tectonic and gravitational forces causes the deformation to propagate away from the plate boundary and for a plateau to grow outward in a smooth fashion (England and Houseman 1988). However, if the entire lithosphere was shortened and thickened by a factor of 2, an appropriate number for the double-thickness crust of the Tibetan Plateau, then the surface of the plateau is predicted to rise no more than 4 km (England and Houseman 1989). A 4 km high plateau is 1.5 km lower than the average surface elevation of the central Tibetan Plateau today (Fielding et al. 1994).

Thickening of the mantle lithosphere produces a dense root compared to the less dense asthenosphere that surrounds it. Under certain circumstances, mantle lithospheric "roots" are not stable and are predicted to be removed by convective processes (e.g., Houseman et al. 1981). Mantle foundering replaces dense mantle lithosphere by less-dense, hotter asthenosphere, which predicts increased lithospheric buoyancy and elevation gain. Removal of the mantle lithosphere beneath the Tibetan Plateau would explain the additional ~ 1-2 km of elevation gain needed to explain the modern height of the plateau (England and Houseman 1989; Molnar et al. 1993). If mantle lithosphere thinning or removal is a common process beneath orogenic belts then regular recycling of continental material to the mantle influences the geochemical and rheologic evolution of continents.

A vertically-averaged model of continuous deformation predicts crustal thickening and elevation to change smoothly away from the compressional boundary (Fig. 2B). Strain migrates away from the plate boundary, causing the plateau to elevate then expand outward. Deformation and elevation increase continues along the length of the eastern and western plateau as the plateau grows northward. Similar to Case 1, early growth of the plateau predicts increases in mean elevation, relief, and erosion rates concentrated on the plateau margins. Unlike the step-wise pattern predicted by Case 1, continuous deformation will predict a spatially smooth pattern of changes as the plateau grows upward and outward. As the plateau expands, reduction of relief across the interior plateau will be limited by rock strength as opposed to erosional lowering. Therefore, the magnitude of erosion of the central plateau required for Case 2 is much less than for Case 1. The abrupt change of 1-3 km mean elevation across the entire plateau due to mantle lithosphere removal will lead to increased relief and acceleration of erosion rates concentrated at all the plateau's modern margins (Molnar et al. 1993).

Case 3: Continuous deformation: vertically partitioned strain. The third case focuses on the fluid behavior of a weak, lower or middle crustal layer using equations that govern fluid flow in a viscous media. While similar in its approach to Case 2, the lower crustal flow model considers a rheologically stratified lithosphere instead of using a vertical average, and considers crustal thickening over a plate-like mantle that does not simultaneously thicken. Laboratory experiments suggest that for regions of elevated geothermal gradient and/or thick crust, quartzose-feldspathic rocks at mid or lower crustal depths may become much weaker than the overlying upper crust and deform by ductile flow (e.g., Goetze and Evans 1979; Brace and Kohlstedt 1980). If such a crustal layer exists, lateral variations in topography or crustal thickness will create horizontal pressure gradients that could drive flow of this low-viscosity material over distances of 10s to even several 100s of kilometers (Bird 1991; Royden 1996; Royden et al. 1997; Clark and Royden 2000). Flow can also be driven by changes in basal stresses (e.g., lateral changes in lid thickness) (e.g., Liu and Shen 1998; Schott and Schmeling 1998; Pysklywec et al. 2002). Long-distance transport of lower crustal rocks allows for differential mass distribution throughout the orogen, possibly even involving nearby regions where surface shortening strain is minimal or absent (Royden et al. 1997; Clark and Royden 2000) (Fig. 1C). Therefore, extrapolation of the strain recorded in the geological record at the surface to depth within orogens may be problematic.

Following Royden (1996), crustal deformation is coupled to mantle motions where the lower crust is strong, but is decoupled where a low-viscosity layer develops. A weak lower crustal layer is unable to support large topographic stresses, which results in a flat-topped (low-relief) plateau. Crustal thickening without mantle thickening may allow a high elevation plateau to develop without the need for later thinning or removal of dense, thickened mantle lithosphere as required by Case 2.

The gross pattern of mean elevation and relief change is similar to the vertically-averaged model in Case 2. Initially, a narrow high plateau develops then expands outward in a smooth fashion away from the plate boundary, leaving a trail of elevation and relief increase (Fig. 2C). This model differs in that the modern high elevation can be achieved early in the orogen's history (because the crust thickens and the mantle does not) and remains constant as deformation progresses away from the plate boundary. No abrupt increase in surface elevation is predicted late in the orogen's history. Like Case 2, mean elevation continues to increase across the length of the eastern and western plateau as the plateau expands northward, and relief is reduced in the interior of the expanding plateau because of the low strength of the middle crust without need for the large amounts of erosion required by Case 1.

Basin and Range Province: widespread extension

In some regions, wide-spread extensional deformation follows a period of contractional deformation (e.g., Dewey 1988). A commonly cited example is the Basin and Range Province of the western United States where contractional deformation during Late Paleozoic to Early Cenozoic orogenesis was followed by large magnitude crustal extension (50 to 200% elongation) (for a review see McQuarrie and Wernicke 2005) during Cenozoic time. Despite large magnitude crustal thinning, the Basin and Range stands anomalously high at 1.75 km elevation (Thompson and Zoback 1979), the Moho is flat, and the crust is regionally ~ 30 km thick (Allmendinger et al. 1987; Hauser et al. 1987; Thompson et al 1989; Smith et al 1989). Attempts to reconcile the basic paradox of a highly extended, yet normal thickness and moderately high elevation terrain have yielded a wealth of models to explain widespread extensional tectonism. These models invoke varied processes such as crustal thinning, mantle thinning or removal, lower crustal flow, and magmatic addition. Differences among models include (1) how strain is partitioned vertically in the lithosphere during orogen-scale extension, and (2) whether mass is conserved in the extending region. Each of these models offers testable predictions about the surface elevation history of the extended region as well as neighboring regions prior to and during extension. Unlike proposed models for Tibet that typically prescribe elevation histories, many models for the Basin and Range address a set of geologic or geophysical phenomena and do not explicitly predict regional elevation change. The predicted elevation histories of these models are my own, and do not necessarily reflect the intentions of the original authors.

In this section, I focus on four commonly-cited models for the Basin and Range Province that illustrate a range of mechanical behaviors during widespread extensional tectonism.

Case 1: Crustal thinning of an orogenic plateau. The first model predicts that a substantially thickened crust, caused by the protracted period of Late Paleozoic to Early Cenozoic orogeny, was later thinned by Cenozoic extension resulting in the present-day crustal thickness and elevation (e.g., Coney and Harms 1984; DeCelles 2004) (Fig. 3A). Upper crustal extension is representative of lower crustal extension, but not the mantle lithosphere. The consequence of this model is that significant gravitational potential energy stored by high, thickened crust drives extensional deformation (Jones et al. 1996; Sonder and Jones 1999). The effect of lithospheric cooling in time is ignored, as is the flexural rift-flank topography that develops across individual high-angle normal faults.

This model assumes that Mesozoic tectonism produced a high standing, thick crust much like the modern day Altiplano, or a smaller version of the modern Tibetan Plateau. While the present day crustal thickness of 30 km occurs with high modern elevations (1.75 km), this elevation is significantly lower than the pre-extensional elevation. In this case, any given location within the extended region would have a high mean elevation prior to extension, decrease during extensional tectonism, and stabilize at a lower mean elevation after extension ceases (Fig. 4A). High relief and erosion rates would be expected at the margins of the paleo-

A. Case 1: Crustal thinning of orogenic plateau

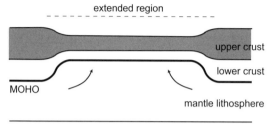

B. Case 2: Lithospheric thinning of an orogenic plateau

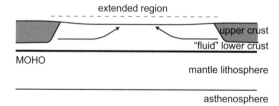

C. Case 3: Regional lower crustal flow

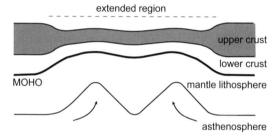

D. Case 4: Magmatic addition

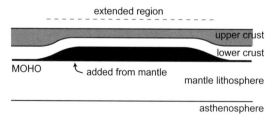

Figure 3. Models of crust and mantle processes during Basin and Range province evolution (continental extension). A) Uniform crustal thinning of a previously thick, high elevation, orogenic plateau (e.g., Coney and Harms 1984). B) Lithospheric thinning of an orogenic plateau where crustal extension is mirrored by significant thinning of the mantle lithosphere. Thinning may occur over a different wavelength in the mantle lithosphere than the crust, which produces shorter wavelength topography in the extended region. Figure modified from Froidevaux (1986). This model may also generally represent elevation histories that involve mantle lithosphere processes such as removal or delamination following gravitational instabilities, slab retreat or mantle upwellings (plume) (e.g., Parsons et al. 1994; Humphreys 1995). C) Regional lower crustal flow. Heterogeneous upper crustal extension is accommodated by infilling of low-viscosity lower crustal material of similar density (e.g., Block and Royden 1990; Bird 1991; Buck 1991). Crustal material may possibly be derived long-distances from beneath stable (undeforming) regions. Figure modified from Wernicke et al. (1996). D) Magmatic addition. Subcrustal magmatic flux provides a source of material to the extending region in order to maintain normal thickness crust during large-magnitude crustal thinning. Figure modified from Gans (1987).

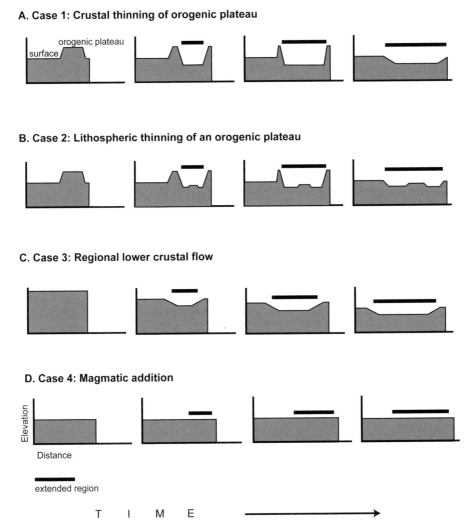

Figure 4. Predicted elevation histories for models in Figure 3. Thick black line indicates extended region. A) Uniform crustal thinning of a previously thick, high elevation, orogenic plateau. Crustal thinning over previously thickened and elevated plateau produces subsidence over the extending region. B) Lithospheric thinning of an orogenic plateau where crustal extension is mirrored by significant thinning of the mantle lithosphere over a different wavelength than the crust. This model also predicts subsidence over the extending region, however the magnitude of subsidence is less where the mantle lithosphere is also thinned. C) Regional lower crustal flow predicts regional lowering of a pre-existing highland as extension progresses. Because material flows from the surrounding stable regions to the extending region, subsidence is regionally averaged and surface lowering is predicted for both the extending region and the surrounding stable regions. Regional slopes between the extending region and the stable region are subdued by flow. D) Magmatic addition predicts little change in elevation over the extending region assuming crustal-density magmatic addition. This model allows for moderate initial elevation because magmatic addition could possibly offset crustal thinning. Elevation change of the extending region is unrelated to elevation change of the surrounding region.

plateau, and as extension progresses, would be concentrated at the margins of the extending region. As the extension migrates, the previously extended region is reduced in relief and erosion rates become lower. This scenario predicts mean elevation change only in areas experiencing extensional tectonism. The mean elevation of the surrounding area would be unaffected.

Case 2: Lithospheric thinning of an orogenic plateau. In the second model, heterogeneous upper crustal thinning during extensional tectonism is accompanied by more regionally distributed thinning of the lower crust and equally significant thinning of the mantle lithosphere. Here, I broadly consider any mechanism by which the mantle lithosphere is thinned or removed, although mantle lithosphere thinning may happen on a different lengthscale, or even different timing, than that of the crust (Crough and Thompson 1977; Froidevaux 1986; Lastowka et al. 2001) (Fig. 3B). Thinning or removal of mantle lithosphere by extension, gravitational instabilities, or roll-back of the Farallon slab (Humphreys 1995) produce similar elevation results and are not distinguished here by separate elevation models. Likewise, rapid thinning of the mantle lithosphere by plume emplacement (Fleitout et al. 1986; Parsons et al. 1994; Saltus and Thompson, 1995) is also broadly considered by this case. Mantle upwellings or slab retreat may cause or contribute to crustal extensional strain by thermally weakening the crust or increasing gravitational potential energy.

As in the case of uniform crustal thinning, lithospheric thinning produces subsidence over the extending region (Fig. 4B). In this case, I consider an example where extension of the mantle lithosphere occurs on a different lengthscale than the crust (Fletcher and Hallet 1983) (Fig. 3B). The difference in lengthscale over which thinning occurs produces mean elevation differences within the extending region that were not observed in Case 1. Replacement of thinned mantle lithosphere by thermally buoyant asthenosphere reduces the elevation decrease and produces intermediate wavelength relief across the extending region. Where relief is high at the margins of the extending region, or between areas of thinned/not thinned mantle lithosphere, greater erosion rates may be expected to drive an increase in local relief. Although not considered here, this model may even predict elevation rise compared to the pre-extension elevation depending on the ratio of crust to mantle lithospheric thinning and the magnitude of thermal buoyancy of the asthenosphere that replaces the thinned lithosphere.

Case 3: Regional lower crustal flow. The third and fourth models neglect the assumption that crustal mass is conserved within the extending region (e.g., McKenzie 1978). The source of new crustal material to the extending region is derived either from the lower crust of adjacent stable regions or magmatic material derived from the mantle. Inflation of the crust by material addition may explain the current normal thickness crust of the Basin and Range, despite extreme extension recorded in the upper crust, without necessarily having to appeal to a thick pre-Cenozoic crust. The difference in density between low-viscosity lower crustal material or emplaced magma and average crustal material make specific predictions about the elevation change of the extending region, as well as the elevation change of the surrounding area.

The third model invokes compensation of upper crustal extension by lateral influx of weak lower crustal material (Block and Royden 1990; Wernicke 1990; Kruse et al. 1991; Bird et al. 1991) (Fig. 3C). The pressure gradient created by the negative load of upper crustal thinning creates a lateral pressure gradient that drives long-distance flow of weak mid or lower crustal rocks (Block and Royden 1990). Upper crustal extension is accommodated by material of similar density (i.e., lower crust and not mantle) and produces no elevation change locally due to isostatically compensated changes to the integrated weight of the lithospheric column. Because the entire region is extending, mean elevation will decrease regionally. The original conception of this model suggested lower crustal flow over lengthscales of 10s of km beneath individual core complexes (Block and Royden 1990; Kruse et al. 1991). More extreme versions of this model suggest that crustal material could be regionally derived from surrounding stable regions, such as the Colorado Plateau and Sierra Nevada, and may compensate the Basin and Range as a whole (Wernicke et al. 1996). This model is also consistent with high GPE driving extension. Most importantly, this model predicts vertical strain partitioning where deformation on the surface may be horizontally removed from deformation at depth. Therefore surface strain in the upper crust may not be indicative of strain directly below it in the lower crust.

The initial elevation and crustal thickness of both the extending region and the surrounding stable regions are higher than at present and lower as extension progresses (Fig. 4C). An elevation difference between the stable and extending regions is maintained to continue to drive crustal flow. Lowering of surrounding regions occurs as lower crustal material is derived from these areas into the extending region. One significant difference between Case 3 and the previous two cases is that relief change between the extended and stable regions will be subdued by crustal flow because the weak layer is unable to support large topographic stresses (e.g., Bird 1991). Subdued relief will predict slower erosion rates than may be expected in Cases 1 and 2. If basins are internally drained, later drainage integration may drive pulses of relief change unrelated to tectonism (e.g., Pederson et al. 2002a).

Case 4: Magmatic addition. Magmatic addition to the crust from the mantle can potentially compensate for crustal extension (Lachenbruch and Sass 1978; Furlong and Fountain 1986; Thompson and McCarthy 1986; Gans 1987; Miller and Gans 1989) (Fig. 3D). To illustrate this process, I assume that extension is entirely compensated by subcrustal magmatic flux. This flux significantly alters volume constant calculations of crustal thickness based on upper crustal extensional strain. This model does not require gravitational potential energy as a driving mechanism to regional extension. Possibly, heat derived from magmatism could significantly "weaken" the crust and induce widespread extensional strain instead.

Depending on the density of the magmatic intrusions, lithospheric buoyancy can either decrease or remain neutral during extension. It is possible that differentiated magmas provide material such that average compositions are silicic or intermediate and would reduce the amount of predicted subsidence. Initial elevations of the extended region are difficult to predict. However, it is important to note that this model is the only one that does not require a high initial elevation of the extending region and allows for moderate conditions instead. In an extreme view, differentiated magmas that are similar in density to the crust are injected and elevation becomes only slightly reduced or remains level (Fig. 4D). As with Case 3, no significant relief change occurs within the extending region or between the extending and stable regions. An absence of relief would also keep erosion rates low.

DISCUSSION

Predicted elevation histories from deformation models alone provide only simple guidance. Tectonically driven elevation change will also drive erosional processes that will alter the resulting topography of the orogen. Greater complexity can also come into play if feedbacks between tectonics, erosion and climate exist. Elevation or relief changes can also be unrelated to tectonism. In the contractional setting discussed here, relief may increase again if closed basins are later breached by headward erosion, possibly related to more erosive climate conditions. Growth of an extensive, high elevation plateau has been suggested to cause intensification of the monsoon climate in southern Asia and aridification of central Asia (Kutzbach et al. 1989, 1993; Manabe and Broccoli 1990; Kroon et al. 1991; Broccoli and Manabe 1992; Prell et al. 1992; An 2001; Dettman et al. 2001). Changing global climate, regional climate due to plateau expansion, or local climate due to relief changes, could further alter relief, erosion rates, temperature and isotopic lapse rates in complicated ways across the orogen that would be unrelated to crustal thickening. Large amounts of erosion also have the potential to alter the thermal structure of the crust by advecting heat toward the surface as rocks are removed, which may alter deformation styles (e.g., Avouac and Burov 1996; Willett 1999). Erosion may also change the topographic slope of the plateau margin, which alone could arrest or propagate deformation in the absence of changes of plate boundary forces (e.g., Dahlen and Suppe 1988).

As in the case for contractional deformation, patterns of erosion and relief change may also occur unrelated to changes in mean elevation in an extensional setting. Relief may increase during a period of time that is not related to significant, tectonically-driven elevation changes if closed basins are breached by headward erosion, possibly related to more erosive climate conditions. Changes in the topography of the Rocky Mountain region, or rise of the Sierra Nevada, may affect regional climate conditions, create orographic barriers and alter atmospheric circulation within the Basin and Range Province. Regional climate effects, in addition to global climate conditions, or local climate conditions due to relief changes, could further alter relief, erosion rates, temperature and isotopic lapse rates.

Looking forward, the techniques discussed in this volume will not only elucidate elevation histories, but applied in combination may also explore how climate, relief, erosion and elevation change independently or interdependently during orogenesis. Such combinations will ultimately advance our understanding of deformation processes on the continents and the interrelationships between climate, erosion and tectonics.

SUMMARY

Determination of precise elevation change offers exciting possibilities to further the field of continental tectonics. The variety of new techniques, combined with existing techniques, promises broad applicability of paleoelevation measurements to common geologic settings. Because deformation in the deep lithosphere can occur without surface deformation directly above, the geologic record alone is insufficient to describe behavior of the lithosphere. As such, elevation histories are an important discriminator between competing models of deformation. The pattern of elevation increase (or subsidence) in a deforming region can indicate thickening or thinning of the deep crust or mantle lithosphere that can otherwise not be directly measured. Elevation change where deformation is absent may indicate mass transfer of crustal material from or to the deforming region. Therefore widespread and long paleoelevation records are necessary to test between suggested processes that act to deform the lithosphere.

Elevation predictions from tectonic models alone only provide simple guidance because topography is the product of complex reactions among climate, erosion and tectonics. Paleoelevation techniques discussed in this volume measure different aspects either directly or indirectly of topography: some techniques are sensitive to local elevation, some to relief, and some rely on proxies of environmental variables. Future directions may include measurement of different variables of the tectonic-climate-erosion system independently from one another and understanding the interaction between them. From these relationships, directly or indirectly, the geodynamic processes may be further constrained.

ACKNOWLEDGMENTS

I thank Craig Jones, Joel Pederson and Matt Kohn (editor) for their constructive and thorough reviews that greatly improved this manuscript. I also thank Carmala Garzione, Nathan Niemi, and Alison Duvall for discussions about paleoaltimetry techniques, Basin and Range tectonics and geomorphology respectively.

REFERENCES

Allmendinger RW, Nelson KD, Potter CJ, Barazangi M, Brown LD, Oliver JE (1987) Deep seismic reflection characteristics of the continental crust. Geology 15:304-310

An Z, Kutzbach JE, Prell WL, Porter SC (2001) Evolution of Asian monsoons and phased uplift of the Himalaya-Tibet Plateau since Late Miocene times. Nature 411(6833):62-66

Avouac J-P, Burov EB (1996) Erosion as a driving mechanism of intracontinental mountain growth. J Geophys Res 101(B8):17,747-17,769

Beaumont C, Jamieson RA, Nguyen MH, Lee B (2001) Himalayan tectonics explained by extrusion of a low-viscosity crustal channel coupled to focused surface denudation. Nature 414(6865):738-742

Block L, Royden LH (1990) Core complex geometries and regional scale flow in the lower crust. Tectonics 9(4):557-567

Bird P (1988) Formation of the Rocky Mountains, Western United States: A continuum computer model. Science 239(4847):1501-1507

Bird P (1991) Lateral extrusion of lower crust from under high topography, in the isostatic limit. J Geophys Res 96(B6):10,275-10,286

Brace WF, Kohlstedt DL (1980) Limits on lithospheric stress imposed by laboratory experiments. J Geophys Res 85:6248-6252

Braun J (2002) Estimating exhumation rate and relief evolution by spectral analysis of age-elevation datasets. Terra Nova 14:210-214

Broccoli AJ, Manabe S (1992) The effects of orography on midlatitude northern-hemisphere dry climates. J Climate 5(11):1181-1201

Buck R (1991) Modes of continental lithospheric extension. J Geophys Res 96(B12):20,161-20,178

Carson MA (1976) Mass-wasting, slope development and climate. *In:* Geomorphology and climate. Derbyshire E (ed) John Wiley and Sons, New York, p 101-136

Clark MK, Royden LH, (2000) Topographic ooze: Building the eastern margin of Tibet by lower crustal flow. Geology 28:703-706

Clark MK, House MA, Royden LH, Whipple KX, Burchfiel BC, Zhang X, Tang W (2005a) Late Cenozoic uplift of southeastern Tibet. Geology 33:525-528

Clark MK, Maheo G, Saleeby J, Farley KA (2005b) The non-equilibrium landscape of the southern Sierra Nevada, California. Geol Soc Am Today 15(9):4-10

Clark MK, Royden LH, Whipple KX, Burchfiel BC, Zhang X, Tang W (2006) Use of a regional, relict landscape to measure vertical deformation of the eastern Tibetan Plateau. J Geophys Res – Earth Surface 111(F3): Art No. F03002

Coney PJ, Harms TA (1984) Cordilleran metamorphic core complexes: Cenozoic extension relics of Mesozoic compression. Geology 12:550-554

Crough ST, Thompson GA (1977) Upper mantle origin of Sierra Nevada uplift. Geology 5:396-399

Dahlen FA, Suppe J (1988) Mechanics, growth, and erosion of mountain belts. Geol Soc Am Sp Paper 218:161-178

DeCelles PG (2004) Late Jurassic to Eocene evolution of the Cordilleran thrust belt and foreland basin system, western U.S.A. Am J Sci 304:105-168

Dettman DL, Kohn ML, Quade J, Ryerson FJ, Ojha TP, Hamidullah S (2001) Seasonal stable isotope evidence for a strong Asian monsoon throughout the past 10.7 m.y. Geology 29:31-34

Dewey JF (1988) Extensional collapse of orogens. Tectonics 7(6):1123-1139

Eaton GP, Wahl RR, Prostka JH, Mabey DR, Kleinkopf MD (1978) Regional gravity and tectonic patterns; their relation to late Cenozoic epeirogeny and lateral spreading in the western Cordillera. Geol Soc Am Mem 152:51-91

England P, McKenzie D (1982) A thin viscous sheet model for continental deformation. Geophys J R Astr Soc 70:295-321

England PC, Houseman GA (1986) Finite strain calculations of continental deformation, 2, Comparison with the India-Asia collision. J Geophys Res 91:3664-3676

England PC, Houseman GA (1988) The mechanics of the Tibetan Plateau. Phil Trans R Soc Lond A 326:301-320

England PC, Houseman GA (1989) Extension during continental convergence, with application to the Tibetan Plateau. J Geophys Res 94:17,561-17,579

England P, Molnar P (1990) Surface uplift, uplift of rocks, and exhumation of rocks. Geology 18:1173-1177

Fielding E, Isacks B, Barazangi M, Duncan C (1994) How flat is Tibet? Geology 22(2):163-167

Fleitout L, Froidevaux C, Yuen D (1986) Active lithospheric thinning. Tectonophysics 132:271-278

Flesch LM, Holt WE, Haines AJ, Shen-Tu BM (2000) Dynamics of the Pacific-North American plate boundary in the western United States. Science 287(5454):834-836

Fletcher RC, Hallet B (1983) Unstable extension of the lithosphere – A mechanical model for basin-and-range structure. J Geophys Res 88(NB9):7457-7466

Forest CE (2007) Paleoaltimetry: a review of thermodynamic methods. Rev Mineral Geochem 66:173-193
Froidevaux C (1986) Basin and Range large-scale tectonics: Constraints from gravity and reflection seismology. J Geophys Res 91(B3):3625-3632
Furlong KP, Fountain DM (1986) Continental crustal underplating – thermal considerations and seismic-petrologic consequences. J Geophys Res 91(B8):8285-8294
Gabet EJ, Pratt-Sitaula BA, Burbank DW (2004) Climatic controls on hillslope angle and relief in the Himalayas. Geology 32(7):629-632
Gans P (1987) An open-system, two-layer crustal stretching model for the eastern Great Basin. Tectonics 6(1):1-12
Goetze C, Evans B (1979) Stress and temperature in the bending lithosphere as constrained by experimental rock mechanics. Geophys J Royal Astr Soc 59(3):463-478
Hager BH, Clayton RW, Richards MA, Comer RP, Dziewonski AM (1985) Lower mantle heterogeneity, dynamic topography and the geoid. Nature 313:541-545
Hauser EC, Gephart J, Latham T, Oliver J, Kaufman S, Brown L, Lucchitta I (1987) COCORP Arizona transect: Strong crustal reflections and offset Moho beneath the transition zone. Geology 15:1103-1106
Hodges KV, Hurtado JM, Whipple KX (2001) Southward extrusion of Tibetan crust and its effect on Himalayan tectonics. Tectonics 20(6):799-809
Hodges KV, Wobus C, Ruhl K, Schildgen T, Whipple K (2004) Quaternary deformation, river steepening, and heavy precipitation at the front of the Higher Himalayan ranges. Earth Planet Sci Lett 220(3-4):379-389
House MA, Wernicke BP, Farley KA (1998) Dating topography of the Sierra Nevada, California, using apatite (U-Th)/He ages. Nature 396:66-69
House MA, Wernicke BP, Farley KA (2001) Paleo-geomorphology of the Sierra Nevada, California, from (U-Th)/He ages in apatite. Am J Sci 301:77-102
Houseman GA, McKenzie DP, Molnar P (1981) Convective instability of a thickened boundary-layer and its relevance for the thermal evolution of continental convergent belts. J Geophys Res 86(NB7):6115-6132
Hovius N (1996) Regular spacing of drainage outlets from linear mountain belts. Basin Res 8:29-44
Humphreys E (1995) Post-Laramide removal of the Farallon slab, western United States. Geology 23(11):987-990
Jin Y, McNutt MK, Zhu YS (1994) Evidence from gravity and topography data for folding of Tibet. Nature 371(6499):669-674
Jones CH, Unruh JR, Sonder LJ (1996) The role of gravitational potential energy in active deformation in the southwestern United States. Nature 381(6577):37-41
Kohn MJ, Dettman DL (2007) Paleoaltimetry from stable isotope compositions of fossils. Rev Mineral Geochem 66:119-154
Koons P (1989) The topographic evolution of collisional mountain belts – A numerical look at the Southern Alps, New-Zealand. Am J Sci 289(9):1041-1069
Kouwenberg LLR, Kürschner WM, McElwain JC (2007) Stomatal frequency change over altitudinal gradients: prospects for paleoaltimetry. Rev Mineral Geochem 66:215-242
Kroon D, Steens T, Troelstra SR (1991) Onset of monsoonal related upwelling in the western Arabian Sea as revealed by planktonic foramifers. *In*: Proceedings of the Ocean Drilling Project, Sci Results 117, Ocean Drilling Program, College Station, Texas 257-263
Kruse S, McNutt M, Phipps-Morgan J, Royden L, Wernicke B (1991) Lithospheric extension near Lake Mead, Nevada – A model for ductile flow in the lower crust. J Geophys Res 96(B8):4435-4456
Kutzbach JE, Guetter PJ, Ruddiman WF, Prell WL (1989) Sensitivity of climate to late Cenozoic uplift in Southern Asia and the American West: Numerical experiments. J Geophys Res 94:18,393-18,407
Kutzbach JE, Prell WL, Ruddiman WF (1993) Sensitivity of Eurasian climate to surface uplift of the Tibetan Plateau. J Geol 101:177-190
Lachenbruch AH, Sass JH (1978) Models of an extended lithosphere and heat flow in the Basin and Range province. Geol Soc Am Mem 152:209-250
Lachenbruch AH, Morgan P (1990) Continental extension, magmatism and elevation; formal relations and rules of thumb. Tectonophysics 174:39-62
Lague D, Hovius N, Davy P (2005) Discharge, discharge variability, and the bedrock channel profile. J Geophys Res 110, doi:10.1029/2004JF000259
Langbein WB, Schumm SA (1958) Yield of sediment in relation to mean annual precipitation. Am Geophys Union Trans 39:1076-1084
Lastowka LA, Sheehan AF, Schneider JM (2001) Seismic evidence for partial lithospheric delamination model of Colorado Plateau uplift. Geophys Res Let 28(7):1319-1322
Lithgow-Bertelloni C, Silver PG (1998) Dynamic topography, plate driving forces and the African Superswell. Nature 395:269-271
Lithgow-Bertelloni C, Guynn JH (2004) Origin of the lithospheric stress field. J Geophys Res 109(B1), Art. No. B01408

Liu M, Shen YQ (1998) Sierra Nevada uplift: A ductile link to mantle upwelling under the basin and range province. Geology 26(4):299-302

Manabe S, Broccoli AJ (1990) Mountains and arid climates of middle latitudes. Science 247(4939):192-194

Mancktelow NS, Grasemann B (1997) Time-dependent effects of heat advection and topography on cooling histories during erosion. Tectonophysics 270:167-195

McKenzie DP (1978) Some remarks on the development of sedimentary basins. Earth Planet Sci Lett 40:25-32

McQuarrie N, Wernicke BP (2005) An animated tectonic reconstruction of southwestern North America since 36 Ma. Geosphere 1(3):147-172

Metivier F, Gaudemer Y, Tapponnier P, Meyer B (1998) Northeastward growth of the Tibet plateau deduced from balanced reconstruction of two depositional areas: The Qaidam and Hexi Corridor, China. Tectonics 17(6):823-842

Meyer B, Tapponnier P, Bourjot L, Metivier F, Gaudemer Y, Peltzer G, Shunmin G, Zhitai C (1998) Crustal thickening in Gansu-Qinghai, lithospheric mantle subduction, and oblique, strike-slip controlled growth of the Tibet plateau. Geophys J Int 135(1):1-47

Meyer HW (2007) A review of paleotemperature–lapse rate methods for estimating paleoelevation from fossil floras. Rev Mineral Geochem 66:155-171

Miller EL, Gans PB (1989) Cretaceous crustal structure and metamorphism in the hinterland of the Sevier thrust belt, western U.S. Cordillera. Geology 17:59-62

Molnar P (1988) Continental tectonics in the aftermath of plate tectonics. Nature 335:131-137

Molnar P, England P (1990) Late Cenozoic uplift of mountain-ranges and global climate change – chicken or egg. Nature 346:29-34

Molnar P, England P, Martinod J (1993) Mantle Dynamics, uplift of the Tibetan Plateau, and the Indian Monsoon. Rev Geophys 31:357-396

Molnar P, Anderson RS, Kier G, Rose J (2006) Relationships among probability distributions of stream discharges in floods, climate, bed load transport, and river incision. J Geophys Res 111, doi:10.1029/2005JF000310

Montgomery DR, Balco G, Willett SD (2001) Climate, tectonics, and the morphology of the Andes. Geology 29(7):579-582

Montgomery DR (2001) Slope distributions, threshold hillslopes, and steady-state topography. Am J Sci 301:432-454

Montgomery DR, Brandon MT (2002) Topographic controls on erosion rates in tetconically active mountain ranges. Earth Planet Sci Lett 201:481-489

Mulch A, Graham SA, Chamberlain CP (2006) Hydrogen isotopes in Eocene river gravels and paleoelevation of the Sierra Nevada. Science 313(5783):87-89

Mulch A, Chamberlain CP (2007) Stable isotope paleoaltimetry in orogenic belts – the silicate record in surface and crustal geological archives. Rev Mineral Geochem 66:89-118

Parsons T, Thompson GA, Sleep NH (1994) Mantle plume influence on the Neogene uplift and extension of the U.S. western Cordillera? Geology 22:83-86

Pederson JL, Mackley RD, Eddleman JL (2002a) Colorado Plateau uplift and erosion evaluated using GIS. GSA Today 12:4-10

Pederson JL, Karlstrom K, Sharp W, McIntosh W (2002b) Differential incision of the Grand Canyon related to Quaternary faulting – Constraints from U-series and Ar/Ar dating. Geology 30(8):739-742

Prell WL, Murray DW, Clemens SC, Anderson DM (1992) Evolution and variability of the Indian ocean summer monsoon: Evidence from the western Arabian Sea drilling program. *In*: Synthesis of Results from Scientific Drilling in the Indian Ocean. Duncan RA, Rea DK, Kidd RB, von Rad U, Weissel JK (eds) Am Geophys Union, Washington D.C., p 447-469

Pysklywec RN, Beaumont C, Fullsack P (2002) Lithospheric deformation during the early stages of continental collision: Numerical experiments and comparison with South-Island, New Zealand. J Geophys Res 107(B7), Art. No. 2133

Quade J, Garzione C, Eiler J (2007) Paleoelevation reconstruction using pedogenic carbonates. Rev Mineral Geochem 66:53-87

Reiners PW, Ehlers TE, Mitchell SG, Montgomery DR (2003) Coupled spatial variations in precipitation and long-term erosion rates across the Washington Cascades. Nature 426(6967):645-647

Reiners PW, McPhillips D, Brandon MT, Mulch A, Chamberlain CP (2006) Thermochronologic approaches to paleotopography. Geochimica et Cosmochimica Acta 70:A525

Reiners PW (2007) Thermochronologic approaches to paleotopography. Rev Mineral Geochem 66:243-267

Riihimaki CA, Libarkin JC (2007) Terrestrial cosmogenic nuclides as paleoaltimetric proxies. Rev Mineral Geochem 66:269-278

Roe G (2005) Orographic precipitation. Annu Rev Earth Planet Sci 33:645-671

Rowley DB (1996) Age of initiation of collision between India and Asia: A review of stratigraphic data. Earth Planet Sci Lett 145(1-4):1-13

Rowley DB, Pierrehumbert RT, Currie BS (2001) A new approach to stable isotope-based paleoaltimetry: implications for paleoaltimetry and paleohypsometry of the High Himalaya since the Late Miocene. Earth Planet Sci Lett 188(1-2):253-268

Rowley DB (2007) Stable isotope-based paleoaltimetry: theory and validation. Rev Mineral Geochem 66:23-52

Roy M, Royden LH (2000) Crustal rheology and faulting at strike-slip plate boundaries 2. Effects of lower crustal flow. J Geophys Res 105(B3):5599-5613

Roy M, MacCarthy JK, Selverstone J (2005) Upper mantle structure beneath the eastern Colorado Plateau and Rio Grande rift revealed by Bouguer gravity, seismic velocities, and xenolith data. Geochem Geophys Geosyst 6(Q10007), doi:10.1029/2005GC001008

Royden LH (1996) Coupling and decoupling of crust and mantle in convergent orogens: Implications for strain partitioning in the crust. J Geophys Res 101(B8):17,679-17,705

Royden LH, Burchfiel BC, King RW, Wang E, Chen Z, Shen F, Liu Y (1997) Surface Deformation and Lower Crustal Flow in Eastern Tibet. Science 276:788-790

Sahagian D, Proussevitch A (2007) Paleoelevation measurement on the basis of vesicular basalts. Rev Mineral Geochem 66:195-213

Saltus RW, Thompson GA (1995) Why is it downhill from Tonopah to Las Vegas? A case for mantle plume support of the high northern Basin and Range. Tectonics 14(6):1235-1244

Schott B, Schmeling H (1998) Delamination and detachment of a lithospheric root. Tectonophysics 296(3-4):225-247

Sklar LS, Dietrich WE (2004) A mechanistic model for river incision into bedrock by saltating bed load. Water Resources Res 40(6), Art No. W06301

Smith RB, Nagy WC, Julander KA, Viveiros JJ, Barker CA, Gants DG (1989) Geophysical and tectonic framework of the eastern Basin and Range-Colorado Plateau-Rocky Mountain transition. *In:* Geophysical Framework of the Continental United States. LC Pakiser, Mooney WD (eds) Geol Soc Am, Boulder, p 205-233

Sobel ER, Hilley GE, Strecker MR (2003) Formation of internally drained contractional basins by aridity-limited bedrock incision. J Geophys Res 108(B7), doi:10.1029/2002JB001883

Sonder LJ, Jones CH (1999) Western United States Extension: How the West was Widened. Annu Rev Earth Planet Sci 27:417-462

Stock GM, Ehlers TA, Farley KA (2006) Where does sediment come from? Quantifying catchment erosion with detrital apatite (U-Th)/He thermochronometry. Geology 34:725-728

Stockli DF (2006) Thermochronometric constraints on paleoaltimetry and paleotopography – Case studies from the Colorado Plateau, Tibet and Labrador. Geochim Cosmochim Acta 70:A617

Stuwe K, White L, Brown R (1994) The influence of eroding topography on steady-state isotherms – Application to fission-track analysis. Earth Planet Sci Lett 124:63-74.

Tapponnier P, Peltzer G, Le Dain AY, Armijo R (1982) Propagating extrusion tectonics in Asia: New insights from simple experiments with plasticine. Geology 10:611-616

Tapponnier P, Meyer B, Avouac JP, Peltzer G, Gaudemer Y, Guo SM, Xiang HF, Yin KL, Chen ZT, Cai SH, Dai HG (1990) Active thrusting and folding in the Qilian Shan, and decoupling between upper crust and mantle in northeastern Tibet. Earth Plan Sci Lett 97(3-4):382-403

Tapponnier P, Xu ZQ, Roger F, Meyer B, Arnaud N, Wittlinger G, Yang JS (2001) Oblique Stepwise Rise and Growth of the Tibet Plateau. Science 294:1671-1677

Thatcher W (2007) Microplate model for the present-day deformation of Tibet. J Geophys Res 112(B1), Art. No. B01401

Thompson GA, Zoback ML (1979) Regional geophysics of the Colorado Plateau. Tectonophysics 61:149-181

Thompson GA, McCarthy J (1986) Geophysical evidence for igneous inflation of the crust in highly extended terrains. EOS Trans 67:1184

Thompson GA, Catchings R, Goodwin E, Holbrook S, Jarchow C, Mann C, McCarthy J, Okaya D (1989) Geophysics of the western Basin and Range Province. *In:* Geophysical Framework of the Continental United States, LC Pakiser, Mooney WD (eds) Geol Soc Am Mem 172:177-203

Wernicke B (1990) The fluid crustal layer and its implications for continental dynamics. *In:* Exposed Cross-Sections of the Continental Crust. Salisbury MH, Fountain DM (eds) NATO ASI Series. Series C: Mathematical and Physical Sciences 317:509-544

Wernicke B, Clayton R, Ducea, M, Jones CH, Park S, Ruppert S, Saleeby J, Snow JK, Squires L, Fliedner M, Jiracek G, Keller R, Klemperer S, Luetgert J, Malin P, Miller K, Mooney W, Oliver H, Phinney R (1996) Origin of high mountains in the continents: The Southern Sierra Nevada. Science 271(5246):190-193

Whipple KX, Kirby E, Brocklehurst SH (1999) Geomorphic limits to climate-induced increases in topographic relief. Nature 401:39-43

Whipple KX, Meade BJ (2006) Orogen response to changes in climatic and tectonic forcing. Earth Planet Sci Lett 243(1-2):218-228

Willett S (1999) Orogeny and orography: The effects of erosion on the structure of mountain belts. J Geophys Res 104(B12):28,957-28,981

Wu S, Bras RL, Barros AP (2006) Sensitivity of channel profiles to precipitation properties in mountain ranges. J Geophys Res 111, doi:10.1029/2204JF000164

Zachos J, Pagani M, Sloan L, Thomas E, Billups K (2001) Trends, rhythms, and aberrations in global climate 65 Ma to present. Science 292:686-693

Zeitler PK, Meltzer AS, Koons PO, Craw D, Hallet B, Chamberlain CP, Kidd WSF, Park SK, Seeber L, Bishop M, Shroder J (2001) Erosion, Himalayan geodynamics, and the geomorphology of metamorphism. GSA Today 11(1):4-9

Zheng PZ, Molnar P, Downs WR (2001) Increased sedimentation rates and grain sizes 2 – 4 Myr ago due to the influence of climate change on erosion rates. Nature 410(6831):891-897

Reviews in Mineralogy & Geochemistry
Vol. 66, pp. 23-52, 2007
Copyright © Mineralogical Society of America

2

Stable Isotope-Based Paleoaltimetry: Theory and Validation

David B. Rowley

Department of the Geophysical Sciences
The University of Chicago
5734 S. Ellis Avenue
Chicago, Illinois, 60637, U.S.A.
rowley@geosci.uchicago.edu

ABSTRACT

Paleoaltimetry is the quantitative estimate of the surface height above mean sea level of ancient landforms. Atmospheric thermodynamic modeling of the behavior of ^{18}O relative to ^{16}O during condensation from water vapor establish the systematic relationship that exists between $\Delta(\delta^{18}O_p)$ and elevation, where $\Delta(\delta^{18}O_p)$ is the difference between a low altitude, preferably sea level $\delta^{18}O_p$ and a potentially high elevation sample. Comparison of model predictions with observations suggests that the model captures the first-order behavior of $\delta^{18}O_p$ during condensation and precipitation in orographic settings. The actual relationship between $\Delta(\delta^{18}O_p)$ and elevation depends sensitively on climate, and specifically starting temperature and to a lesser degree relative humidity. Thermodynamic modeling allows the $\Delta(\delta^{18}O_p)$ and elevation relationship to be explicitly calculated for any given starting climate state. This makes the theoretical approach significantly more appropriate than empirically calibrated approaches based typically on quite limited samples presently available in most orographic settings today.

Paleoaltimetry archives derive their isotopic compositions from surface and or ground water, and hence it is important to understand the systematic differences between these reservoirs and precipitation. Surface waters and ground waters integrate not just the change in isotopic composition with altitude, but also variations in hypsometry within the drainage basin and precipitation amount as functions of elevation. Thus these archives should reflect the precipitation amount weighted hypsometric mean elevation of the (paleo)-drainage basin from which they derive their waters. Analysis of modern data from the Himalayan region supports this expectation. Foreland basin rivers are such integrators and it is shown, using an example from the Siwaliks that the rivers draining the front of the Himalayas in the past had isotopic compositions comparable with modern rivers draining the Himalayan front suggesting little net change in Himalayan hypsometry over the past 11 million years.

INTRODUCTION

Paleoaltimetry is the quantitative estimate of the surface height above mean sea level of features in the past. Unlike quantitative paleobathymetry, which has a long history of development and application, quantitative paleoaltimetry is a recently emerging area of investigation and one still in its infancy of development and validation. Nonetheless, there has been an explosion of interest in paleoaltimetry in the past 10 years and considerable progress since the publication of the benchmark review of this field by Chase et al. (1998). The recent review by Rowley and Garzione (2007) summarizes much of this progress. The focus of the present discussion will be on the theoretical underpinnings of $\delta^{18}O$ isotope-based

     DOI: 10.2138/rmg.2007.66.2

approaches to paleoaltimetry and its validation in the modern. Modeling of δ^2H is included for completeness, but is not discussed beyond that. This particular review is not intended to be comprehensive of work in the field of either paleoaltimetry in general or stable isotope paleoaltimetry in particular. Rather the review will address a particular modeling approach to paleoaltimetry, some of the rationale for adopting a theoretical approach as opposed to empirical calibration approaches, and continued validation of this modeling approach using a spectrum of data from the present. In addition, one new application of this approach to the past is outlined exploring paleo-hypsometric relations in paleodrainages feeding the Siwalik foreland basin of the Himalayas. For broader reviews of stable isotope-based paleoaltimetry the reader is directed to Rowley and Garzione (2007), Blisnuik and Stern (2005), as well as important contributions in the present volume, including Quade et al. (2007), Mulch and Chamberlain (2007) and Kohn and Dettman (2007).

The primary foci of paleoaltimetry are regions of the Earth with elevations typically 2 to 3 km and higher. Although these are clearly tectonically interesting and important, it is important to recognize that regions above 2 km represent only about 11% of the surface of the continents and a bit more than 3% of the surface of the Earth as a whole. Much of the interest in paleoaltimetry to date is focused on the Himalayas (Rowley et al. 1999; Garzione et al. 2000a,b; Rowley et al. 2001), Tibetan Plateau (Currie et al. 2005; Cyr et al. 2005; Rowley and Currie 2006) and the Andes (Garzione et al. 2006; Ghosh et al. 2006), regions with current elevations greater than 4 km. Together these represent less than 2% of the surface area of the continents (Fig. 1). Regions such as the Sierra Nevada (Mulch et al. 2006), Cascades (Kohn et al. 2002), western United States (Horton and Chamberlain 2006; Kent-Corson et al. 2006), among other, less highly elevated orographic entities have also been investigated with stable isotope-based paleoaltimetric approaches. It is important to recognize that all such regions represent but a small fraction of the Earth's surface environment.

Paleoaltimetry, at least in its present form, is primarily directed at developing an understanding of the elevation history of what are currently highly elevated regions that represents a small but intriguing fraction of Earth's hypsometric development. It seems clear that as confidence is gained in the ability to discern paleoelevation histories in these

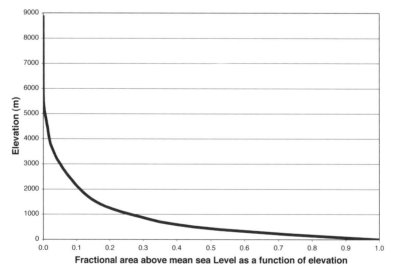

Figure 1. Fractional area of the continents with elevations above mean sea level based on Globe 30 second topography (GLOBE Task Team and others (Hastings 1999).

regions, where the signals are large, there will be increasing interest in the application of these techniques to deeper time and to regions no longer at high elevations.

The majority of continental topography of Earth is, to first order, in simple isostatic equilibrium, and thus there exists a good correlation between topography and crustal thickness. This suggests that estimates of paleoelevation can be derived from dating times of change of crustal thickness as a proxy for elevation. Prior to the ability to independently estimate the elevation of the surface as a function of time, the generally held view was that elevation change was inherently correlated with times of crustal deformation, and in the case of mountain building primarily times of shortening of the crust. The assumption being that if one can date and estimate the magnitude of crustal shortening then the timing and amplitude of surface elevation change can be directly determined, taking appropriate account of isostasy. This obviously does not pertain to regions, such as South Africa where the surface rocks are undeformed, including interbedded marine strata, nor specifically in the cases of rift shoulder uplift, where crustal thickening clearly is not the underlying driver of increases in elevation. However, in orogenic settings this assumption appeared to be an obvious corollary of the process.

Structural analysis of highland areas is, at least in part, motivated to determine the magnitude and timing of shortening from which to derive estimates of the timing of surface uplift. Argand's (1924) insightful surmise of the doubled thickness of Tibetan crust provided considerable impetus to field investigations of the magnitude and timing of crustal deformation as a function of location within the Himalayas and Tibet (see Coward et al. 1988; Dewey et al. 1988) among many other regions. The absence of simple, unequivocal and pervasive evidence of ~50% surface shortening since the late Early Eocene initiation of India-Asia collision (Zhu et al. 2005), in Tibet raises questions regarding this assumption. The relative constancy of elevation of the Tibetan Plateau has been cogently argued to reflect flow in the lower or middle crustal levels in response to potential changes in surface loading (Zhao and Morgan 1987). This point was further emphasized by Royden et al. (1997) in their analysis of eastern Tibet. In eastern Tibet there exist quite clear geologic relations indicating limited Tertiary crustal shortening, and yet modern surface elevations above 4 km argued that relatively recent surface uplift had indeed taken place (Clark et al. 2005). Royden et al (1997) among many others invoke flow of continental crust at lower or mid-crustal depths driven by topographic gradients away from regions of southern and central Tibet elevated earlier in this history.

The potential role of the mantle in the evolution of topography in orogenic settings has been analyzed by Bird (1979), England and Houseman (1986), Molnar et al. (1993) and Garzione et al.(2006) among others. England and Houseman (1986) and Molnar et al. (1993) brought forth arguments, both based on numerical modeling and geological observations that mantle lithospheric thickening and subsequent convective destabilization and removal may have played a significant role in the evolution of Tibetan topography. Pyskylwec et al. (2000) demonstrated that the growth and evolution of Rayleigh-Taylor instabilities in convergence systems depends on a number of parameters, including the rate of convergence, the density field, and the rheology of the mantle lithosphere. These types of instabilities may or may not develop depending on relationships among the various controlling parameters and therefore should not be expected to be an intrinsic contributor in the evolution of every collisional domain.

Both lower or mid-crustal flow and lithospheric mantle contributions decouple surface geology, and particularly evidence of crustal shortening, from elevation history at least in some regions. That is to say that some fraction of the topographic development may not be correlated with or controlled by observable evidence at the surface of crustal thickening. In order to sort out various potential contributions to the elevation history in a given orogenic system it is therefore necessary to have techniques that allow independent estimates of the surface elevation history. One such technique that has been developed that potentially allows for this is stable isotope-based paleoaltimetry. The theoretical basis for this paleoaltimeter has been discussed by Rowley et al. (2001) and Rowley and Garzione (2007) and will be briefly reviewed and updated here. In order to validate the model, modern data primarily derived from surface waters are discussed with an emphasis on the Himalayas and southern Tibet.

STABLE ISOTOPE-BASED PALEOALTIMETRY

Measurement of $\delta^{18}O$ and δ^2H in precipitation quickly led to the recognition of various factors, often referred to as effects, controlling the spatial variation of isotopic compositions. Among the various effects are temperature, latitude, amount, continentality, and elevation. The relative importance of these various effects has generally been assessed by employing multiple regression techniques to the global network of isotopes in precipitation (GNIP) datasets (Craig 1961; Dansgaard 1964; Rozanski et al. 1993), a product of the Isotope Hydrology Section of the International Atomic Energy Agency, located in Vienna, Austria (IAEA-Yearly 2004; IAEA-Monthly 2004). The GNIP dataset contains relatively fewer high elevation stations and so the altitude effect is relegated typically to 4th or less significance in controlling the isotopic composition of precipitation (Rozanski et al. 1993). This might be taken to imply stable isotope-based paleoaltimetry has a quite limited potential. However, $\delta^{18}O$ in precipitation exceeding -10 to $-15\%_0$ relative to low elevation starting compositions within <100 km in orographic settings are comparable to $75°$ of the so-called latitudinal effect, thus testifying to the potential significance of elevation in controlling isotopic compositions. More recently, Bowen and Revenaugh (2003) model the spatial pattern of modern isotopic of precipitation using only two parameters the latitude and elevation quite successfully. This approach uses a simple linear scaling of isotopic composition to elevation (Bowen and Revenaugh 2003; Dutton et al. 2005). This approach is quite good at fitting the modern isotopic composition of precipitation but is obviously not applicable to the determination of paleoelevations. In order to investigate the potential for employing oxygen and or hydrogen isotopes as paleoaltimeters, it is important to have a first-order theoretical understanding of why correlation should exist between the isotopic composition of precipitation and elevation. In the following section the theory behind this stable isotope paleoaltimeter is explored, followed by various tests of this model using modern day data.

ATMOSPHERIC THERMODYNAMICS OF OXYGEN AND HYDROGEN ISOTOPE-BASED ESTIMATES OF ELEVATION FROM OROGRAPHIC PRECIPITATION

Rowley et al. (2001) presented a model that theoretically predicts the expected relationship between stable isotopic composition of the condensed water phase and elevation. The model tracks the moist static energy, water vapor content, and water vapor and condensate isotopic composition along ascending, precipitating trajectories and is summarized here. An empirical fit, that is re-examined and updated from Rowley et al. (2001) here, between the condensed phase isotopic composition and precipitation extends the theory such that the output of the model is the expected systematic behavior of oxygen and hydrogen isotopic composition of precipitation as a function of elevation. In this discussion we employ $\Delta(\delta^{18}O)$ as introduced by Ambach et al. (1968) and $\Delta(\delta^2H)$, the difference in isotopic composition between a low, preferably near sea level composition and a potentially elevated sample as the monitor of elevation recognizing that this difference rather than the absolute isotopic composition is the measure of elevation. It should be stressed that positive values of $\Delta(\delta^{18}O)$ or $\Delta(\delta^2H)$ have no paleoelevation significance, but instead primarily indicate (1) error in the estimate of the mean low altitude isotopic composition used in the normalization, or (2) typical scatter in isotopic data, or (3) significant evaporation of precipitation or surface waters in the unknown site relative to the low elevation normalizing value. Positive values, which if used to estimate paleoelevation using equations described in this text or comparable ones in Currie et al. (2005) or Rowley and Garzione (2007) yield negative, i.e., below sea level, heights that would in turn imply that the precipitation was enriched beyond its initial values by descending in altitude. This is impossible and hence any such values should simply be ignored in terms of implications for stable isotope-based (paleo)altimetry.

THE MODEL

At equilibrium, there is a fractionation of ^{18}O relative to ^{16}O and ^2H relative to ^1H that occurs as water vapor condenses to form condensate (water or ice). The magnitude of fractionation, at least according to the model, is determined by the equilibrium fractionation factor, α, that for oxygen is defined as:

$$\alpha_O = \frac{R_p}{R_v} = \frac{\left(\delta^{18}O_p + 1000\right)}{\left(\delta^{18}O_v + 1000\right)}$$

where R_p is the ratio of $^{18}O/^{16}O$ in condensate and R_v is the ratio of $^{18}O/^{16}O$ in water vapor. The quantities $\delta^{18}O_p$, $\delta^{18}O_v$ are the ratios in condensate and vapor, respectively, relative to a standard, (SMOW) expressed as per mil (‰), such that, for oxygen:

$$\delta^{18}O_p = \left(\frac{R_p}{R_{SMOW}} - 1\right) \cdot 1000$$

Substitution of ^2H and ^1H for ^{18}O and ^{16}O, respectively, in the above results in the identical relations for $^2H/^1H$ fractionation. The fractionation factor, α, is a function of the temperature at which condensation takes place and the phases involved. In the atmosphere, fractionation occurs between water vapor and liquid water or between water vapor and water ice. The temperature dependence of $\alpha(T)$ has been determined experimentally for liquid-vapor equilibrium (Majoube 1971b; Horita and Wesolowski 1994), and for ice-vapor equilibrium (Merlivat and Nief 1967; Majoube 1971a). Existing experimental results are in quite close agreement and we use these relations in the model (Fig. 2).

Simple application of the empirical fits would imply a greater than 3.4‰ difference between water vapor condensing as liquid water and water ice just above and below 0 °C. However, water is well known to cool below its freezing temperature by 20 K or more and thus we adopt a linear mixing model such that there is not an abrupt step in α_O or α_H at 273.15 K. Thus for temperatures between 273.15 K and 253.15 K we mix between ice and liquid water fractionation (Fig. 2).

At any given T the equilibrium isotopic compositions of oxygen and hydrogen in condensate ($\delta^{18}O_p$ or δ^2H_p) and vapor ($\delta^{18}O_v$ or δ^2H_v) are described by the relation:

$$\delta^{18}O_p = \alpha_O(T) \cdot \left(\delta^{18}O_v + 1000\right) - 1000$$

and

$$\delta^2H_p = \alpha_H(T) \cdot \left(\delta^2H_v + 1000\right) - 1000$$

The isotopic composition of the vapor and condensate are simply offset at any given height in the atmosphere by the appropriate equilibrium fractionation factor and hence the derivative of the isotopic composition for the vapor and condensate with respect to elevation are identical as described below by Equation (1). Open system distillation, as modeled by Rayleigh condensation, removes the condensate as it condenses from the vapor leaving the isotopic composition of the residual vapor progressively depleted in ^{18}O and ^2H. We use $\zeta = -\ln(p/p_s)$

as our vertical coordinate, where p is ambient pressure and p_s is the surface pressure, such that ζ represents a scale height in the atmosphere. The distillation process can then be expressed by the differential equation:

$$\frac{dR_v}{d\zeta} = \frac{dR_p}{d\zeta} = R_v\left[a(T)-1\right]\frac{1}{q}\frac{dq}{d\zeta} \tag{1}$$

where R_p and R_v are the isotopic ratios in the incremental condensate and the vapor, respectively, q is the mass mixing ratio of water, and $dq/d\zeta$ is the amount of water condensed from the air parcel in order to maintain saturation as a consequence of adiabatic ascent. From atmospheric thermodynamics it is possible to determine $dq/d\zeta$ with three basic equations. These are:

$$\frac{dz}{d\zeta} = \frac{RT}{g} \tag{2}$$

where z is altitude in meters, and R (without any subscripts) is the gas constant for air ($R = 287$ J kg^{-1} K^{-1}). The change in temperature with height depends on whether or not condensation is occurring, and for rapidly ascending, thermally isolated parcels is described by the relations

$$\frac{dT}{d\zeta} = -\frac{RT + Lq_s}{C_p + Lq_s\left(\ln(e_s)\right)'}, \qquad \text{if moisture is condensing} \tag{3a}$$

$$\frac{dT}{d\zeta} = -\left(\frac{R}{C_p}\right)T \qquad \text{if non-condensing}(q < q_s) \tag{3b}$$

where C_p is the heat capacity of air (1004.0 J kg^{-1} K^{-1}), q_s is the saturation mass mixing ratio of water ($\approx .622\, e_s/p$), e_s is $e_s(T)$, which is the saturation vapor pressure of water as a function of T (Fig. 3), and L is the latent heat contribution due to condensation, which also varies as a function of T. Equation (3b) is the formula for the dry adiabat. Equation (3a) incorporates the change in saturation vapor pressure with temperature through the expression,

$$\left(\ln(e_s)\right)' = \frac{1}{e_s}\frac{de_s}{dT}$$

Finally, the amount of water condensed from the air parcel as ice or liquid in order to maintain saturation is determined through the relation:

$$\frac{1}{q}\frac{dq}{d\zeta} = 1 + \left(\ln(e_s)\right)'\frac{dT}{d\zeta} \qquad \text{if condensing } (q \geq q_s) \tag{4a}$$

$$\frac{1}{q}\frac{dq}{d\zeta} = 0 \qquad \text{if noncondensing}(q < q_s) \tag{4b}$$

If there is no condensation, q is conserved following the air parcel.

The above calculation takes as initial conditions the temperature (T) and relative humidity (RH) that determines the water vapor concentration of the starting air mass. Air starting at the ground with a specific T and RH is lifted along the dry adiabat following (Eqn. 3b) and (Eqn. 4b) with the vapor fraction equal 1.0 until condensation starts at the cloud condensation level when $q = q_s$ (Fig. 4). Condensation then occurs at all levels above the cloud condensation level as described by (Eqn. 3a) and (Eqn. 4a) resulting in progressive decrease in the remaining vapor fraction as a function of adiabatic ascent. Latent heat release associated with condensation changes the temperature lapse rate to a moist adiabat. It is this temperature and the associated phase(s) that controls the equilibrium fractionation between the remaining vapor and condensate as represented by the corresponding α's (Fig. 4). This thermodynamically

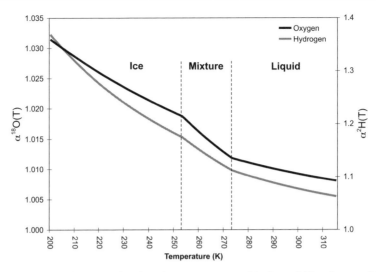

Figure 2. Fractionation factors as a function of temperature as used in the model based on empirical fits determined by (Majoube 1971b; Horita and Wesolowski 1994) and (Merlivat and Nief 1967; Majoube 1971a) See Rowley et al. (2001), Rowley and Garzione (2007) for original publications for the equations graphed in Figure 2.

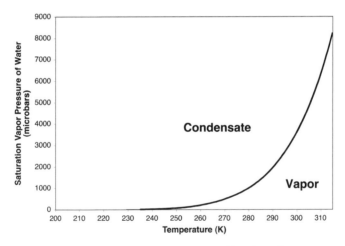

Figure 3. Saturation vapor pressure curved based on a fit to the Smithsonian Tables as a function of temperature.

determined adiabatic lapse rate depends solely on the starting T and RH. Therefore, different starting air mass conditions will yield different rates of condensation with elevation and hence $\delta^{18}O$ (δD) vs. altitude relationships. The decreasing water vapor fraction, and hence decreasing ratio of initial to remaining water vapor, with height results in a decrease in the latent heat contribution that together drives the Rayleigh distillation process resulting in the progressive isotopic depletion of the remaining reservoir from which subsequent condensation occurs.

The global mean isotopic lapse rate of precipitation in low latitudes depends sensitively on the initial sea level T and RH frequency distribution (Rowley et al. 2001). Coupled monthly mean T and RH data derived from NCEP reanalysis output (Kalnay et al. 1996) of 40 years

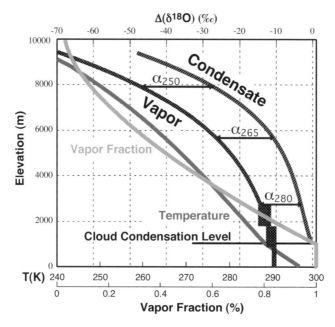

Figure 4. Graph of temperature, vapor fraction, and isotopic composition of water vapor and condensate as a function of elevation for an air mass with a starting T and RH of 295 K and 80 % (modified after Rowley and Garzione 2007). Note that the vapor and condensate are the isotopic composition reflected in the scale along the top of the graph with $\Delta(\delta^{18}O)$ normalized relative to the initial isotopic composition of the condensate. The isotopic composition of modeled condensate is not the same as precipitation as is discussed below.

data were extracted from the monthly mean T and RH data from all low latitude ($\leq 35°$N and S), entirely oceanic, $2° \times 2°$ grid cells in the reanalysis product. A total of 614 unique pairs of T and RH values, corresponding to 33,168 instances that are taken to represent the likelihood of starting air mass conditions for the model. These define the probability density function of these parameters (see Fig. 2 of Rowley et al. 2001) used to model global mean isotopic lapse rate of precipitation. Rowley et al. (2001) and Rowley and Garzione (2007) used Monte Carlo simulation of 1,000 to 5,000 random pairs to estimate the probability density functions of modeled vertical profiles of temperature and vapor fraction and $\Delta(\delta^{18}Op)$ with respect to elevation. That analysis is updated by using all 33,168 instances to estimate the probability density functions of modeled vertical profiles of temperature and vapor fraction (Fig. 5).

Rowley et al. (2001) derived an empirically-based scheme to convert the isotopic composition of condensate that the model calculates to the isotopic composition of precipitation, which is what is observed. The scheme devised by Rowley et al. (2001) was based on GNIP weighted mean annual isotopic composition of precipitation as a function of elevation in the Alps. The scheme weights the isotopic composition of the condensate as a function of elevation by the condensation amount as a function of elevation within a 1,000 m thick parcel of air between 1,000 and 2,000 m above the ground surface. The precipitation thus calculated represents the condensation amount weighted mean isotopic composition of the modeled condensate extracted from within this kilometer thick parcel of air. This scheme is re-investigated here using data from tropical regions, collected by either Gonfiantini et al. (2001) or released by IAEA in 2004 for the period up to 2001 (IAEA-Monthly 2004; IAEA-Yearly 2004). The observed isotopic compositions are normalized to $\Delta(\delta^{18}Op)$ by subtracting

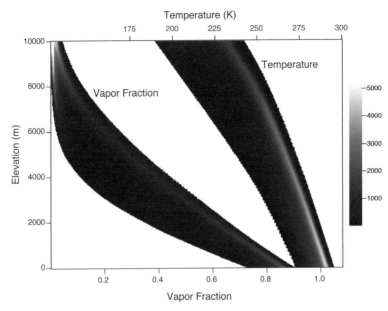

Figure 5. Frequency distribution of modeled condensation weighted mean vapor fraction and temperature relative to elevation. Elevation represents the surface elevation with the frequency representing the modeled parcel 2,000 ± 500 m above the surface from which precipitation falling to the ground is derived. Gray scale color scaled to the number of times a given value of temperature and vapor fraction as a function of elevation is modeled.

an estimate of the low altitude starting weighted mean annual isotopic composition from each of the remaining weighted mean annual isotopic values as listed in Table 1. The starting surface T of 299 K and RH of 80% is appropriate for each of the regions so no correction needs to be applied for the analysis. The model is run for this starting T and RH and the spectrum of curves for various sampling elevations is computed (Fig. 5, inset). The best fit sampling elevation is determined by minimizing the root mean squared difference between the model-derived predicted isotopic composition and the observed $\Delta(\delta^{18}O_p)$ at each of the station elevations listed in Table 1. Based on these data the best-fit sampling elevation for a 1000m thick parcel is located between 1500 m and 2500 m higher than the land surface, that is with a mean elevation of 2000 m higher than the land surface (Fig. 6).

As is clear from the inset figure (Fig. 6) there is little difference at low elevations and so the main controllers of the best-fit solution are the higher altitude stations. This solution is preferable to that used by Rowley et al. (2001) in that the highest sampling station in the Alpine dataset was Grimsel at just over 2000 m, hence not able to tightly constrain the curve, as well as the fact that the sites used here are well within the ±40° latitude band where the model is most appropriately applied.

In the previous analyses Rowley et al. (2001), and derivative papers including Rowley and Currie (2006) and Rowley and Garzione (2007) quoted model results for the weighted mean $\Delta(\delta^{18}O_p)$ as a function of elevation. It turns out that the relationship between $\Delta(\delta^{18}O_p)$ and elevation calculated in Rowley et al. (2001) was the average of $\Delta(\delta^{18}O_p)$ at each elevation, as opposed to the average elevation at each $\Delta(\delta^{18}O_p)$. The confidence intervals, however, were correctly computed in terms of frequency distribution of elevation as a function of $\Delta(\delta^{18}O_p)$. The result of this difference in computation of the median and weighted mean was a significant difference in the predicted elevation based on the weighted mean and the median (see Rowley

Table 1. Weighted mean $\Delta(\delta^{18}O_p)$ versus elevation used to calibrate the sampling elevation of the calculated isotopic composition of condensation versus elevation.

Station	Station Altitude	Weighted Mean $\Delta(\delta^{18}O_p)$	Location	Ref.
Bakingele	10	0.2	Mt. Cameroon	1
Debundscha	20	0.1	Mt. Cameroon	1
Idenau	30	−0.4	Mt. Cameroon	1
Batoki	50	0.1	Mt. Cameroon	1
Brasseries	180	−0.4	Mt. Cameroon	1
Bomana	460	−0.6	Mt. Cameroon	1
Bonakanda	860	−1.4	Mt. Cameroon	1
Foret SW	1000	−1.0	Mt. Cameroon	1
Upper Farm	1100	−1.7	Mt. Cameroon	1
Route VHF	1610	−2.4	Mt. Cameroon	1
Limite Foret SW	2320	−3.9	Mt. Cameroon	1
Station VHF	2460	−3.7	Mt. Cameroon	1
Versant Nord	2475	−3.2	Mt. Cameroon	1
Versant Nord	2500	−3.8	Mt. Cameroon	1
Limite BUEA	2500	−3.4	Mt. Cameroon	1
Hutte 2	2925	−4.4	Mt. Cameroon	1
Versant SW	3000	−4.7	Mt. Cameroon	1
Nord	3050	−4.8	Mt. Cameroon	1
Southwest	3300	−5.4	Mt. Cameroon	1
Sommet Bottle Peak	4050	−6.3	Mt. Cameroon	1
Trinidad	200	0.0	Bolivian Andes	1
Rurrenabaque	300	−0.8	Bolivian Andes	1
Coroico	1700	−3.4	Bolivian Andes	1
Chacaltaya	5200	−10.3	Bolivian Andes	1
El Alto	4080	−9.9	Bolivian Andes	1
Izobamba-Sao Gabriel	3058	−7.2	GNIP Brazil-Peru	2
Bogota-Sao Gabriel	2547	−5.0	GNIP Brazil-Colombia	2

(1) Gonfiantini et al. (2001), (2) (IAEA-Yearly 2004). Mount Cameroon normalized relative to the precipitation amount weighted average isotopic composition of −3.2‰ of the stations below 100 m elevation. Bolivian Andes data normalized to precipitation amount weighted average isotopic composition of −5.2‰ at Trinidad. Izobamba and Bogota are normalized to weighted mean isotopic composition measured at Sao Gabriel of −4.4‰.

et al. 2001-Fig. 8). Figure 7 shows the revised model results in which both the weighted mean and the confidence intervals are computed as averages of elevation at each $\Delta(\delta^{18}O_p)$, and with the revision of the sampling height of the condensate to 2,000±500 m. Included with this figure is the relationship between $\Delta(\delta^{18}O_p)$ and elevation from Rowley et al. (2001) computed with the polynomial fit reported by Currie et al. (2005). These two revisions effectively cancel each other such that the best fit curve with the current scheme is nearly identical with that of Rowley et al. (2001) and Currie et al. (2005). Increasing the mean sampling height effectively systematically lowers the median and ±1σ and ±2σ distributions to correspond with the revised weighted mean curve.

Given that this revised calibration scheme together with the correction of the weighted mean calculation are nearly identical (< ±150 m difference) to the previous model results all of the comparisons of the modern isotopic compositions of precipitation from other regions are effectively unchanged and hence these comparisons demonstrates that the model yields quite reasonable fits without adjustment (see Rowley et al. 2001 and Rowley and Garzione 2007). However it should always be made clear that the empirical scheme to model precipitation from condensate does not represent the microphysics of water droplet formation, coalescence,

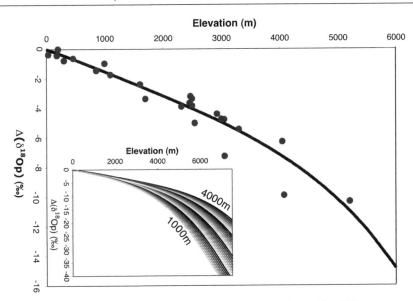

Figure 6. Empirical fit of isotopic composition of condensate sampled from a 1 km thick parcel at a mean elevation 2,000 m above the ground surface. Dots represent data from Table 1. Inset-family of curves representing a range of mean sampling elevations from 1000 to 4000 m above the surface. The best fit curve is at 2000 m elevation above the surface.

turbulent mixing, and eventual fall or re-evaporation within clouds, but rather to simply derive a means of providing an empirical match between condensate and precipitate in orographic settings. Further, many regions, particularly those where significant evaporation occurs during precipitation descent from the level of condensation to the ground will not be fit by this relationship. In the modern world this is readily tested using deviations of precipitation from the global meteoric water line (GMWL) (Craig 1961; Dansgaard 1964) and particular care needs to be taken in applying this or any Rayleigh distillation-based approach in such regions. Evaporation enriches the precipitation in ^{18}O and to a lesser degree ^{2}H making it appear to have condensed at lower elevations. For example, the GNIP station at Addis Abba at 2360 m has an amount weighted mean annual $\Delta(\delta^{18}O_p)$ of about -1.0 ‰, that is more enriched than the mean $\Delta(\delta^{18}O_p)$ of -3.6 ± 1.6‰ for low elevation (≤ 100 m) and low latitude ($\leq \pm35°$) stations in the GNIP database (IAEA-Yearly 2004). Obviously a positive $\Delta(\delta^{18}O_p)$ has no meaning in terms of elevation but can be an indication of evaporative enrichment of ^{18}O.

Polynomial regression of the relationship of $\Delta(\delta^{18}O_p)$ versus elevation (z) measured in meters derived from modeling all possible modern starting T and RH pairs for values of $\Delta(\delta^{18}O_p)$ between 0‰ and -25‰ results in a curve that is only slightly different from the equation reported by Currie et al. (2005) and used by Rowley and Currie (2006), as well as that of Rowley and Garzione (2007). The revision is Equation (5):

$$z_{weighted\ mean} = -0.0129\ \Delta(\delta^{18}O_p)^4 - 1.121\ \Delta(\delta^{18}O_p)^3 \tag{5}$$
$$- 38.214\ \Delta(\delta^{18}O_p)^2 - 715.22\ \Delta(\delta^{18}O_p)$$

The difference in elevation computed using the equation in Currie et al (2005) and Equation (5) ranges from about $+150$ to -84 m in the range of the fit. Polynomial regression constrained to pass through the origin of the median of the distribution yields the relationship

$$z_{median} = -0.0168\Delta(\delta^{18}O_p)^4 - 1.368\ \Delta(\delta^{18}O_p)^3 - 43.75\ \Delta(\delta^{18}O_p)^2 - 771.11\Delta(\delta^{18}O_p)$$

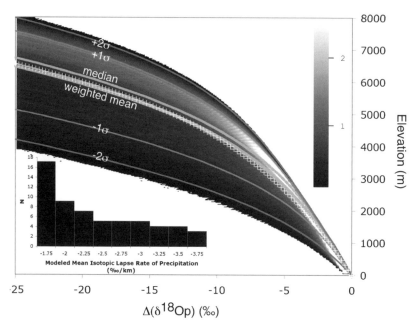

Figure 7. Frequency distribution of $\Delta(\delta^{18}O_p)$ versus elevation derived by weighting all possible pairs of starting T and RH and corresponding computed vertical profiles of T and vapor fraction shown in Figure 6 by their frequency of occurrence in the Rowley et al (2001) probability density function of T and RH. Gray curves highlight the weighted mean (heavy), median, and $\pm1\sigma$ and $\pm2\sigma$ deviations resulting from the probability density distribution of corresponding starting T and RH. White pluses are the Rowley et al. (2001) and Currie et al. (2005) values of the $\Delta(\delta^{18}O_p)$ versus weighted mean elevation. Gray scale shading in percent, saturating to white above 3%. Note that warmer starting temperatures plot in the $+\sigma$ direction whereas colder starting temperatures plot in the $-\sigma$ direction. Any change in mean global climate, and particularly in low latitude mean and variation of sea level temperature will affect the mean lapse rate of isotopic composition in a similar direction. Inset histogram - modeled isotopic lapse rates of $\Delta(\delta^{18}O_p)$ derived from the model for elevations between 1500 m and 6500 m.

It has long been understood that starting temperature has a much larger effect on the isotopic lapse rate than does relative humidity, hence the wide utilization of isotopic composition of snow and ice to derive estimates of temperature variation (Dansgaard 1964). Given that the lapse rate of isotopic compositions is, according to the model, primarily dependent upon starting temperature and secondarily relative humidity (see Rowley and Garzione 2007), and given that the mean starting temperature and relative humidity of air masses in the past are not known, a reasonable estimate of elevation uncertainty is provided by the frequency distribution of $\Delta(\delta^{18}O_p)$ versus elevation values (Fig. 7). Accordingly estimates of the elevation uncertainty relative to the weighted mean elevation are shown in Figure 8 and given by the following fits to the model frequency distribution:

$$z_{+2\sigma\ uncertainty} = 0.0228\Delta(\delta^{18}O_p)^4 + 1.132\Delta(\delta^{18}O_p)^3 + 14.276\Delta(\delta^{18}O_p)^2 - 57.547\Delta(\delta^{18}O_p)$$

$$z_{+1\sigma\ uncertainty} = 0.0150\Delta(\delta^{18}O_p)^4 + 0.738\Delta(\delta^{18}O_p)^3 + 9.031\Delta(\delta^{18}O_p)^2 - 47.186\Delta(\delta^{18}O_p)$$

$$z_{-1\sigma\ uncertainty} = -0.0126\Delta(\delta^{18}O_p)^4 - 0.580\Delta(\delta^{18}O_p)^3 - 5.262\Delta(\delta^{18}O_p)^2 + 89.212\Delta(\delta^{18}O_p)$$

$$z_{-2\sigma\ uncertainty} = -0.0023\Delta(\delta^{18}O_p)^4 + 0.107\Delta(\delta^{18}O_p)^3 + 11.611\Delta(\delta^{18}O_p)^2 + 280.09\Delta(\delta^{18}O_p)$$

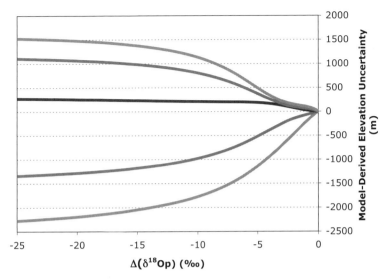

Figure 8. Model-derived elevation uncertainty relative to the weighted mean elevation as a function of $\Delta(\delta^{18}O_p)$ resulting from uncertainty in the starting T and RH of orographically forced ascent of airmasses. Curves are $\pm 1\sigma$ (light gray), $\pm 2\sigma$ (medium gray). The difference between the median (dark gray) and weighted mean is also shown.

Why are these equations represented by 4th order polynomials and not 2nd order curves given that the vertical variation of temperature and vapor fraction are well approximated by second order functions? The simple answer is that the transition from condensing water vapor to liquid water above 0 °C to condensing water ice below −20 °C, and the attendant affect on the fractionation factor (Fig. 2), results in additional structure not captured by 2nd or 3rd order curves. Each of the equations fit their respective model output with an $R^2 > 0.9997$. The lack of symmetry of the modeled uncertainty reflects asymmetry in the probability density function and particularly the long tail toward lower values of T relative to the mean (see Fig. 2 of Rowley et al. 2001). The effect of this long tail is well displayed in both Figure 5 and 7.

The thermodynamic model that determines the theoretical $\delta^{18}O$ (δ^2H) vs. altitude relationship is mathematically one-dimensional, in that the equations need only be integrated with respect to ζ. The vertical trajectories themselves can wander horizontally in an arbitrarily complex way as the parcel ascends. The chief physical assumption is that the air parcel remains relatively isolated from the surrounding air. Although turbulence, among other processes, no doubt contribute to isotopic lapse rates of precipitation in real world orographic settings, the fit of observed isotopic lapse rates with model predictions implies that the model captures the main features determining the relationship between elevation and isotopic composition in many low latitude settings (Rowley et al. 2001; Rowley and Garzione 2007).

Climate change, and particularly changes in mean sea level temperatures have the potential to significantly affect isotopic lapse rates in the past. Figure 9 compares results using the modern low latitude distribution of climate parameters with those derived from GCM model output of Eocene (Huber and Caballero 2003) conditions to demonstrate this point. On average most estimates of past climate suggest that the Present is more toward the colder end of the climate spectrum and thus most times in the past, such as the Eocene would have been warmer with potentially slightly higher sea surface relative humidity, resulting in lower isotopic lapse rates in the past. This illustrates an important aspect of having a model rather than simply employing an empirical calibration of the isotopic lapse rate (see for example Chamberlain

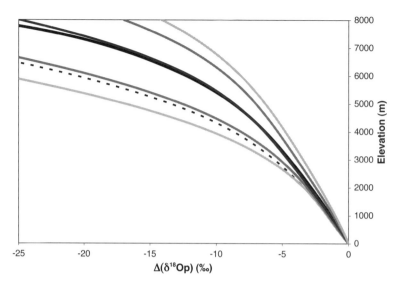

Figure 9. Comparison of modeled Eocene (heavy black curve) and modern (dashed black curve) weighted mean $\Delta(\delta^{18}O_p)$ versus elevation relationships. The warmer (T_{mean}=300 K ± 3 K (1σ), RH $_{mean}$= 80% ± 3% (1σ)) climate of the Eocene modeled by Huber and Caballero (2003) results in a lower isotopic lapse rate predicted for this time. The median (dark gray), and ±1σ (medium gray), and ±2σ curves (light gray) appropriate for the Eocene low latitude climate are also shown. For comparison the modern +2σ curve (Fig. 7) and the Eocene median curve are essentially identical. Comparable predicted isotopic lapse rates can be calculated for any given time for which GCM-derived climate parameters are available. Note the isotopic lapse rate is ‰/m and thus is the inverse of this graph.

and Poage 2000; Garzione et al. 2000a,b, 2006; Poage and Chamberlain 2001). There is no quantitative way to derive the magnitude of the effect of climate change on empirical isotopic lapse rates. One of the nice aspects of having a model is that it is possible to recalculate the relationship between $\Delta(\delta^{18}O_p)$ and elevation, if appropriate GCM derived model results are available. Since the required data are sea level T and RH over oceanic regions rather than continental interiors these data are probably less affected by details of elevation reconstructions within continental interiors and hence less likely to circularly affect the model results for 1D predictions of the isotopic lapse rate.

Figure 9 very nicely demonstrates the effects of global climate change on the isotopic lapse rate. Warmer climates yield shallower isotopic lapse rates while colder, drier climates would be expected to be associated with steeper isotopic lapse rates. On average much of the geologic past is generally assumed to have been warmer and hence application of the mean modern lapse rate will underestimate paleoelevations in these cases.

MODEL VERSUS EMPIRICAL FITTING OF DATA

The model suggests that there should be considerable sensitivity of the isotopic lapse rate to temperature of low elevation air mass from which precipitation is derived. Rowley and Garzione (2007) illustrated this by comparing profiles of $\Delta(\delta^{18}O_p)$ versus elevation for Mount Cameroon based on the data reported by Gonfiantini et al. (2001). The 4 K change in T between the mean of the probability density function of T and RH at 295 K and locally appropriate value of 299 K results in a significant improvement in the correlation between data and model predictions (See Fig. 5 of Rowley and Garzione 2007). Empirical calibration of the isotopic lapse rate using temporally limited sampling will almost invariably underestimate the potential range of variability in isotopic composition as a function of elevation, reflecting the limited range of climate variability represented by those samples. Temporally limited sampling can refer to a single suite of samples collected within a month or so from relatively short streams (e.g., Garzione et al. 2000a,b), or precipitation data collected over less than 28 months (Gonfiantini et al. 2001) of which only one year of the data were used by Garzione et al. (2006), to compilations of river data collected over a range of time scales by different sources as in Poage and Chamberlain (2001). Even the GNIP data (IAEA-yearly 2004) with up to several decades of measurements are potentially temporally limited relative to climatological means. Thus, although a good fit between isotopic composition and elevation may be apparent in each of these data sets, it is not clear how well any of these data sets captures realistic estimates of their respective climatological means. This comment extends to estimates of elevation uncertainty that are typically based on bootstrapping of deviations of observations about the best-fit regression relationship (see Rowley and Garzione 2007). The model, in contrast, represents an ensemble mean that captures the sort of variability expected in low latitude situations. The climatic conditions represented in the model can easily be either perturbed by specifying a constant offset in ΔT and or ΔRH, or use a new probability distribution function of T and RH, based, for example, on GCM output to recalibrate the $\Delta(\delta^{18}O_p)$-elevation relationship (Fig. 9). This cannot be done rigorously with empirical calibrations.

To make this point clearer, the data from Gonfiantini et al. (2001) provides an instructive data set for thinking about issues related to limited temporal sampling and empirical fitting. Gonfiantini et al. (2001) report station elevation and $\delta^{18}O_p$ on a monthly basis for some number of months between December 1982 to April 1986. They also report temperature and precipitation amount for some subset of these months and stations, along a transect from the Bolvian foreland to the Altiplano, from Trinadad to El Alto (see their Table 6). This data set essentially provides six different, although not independent, measures of the relationship between elevation and isotopic composition (Fig. 10). These are unweighted and precipitation amount weighted mean $\delta^{18}O_p$ that are combined into a 1983, 1984, and December 1982 through part of April 1986 average data sets. The 1984 unweighted means, listed incorrectly in Gonfiantini et al. (2001) Table 5 as weighted means, were regressed by Garzione et al. (2006) as corrected in Garzione et al. (2007) in order to constrain their estimate of the paleoelevation history of this part of the Bolivian Altiplano as shown by the dashed curve in Figure 10. Several things are immediately apparent. There are significant differences depending on how the data are combined. The 1984 isotopic compositions, representing up to 11 months of data are significantly more depleted than either 1983, based on between 9 and 12 months of data, or the combined 1982 to 1986 average isotopic compositions, based on between 15 and 27 months of data. Amount weighted means are, on average, more depleted than unweighted means by as much as several per mil. On the basis of these data there is no basis for judging whether any of these data represent the climatological mean relationship. Garzione et al. (2007) prefer regressing the weighted mean of all existing data (bold line in Fig. 10) because it more closely matches isotopic compositions of tributaries collected in 2004 and 2005, but additional data may suggest some other relationship within or beyond the existing data is more representative. Thus my preference for comparing observation against the model, and to use the uncertainties in elevation represented by climate variation embedded in the model to estimate paleoaltitude uncertainties in the past.

One final point regarding Figure 10 is that each of the data sets internally yield comparable isotopic lapse rates, i.e., they are characterized by similar slopes, but significantly different 0 m $\delta^{18}O_p$ intercept values. This emphasizes the important point that it is not the absolute value of the $\delta^{18}O_p$ that correlates with elevation, but rather the difference relative to low elevation starting isotopic composition, and hence the importance of normalizing in terms of $\Delta(\delta^{18}O_p)$, rather than modeling in terms of measured $\delta^{18}O_p$.

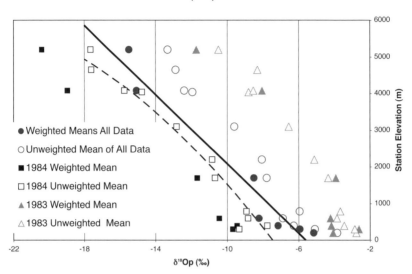

Figure 10. $\delta^{18}O_p$ versus elevation for various groupings of the data presented by Gondiantini et al. (2001) based on data in their Table 6. Black dashed curve is the polynomial regression curve derived from the unweighted mean isotopic composition as a function of elevation used by Garzione et al. (2006) as corrected by Garzione et al. (2007) and the bold line is the linear regression relationship preferred by Garzione et al. (2007).

How does the model compare with various previous empirically derived estimates of the relationship between isotopic composition and elevation, either globally (Poage and Chamberlain 2001), or locally in the Himalayas (Garzione et al. 2000a,b), or in the eastern Andes (Garzione et al. 2006, 2007) (Fig. 11). Garzione et al. (2000b) already compared their fit with an early version of the model, later published as Rowley et al. (2001), and noted the similarity. It is clear from the comparison presented in Figure 11 that there is first-order agreement among these different approaches. The necessary underlying assumption for the empirical fits is that the existing samples represent the long-term climatological means, but as demonstrated above there is no basis for judging how true this is and hence bootstrapping of deviations from best fit relations also likely significantly underestimates uncertainties. The model provides an independent basis for estimating the likely mean and variations in this relationship, and hence provides a more robust estimate of inherent uncertainty than estimated by bootstrapping of a given data set alone.

DATA-MODEL COMPARISONS

Rowley et al. (2001), Currie et al. (2005), Rowley and Currie (2006), and Rowley and Garzione (2007) have reviewed various comparisons of modern day observed isotopic compositions with those predicted by application of this model. For the most part these comparisons have been quite favorable with a fairly close (± ~500 m) match between known and predicted elevations. In the next section additional comparisons are made between various model predictions relative to observed data from the modern world where there is essentially no uncertainty in either the measured isotopic composition or the elevations of various samples.

The discussion below specifically focuses on application of the model to low latitude (<35°N or S) examples, in order to further validate the global model. There are several reasons for emphasizing this latitude range and not higher latitudes. First, tropical latitudes are more

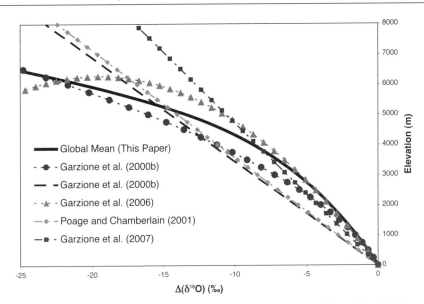

Figure 11. Comparison among a variety of empirically derived estimates of the relationship between $\Delta(\delta^{18}O)$ and elevation. Note that Poage and Chamberlain (2001) and Garzone et al. (2000b) are fits to rivers, not precipitation, hence it is not surprising that in the range of $\Delta(\delta^{18}O)$ values that constrain their estimates that the slopes are less than that predicted for precipitation. The mean climate in the Bolivian Andes transect is about 4 °C warmer than the global mean of the T and RH probability density function, giving rise to the steeper slope of the Garzione et al. (2006) and Garzione et al. (2007) curve than the global mean curve, as shown by Rowley and Garzione (2007).

likely to derive moisture from within these latitudes and hence are well represented by the T and RH probability density distribution used in the model to estimate isotopic lapse rates. Second, strong latitudinal temperature gradients in oceanic source areas at latitudes poleward of about 40° result in significant increase in T variability and hence in variability of estimated isotopic lapse rates at higher latitudes (Rowley and Garzione 2007). Third, mid-latitude and temperate systems are typically characterized by complex frontal systems and air mass mixing with potentially multiple independent moisture sources each with different isotopic, T and RH characteristics, resulting in much more complex initial starting conditions than is captured in our simple one-dimensional model. We thus restrict our discussion to regions within ±35° of the equator. Even within this latitude range, complex vapor trajectories exist giving rise to complex patterns in the isotopic compositions (Friedman et al. 2002a,b).

It is critical to assess the fit of the model to modern systems for which the first-order isotopic composition as a function of elevation has been determined empirically, before any confidence is warranted in the application of this model or for that matter other empirical approaches predicated on application of the same thermodynamic rationale (Garzione et al. 2000a,b) for paleoaltimetry studies. The most robust statistic of the isotopic composition at any given station is the precipitation amount weighted annual mean. This reflects the fact that there is considerable variability in isotopic composition as a function of hour, day, month or season at most stations and hence the comparison that needs to be made is preferably with a multi-decadal record or at least with multi-year records wherever possible. Similarly, surface waters generally integrate the isotopic composition over the mean residence time within the drainage basin and would also be expected to reflect the amounts of precipitation as a function of time rather than just its average isotopic composition. In the following discussion attention will be focused entirely on amount weighted isotopic compositions, unless data on this quantity are not available. Rowley and Garzione (2007) presented comparisons for precipitation collected from altitude transects up Mount Cameroon and Bolivian Andes from Gonfiantini et al. (2001), and Himalaya-Southern Tibet. In this discussion additional data from the Himalayan region will be reviewed. The primary focus will be Himalayan rivers.

However, in order to bring some closure in regards the above discussion regarding the Bolivian Andes data (Gonfiantini et al. 2001) a comparison is presented of these various data sets in relation to the model (Fig. 12). Each of the data sets, the weighted mean isotopic compositions for 1983, 1984, and the average for the interval from 1982 to 1986 as derived from Table 6, rather than Table 5, of Gonfiantini et al. (2001) are normalized relative to the 0 m intercept of linear regressions of each subset of station elevation against $\delta^{18}O_p$, so that they are plotted as $\Delta(\delta^{18}O_p)$. Also included are the isotopic compositions of small tributaries with samples from 2004 and 2005 as a function of elevation from Garzione et al. (2007). The small tributaries are normalized relative to the 0 m intercept of a linear regression of sample elevation against $\delta^{18}O_{mw}$. Figure 12 demonstrates that the model captures the first-order relationships reflected in these various data sets, tends to underestimate the elevations of the highest stream samples, as expected given the 4 °C difference between the global climatological mean and the best estimate of the local climatological mean. In addition, it would appear that 1983 was indeed an anomalous year as suggested by Gonfiantini et al. (2001).

SURFACE WATERS

All archives that might be used for paleoaltimetric purposes derive their isotopic signatures either from surface or ground waters rather than precipitation directly. The isotopic composition of surface waters in particular, and ground waters to a lesser degree can differ significantly from

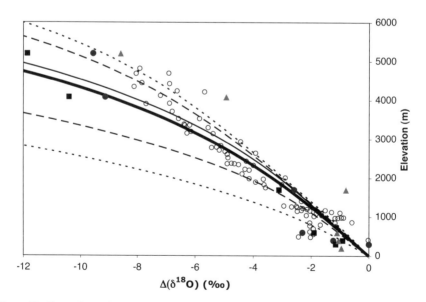

Figure 12. Comparison of normalized isotopic composition of weighted means versus elevation of Bolivian stations from Gonfiantini et al. (2001) for 1983 (triangles), 1984 (squares), and 1982 to 1986 averages (filled circles), and normalized $\delta^{18}O_{sw}$ from small tributaries from Garzione et al. (2007) plotted at sample elevations. Model curves are weighted mean (bold), median (fine), ± 1σ (coarse dashed), and ±2σ (fine dashed). Note that detailed location data needed to compute various hypsometric weighted means as discussed below are not yet published for these surface waters.

that of precipitation. The most important difference between surface or ground waters and precipitation is that rivers and streams integrate precipitation in the drainage basins above the point that a sample is taken (Ramesh and Sarin 1995). Surface waters thus integrate (1) variation of isotopic composition with elevation, (2) area as a function of elevation (i.e., hypsometry), as well as (3) variation in precipitation amount as a function of elevation, in addition to seasonal variations. Hypsometry in drainage systems is not a simple linear function of elevation and hence cannot be represented by the average of the maximum and sample elevation (Fig. 10). Rather the hypsometry of each drainage system needs to be computed individually. One consequence of this hypsometric effect is that isotopic compositions along rivers and streams should not be expected to vary in a simple linear fashion with elevation. This is nicely demonstrated by the individual profiles in the compilation of (Poage and Chamberlain 2001), even though those authors limited their treatment of these data to determining the best fit linear slope.

Precipitation amount also varies as a function of elevation, sometimes with strong gradients in orographic systems, particularly at relatively low elevations (< 3-4 km) (Burbank et al. 2003; Putkonen 2004). Anders et al. (2006) and Roe (2005) have provided a large-scale mapping of precipitation rate as a function of orography in the Himalaya and southernmost Tibet using Tropical Rainfall Measurement Mission (TRMM) satellite data. They model the precipitation as a function of elevation as a combination of the change in saturation vapor pressure as a function of temperature (and hence elevation) and surface slope (Roe et al. 2002; Anders et al. 2006). Rowley and Garzione (2007) show that both the weighted mean annual precipitation and median annual precipitation amount decreases linearly with increasing elevation above about 1,000 m. Regression of the TRMM data from Anders et al. (2006) results in Equation (6) describing the relationship between mean annual precipitation amount (P_z in mm/yr) and elevation (z) in meters up to 4,600 m:

$$P_z = -0.172 \pm 0.006z + 869.7 \pm 22.6 \text{ with an } R^2 = 0.9684 \qquad (6)$$

as shown by Rowley and Garzione (2007). Above about 4600 m, MAP is approximately constant at about 74 mm/yr. Anders et al. (2006) demonstrate that this relationship accords with a dominant correlation to the rate of condensation as a function of temperature largely controlled by the shape of the water saturation vapor pressure curve (Roe et al. 2002; Anders et al. 2006).

Data from an array of drainages within the Himalayas and associated Indo-Gangetic plain are analyzed in order to gain some appreciation of the relationships among oxygen isotopic composition, basin hypsometry, and precipitation amount all of which vary as a function of elevation. Locations and sampling elevations of isotopic compositions of rivers and streams within this region have been reported by a number of studies (Ramesh and Sarin, 1995; Gajurel et al. 2006, among others) that provide a basis for examination of these relationships (Fig. 13). For each sampling location, GIS-based hydrologic tools are used to compute the drainage basin area, maximum elevation, and hypsometry (i.e., area as a function of elevation) above the reported sample elevation based on Asian Hydro1K digital elevation data (Hydro1k 2005), a USGS product, produced by the EROS Data Center. The hypsometric mean elevation (z_{hm}) of the drainage basin is simply the area weighted mean elevation as a function (z) of elevation as given by Equation (7),

$$z_{hm} = \frac{\sum A_z z}{\sum A_z} \qquad (7)$$

with the summations extending from the sample elevation (z_{min}) to the maximum elevation (z_{max}) in the drainage basin. The precipitation weighted mean elevation (Eqn. 8) is computed by weighting the area as a function of elevation (A_z) within the drainage basin by the precipitation

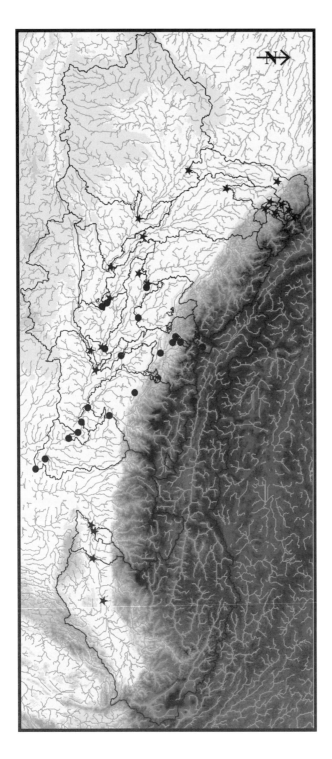

Figure 13. Selection of oxygen isotopic sample localities from the Himalayas and Indo-Gangetic Plain together with outlines of selected associated drainage basins. Stars are primarily from Ranesh and Sarin (1995), filled circles from Gajurel et al. (2006). Rivers and streams and underlying topography derived from Hydro1k digital elevation dataset (Hydro1k 2005). North is to the right.

amount as a function of elevation (P_z), again with the summations extending from z_{min} to z_{max} in the drainage basin.

$$z_{pwm} = \frac{\sum P_z A_z z}{\sum P_z A_z} \tag{8}$$

Figure 14 plots these for a selection of rivers draining the Himalaya and Indo-Gangetic Plain. The important point to emphasize is that the hypsometric mean elevation of each of the drainages plots to the left of the line of 1:1 correlation. The drainage basin area weighted mean difference between z_{pwm} and z_{hm} exceeds 800 m. Further, for low elevation sampling locations there is typically a marked difference between both the hypsometric mean and precipitation weighted mean elevations and the average (i.e., (maximum + sample)/2) elevation of the drainage basin. This difference decreases as basin area decreases and sampling elevation within the larger drainages increases.

The isotopic composition of surface water, $\Delta(\delta^{18}O_{sw})$, at any given height within a drainage basin should reflect the hypsometry of the drainage basin above the sampling site (A_z) integrated with the amount of precipitation falling as a function of elevation (P_z) on that hypsometry and the progressive decrease in isotopic composition of the precipitation $\Delta(\delta^{18}O_p)_z$ with elevation as given by Equation (9),

$$\Delta(\delta^{18}O_{sw}) = \frac{\sum P_z A_z \Delta(\delta^{18}O_p)_z}{\sum P_z A_z} \tag{9}$$

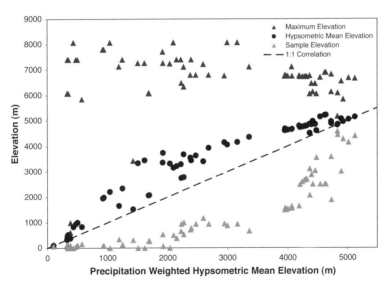

Figure 14. Comparison of hypsometric mean and precipitation amount weighted hypsometric mean elevation for a selection of drainage basins with areas greater than 1,000 km² and up to 610,891 km² in the Himalaya and southern Tibetan region. Also shown are the sample elevation and maximum elevation within that drainage basin. The sample, hypsometric mean, and maximum elevation of each of the drainages is plotted at the computed precipitation amount weighted hypsometric mean elevation of that drainage basin. The straight dashed line shows the line of 1:1 correlation of the hypsometric mean and precipitation weighted hypsometric mean elevation. The horizontal offset of dots from this line measures the mismatch between these two measures of basin hysypsometry. The mismatch between the precipitation weighted hypsometric mean and the average of the maximum and sample elevation would be even larger (not plotted), particularly for the large rivers sampled in Indo-Gangetic plain that plot on the left side of the graph.

where

$$\Delta(\delta^{18}O_p)_z = -7.293\times10^{-15}z^4 - 6.906\times10^{-12}z^3 - 5.517\times10^{-9}z^2 - 1.577\times10^{-3}z$$

is the polynomial fit of the weighted mean $\Delta(\delta^{18}O_p)_z$ with elevation. An alternative way of looking at this is that the isotopic composition of a river or stream sample should record the z_{pwm} of the drainage basin. For our purposes we compare surface data with various measures of the hypsometry. For the Himalayas, surface water isotopic compositions are compared with precipitation amount modeled with Equation (6). In other areas where a mapping of precipitation amount as a function of elevation is lacking this can be replaced with the condensation weighted hypsometric mean elevation z_{cwm} given by

$$z_{cwm} = \frac{\sum C_z A_z z}{\sum C_z A_z} \tag{10}$$

where the condensation rate C_z as a function of elevation based on the model can reasonably be approximated with:

$$C_z = 1.699\times10^{-8}z + 1.599 \tag{10a}$$

Comparison of z_{pwm} and z_{cwm} is shown in Figure 15. Reiterating that one consequence of this hypsometric effect is that isotopic compositions along rivers and streams should not be expected to vary in a simple linear fashion nor, as pointed out by (Ramesh and Sarin 1995), should the isotopic lapse rate of precipitation and the isotopic lapse rate determined from surface waters be the same. Thus the early analysis of Chamberlain and Poage (2000) trying to constrain the "global" isotopic lapse rate by combining data from precipitation, surface water, and groundwater samples is not appropriate.

Quite extensive data sets exist for the Himalaya–Southern Tibet region. Below we summarize findings derived from analysis of some of these data in the context of the model presented above. The approach taken here is different from that adopted by (Garzione et al. 2000a,b) who derived the isotopic lapse rate by empirically fitting a curve to observed surface

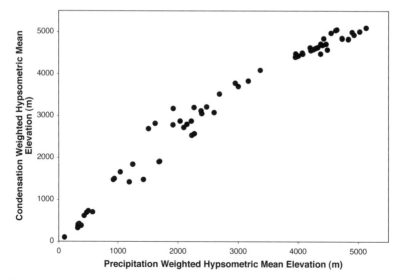

Figure 15. Comparison of the precipitation (z_{pwm}) and condensation (z_{cwm}) weighted hypsometric mean elevations of a selection of Himalayan and southern Tibet drainage basins.

water data. Here the model is used to predict z_{pwm} from the measured $\delta^{18}O_{sw}$ of each sample normalized relative to the amount weighted mean isotopic composition of New Delhi (−5.59‰) to yield $\Delta(\delta^{18}O)$. Predicted z_{pwm} is compared directly with z_{hm} derived from digital elevation data of these drainages (Fig. 16). The close correlation shown by Figure 16 demonstrates that the isotopic compositions of these small drainages are behaving as predicted and that apparent differences between the Seti (Garzione et al. 2000a) and Kali Gandaki (Garzione et al. 2000b) derived lapse rates potentially reflects differences in the precipitation weighted hypsometries of these different drainages as shown by the equally close fit of these two data sets.

One further factor that needs to be taken into account when examining modern surface water data is the seasonal variability in precipitation amount and its isotopic composition. For example, in the case of New Delhi, where, as with much of the Himalayas and southern Tibet, the precipitation regime is dominated by the summer monsoon, there is almost a factor of 40 difference in precipitation amount of the rainiest and driest months of the year, while the isotopic composition varies from around −1‰ during non-monsoon months to as low as −9.3‰ during the monsoon, based on monthly means calculated from 34 years of data for the 1961 to 2001 interval of precipitation records from New Delhi (IAEA-Monthly 2004) (Fig. 17). New Delhi's amount weighted mean annual isotopic composition during this interval is −5.59 ± 0.76‰ (2σ) based on weighting the annual means (IAEA-Yearly 2004). Comparable monthly variability is evident in one year of data from Katmandu (Gajurel et al. 2006). The monsoon season weighted mean isotopic composition at New Delhi is −5.89‰ based on the monthly summary (IAEA-Monthly 2004), hence not different from the annual weighted mean isotopic composition. The effect of significant seasonal variability is potentially most significant in small, short-residence time, drainage basins relative to larger drainages with longer mean residence times.

Rivers and streams not only integrate the precipitation amount weighted hypsometric mean elevation of the drainage basin above the sampling site, but also the seasonal variation in the isotopic composition of the precipitation falling on that watershed. This will be most

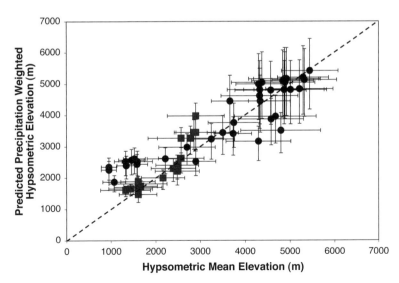

Figure 16. Predicted z_{pwm} vs. Hydro1K-derived (Hydro1k 2005) z_{hm} for the small Himalayan tributary streams of the Seti (Gray squares) and Kali Gandaki (black filled circles) from Garzione et al. (2000a,b). Predicted z_{pwm} is estimated from the measured isotopic composition with ±1σ model uncertainties, and ±1σ deviations about the hypsometric mean elevation. Note from Figure 14 that as the drainage basin decreases in size and as the sample elevation increases that z_{pwm} and z_{hm} become essentially identical.

New Delhi (1961-2001)

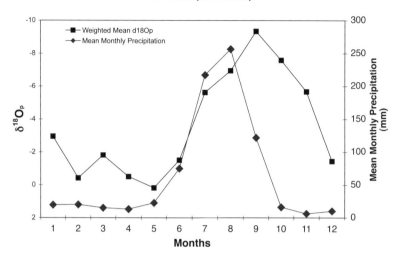

Figure 17. Precipitation amount and amount weighted monthly mean isotopic composition of rainfall in New Delhi from 1961 to 2001 (IAEA-Monthly 2004).

apparent in rivers and streams with short residence times. Given that there are typically only one or at most a few samples from a given sampling location, existing isotopic data from rivers may, but more likely will not reflect the climatological mean isotopic composition of precipitation falling on that watershed. The expected result based on the limited sampling currently available will be a much larger scatter of modern data than would presumably be the case with samples representative of the long-term means.

The discussion below focuses on results summarized from a range of major studies of Himalayan and Ganges rivers by Gajurel et al. (2006). Figure 13 shows sample locations and a subset of the associated drainage basins derived from the Hydro1k hydrologically corrected digital elevation dataset (Hydro1k 2005). In all discussions the measured isotopic compositions are normalized with the 40 year amount weighted mean isotopic composition of precipitation falling in New Delhi to provide an estimate of $\Delta(\delta^{18}O_{mw})$.

Gajurel et al. (2006) summarize isotopic compositions from an array of locations ranging from small drainages well within the Himalayas to samples collected at various locations along the courses of the major rivers within the Ganges plain. These rivers sample a broad spectrum of the hypsometry of this region as reflected in Figure 18, where data on the hypsometric mean elevation, sample elevation, maximum drainage basin elevation are plotted for each sample. Also plotted are the computed z_{pwm} and z_{cwm} for each watershed above the sample elevation. Finally for each sample a model based prediction z_{pwm} based on the reported isotopic composition normalized to New Delhi is plotted.

Rowley and Garzione (2007) presented a comparison of the hypsometry of several large rivers draining the Himalayas with that predicted by measured isotopic compositions reported by Ramesh and Sarin (1995). That comparison was quite favorable, implying that the isotopic composition of foreland basin rivers can indeed record the precipitation weighted hypsometric mean elevation of their drainages. This suggests that it may be possible to discern aspects of orography from foreland basin records in ancient orogenic systems. Rowley and Garzione (2007) noted that due to the strong influence of the precipitation weighting that the isotopic compositions can not be used to say much more than that elevated topography exists, but that

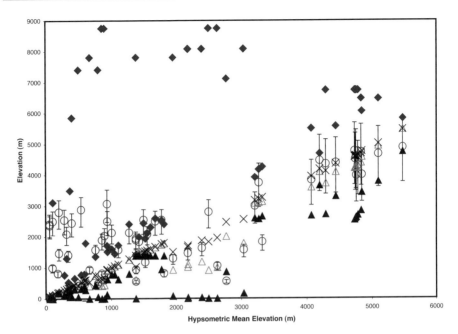

Figure 18. Comparison of various measures of drainage basin hypsometry with the predicted elevation of the precipitation weighted hypsometric mean elevation (black circles) for a selection of river compositions reported by Gajurel et al. (2006). The vertical bars are model-derived ±1σ confidences in the predicted elevation based on measured δ^{18}O$_w$ normalized to New Delhi.

the nature of the hypsometry within that drainage is not discernable. That analysis is extended here using the recently published compilation of Gajurel et al. (2006) (Fig. 18). Again there is a correlation between observation and prediction, but with a larger degree of scatter than evident in the Rowley and Garzione (2007) analysis. Aspects of this comparison have not been analyzed in detail to determine potential contributions from sampling season or temporal variability of isotopic compositions and their correlations with, for example, know variability recorded at New Delhi, so the significance of the differences are hard to fully assess at this time.

An important point that needs to be emphasized is that the isotopic composition of precipitation is variable on all time scales and so the most robust comparisons should be provided by multi-year, preferably multi-decadal or longer isotopic compositions. The fact that what are essentially instantaneous random grab samples that integrate some uncertain aspect of the hydrography of each of these river systems yield any correlation between isotopic composition and hypsometry is pretty remarkable and provides some confidence in estimates based on archival records that integrate isotopic compositions on a very wide range of time scales (Rowley and Garzione 2007).

The comparison reflected by the data in Figure 18 raises the question-is it possible to measure, for example, the isotopic compositions of bivalve shells, or fish, reptilian or aquatic mammalian teeth preserved within fluvial sediments to determine how high the adjacent mountains were? The implication from Figure 18 is that indeed it should be possible to estimate the precipitation weighted hysometric mean elevation of drainages sampled by such materials. Application of this approach to a paleo-case is presented below.

Figures 19 and 20 present tentative results that show the application of this approach to the Himalayan foreland basin. Soil carbonate-derived oxygen isotopic compositions from

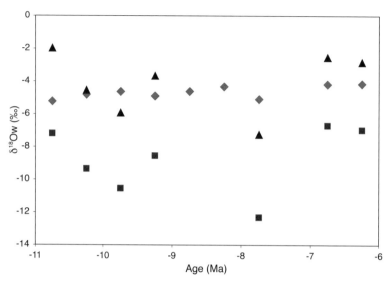

Figure 19. Comparison of Miocene wet season average bivalve isotopic compositions from Dettman et al. (2001) (dark gray squares) with averaged soil carbonate compositions from Bakia Khola (gray diamonds) reported by Harrison et al. (1993) from the Siwaliks. The difference between these is taken as a measure of $\Delta(\delta^{18}O_{mw})$ and hence can be used to estimate the precipitation weighted hypsometric mean elevations of drainages sampled by the bivalve shells. Note that in modern world the wet season amount weighted mean isotopic composition is not significantly from the amount weighted mean annual $\delta^{18}O_p$.

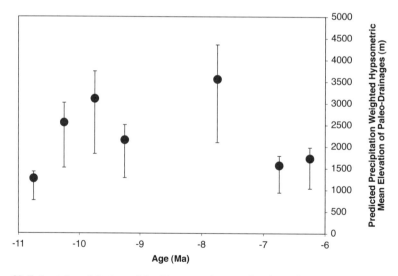

Figure 20. Estimated precipitation weighted hypsometric mean elevations of paleo-drainages sampled by bivalves in the Siwaliks based on estimated $\Delta(\delta^{18}O_{mw})$ shown in Figure 19 as a function of age. Error bars reflect model uncertainty at 2σ based on $\Delta(\delta^{18}O_{mw})$ of the estimated precipitation weighted hypsometric mean elevations.

Bakia Khola reported by Harrison et al. (1993) are used to estimate the isotopic composition of soil waters which are in turn assumed to measure the isotopic composition of low elevation precipitation (Fig. 19). Modern ground waters from the Indo-Gangetic plain are comparable to the weighted mean annual isotopic composition of precipitation measured at New Delhi (Krisnamurthy and Bhattacharya 1991) supporting this assumption. Detailed seasonal isotopic compositions of riverine bivalves preserved within the Siwaliks reported by Dettman et al. (2001) are assumed to provide an estimate of the precipitation weighted hypsometric mean isotopic composition of the drainage basin sampled by a given river system at a given time providing sediments and water to the Siwaliks. The difference between the soil carbonates and bivalve shell compositions are taken as an estimate of $\Delta(\delta^{18}O_{mw})$ that in turn allows an estimate of precipitation weighted hypsometric mean elevation of each of the drainage basins sampled by these bivalves. Note that Figure 19 plots the wet season mean isotopic composition reported by Dettman et al.(2001). In the above discussion of the seasonal variation in isotopic composition of New Delhi it was pointed out that the amount weighted mean monsoon season isotopic composition of precipitation is not significantly different from New Delhi's annual amount weighted mean isotopic composition. Figure 20 implies some temporal variation in z_{pwm} of the drainage basins sampled by these sediments. The variation in time shown on this plot should not be taken to imply changes in Himalayan hypsometry as a whole, but would presumably simply reflect variations in the nature of drainage basin hypsometry being sampled as a function of time. Note also that the paleoelevation estimates are completely compatible with persistence of High Himalayan topography since at least10 Ma (Rowley et al. 1999; Garzione et al. 2000a,b; Rowley et al. 2001) to 20 Ma (France-Lanord et al. 1988; Rowley and Garzione 2007).

CONCLUSIONS

Considerable progress has been made in paleoaltimetry since 1998 when Chase et al. (1998) reviewed the field. The simple 1D thermodynamic model of Rowley et al. (2001) provides a framework for understanding the relationship between stable isotopes and elevation and potential effects of long-term climate change on this relationship. Comparison of transects of precipitation with elevation demonstrate that the model yields good fits to observations, particularly if local temperature and relative humidity conditions are employed (Rowley and Garzione 2007). Existing empirical calibrations are based on temporally limited sampling that may not be representative of the climatological mean conditions, particularly with respect to temperature, to which the modeled relationship is particularly sensitive. Hence their application to the past is uncertain. The global, low latitude, weighted mean model of $\Delta(\delta^{18}O)$ versus elevation represents an ensemble average with variations that, based on comparisons with modern data, appear to robustly capture the characteristic of this relationship. Model based calibrations of $\Delta(\delta^{18}O)$ versus elevation can be computed for any given climatology and hence effects associated with global climate change can be explicitly addressed, which is not possible with empirical calibrations. The model provides a more reasonable estimate of elevation uncertainty than provided by bootstrapping of deviations from limited empirical datasets, since the empirical observations typically capture only a snapshot of climatology that is not likely to be representative of the climatological mean and its variability.

Surface water isotopic compositions do not reflect the local isotopic composition of precipitation, but rather the combined influences of variations in isotopic composition of precipitation as a function of elevation, amount of precipitation as a function of elevation, and drainage basin hypsometry above a given sample elevation. Surface waters thus integrate over the catchment and should be thought of as recording the precipitation amount weighted hypsometric mean elevation of the catchment. Because both the amount of precipitation and typically area decreases with elevation, surface water isotopic compositions are strongly weighted toward recording relatively low elevations. It is difficult to see how to robustly interpret these data in terms of estimating the full hypsometry of such catchments, and thus deriving estimates of paleoelevations of adjacent mountains from measurement of the isotopic compositions derived from, for example, foreland basin fluvial or lacustrine sedimentary sequences. Evaporation of surface water samples, particularly in lacustrine settings can be significant with the consequent enrichment of the waters, resulting in an underestimate of paleoelevations.

Application of stable isotope-based paleoaltimetry to the Himalayas, Tibet and Andes are beginning to elucidate the evolution of paleotopography in these important orographic systems. These and associated approaches have been applied as well to other regions, including the western Cordillera of North America, Southern Alps of New Zealand and Patagonia. These data are beginning to yield significant insights into the process controlling orogenesis (*sensu stricto*). It is still very early in the development and application of these new techniques, but given the recent progress I look forward to what I suspect will be an increasing number of data sets that will provide insights into the paleoelevation histories of mountain belts. These in turn should provide the basis to test existing ideas and hopefully spur new thinking into the underlying dynamic coupling of tectonics, elevation, surface processes and climate.

ACKNOWLEDGMENTS

First and foremost I want to thank Pratigya Polissar, a Canadian Institute for Advanced Research (CIFAR) Post-Doctoral Fellow shared between Kate Freeman at Pennsylvania State University and myself, for carefully going through the previous version of the model and pointing out several issues with the code. This led to the correction in the computation of the weighted mean curve, and, the reappraisal of the mean elevation at which the condensation is sampled. I also want to thank Matt Kohn for enormous patience as I delayed revisions while recoding and re-validation of these modifications in program. Andreas Mulch, an anonymous reviewer, and particularly Matt Kohn helped clarify the manuscript. This work has been supported by the National Science Foundation through grants EAR-9973222 and EAR-0609782, and the CIFAR-Earth System Evolution Program.

REFERENCES

Ambach W, Dansgaard W, Eisner H, Mooler J (1968) The altitude effect on the isotopic composition of precipitation and glacier ice in the Alps. Tellus 20:595-600
Anders AM, Roe GH, Hallet B, Montgomery DR, Finnegan NJ, Putkonen JK (2006) Spatial Patterns of Precipitation and Topography in the Himalaya. *In:* Tectonics, Climate, and Landscape Evolution, Willet SD, Hovius N, Brandon M, Fisher DM (eds), Geological Society of America, Boulder, CO, p 39-53
Argand E (1924) La Tectonique de L'Asie. Presented at Proceedings of the XIIIth International Geological Congress, Brussels
Bird P (1979) Continental delamination and the Colorado Plateau. J Geophys Res 84:7561-7571
Blisnuik P, Stern L (2005) Stable isotope paleoaltimetry-a critical review. Am J Sci 305:1033-1074
Bowen GJ, Revenaugh J (2003) Interpolating the isotopic composition of modern meteoric precipitation. Water Resour Res 39:9-1–9-13
Burbank DW, Blythe AE, Putkonen J, Pratt-Sitaula B, Gabet E, Oskin M, Barros A, Ojha TP (2003) Decoupling of erosion and precipitation in the Himalayas. Nature 426:652-655
Chamberlain CP, Poage MA (2000) Reconstructing the paleotopography of mountain belts from the isotopic composition of authigenic minerals. Geology 28:115-118
Chase CG, Gregory-Wodzicki KM, Parrish JT, DeCelles PG (1998) Topographic history of the western Cordillera of North America and controls on climate. *In:* Tectonic Boundary Conditions for Climate Reconstruction, Crowley TJ, Burke KC (eds), Oxford University Press, New York, p 73-97
Clark MK, House MA, Royden LH, Whipple KX, Burchfiel BC, Zhang X, Tang W (2005) Late Cenozoic uplift of southeastern Tibet. Geology 33:525-528
Coward MP, Kidd WSF, Pan Y, Shackleton RM, Hu Z (1988) The structure of the 1985 Tibet Geotraverse, Lhasa to Golmud. Phil Trans Roy Soc London 327:307-336
Craig H (1961) Isotopic variations in meteoric waters. Science 133:1702-1708

Currie BS, Rowley DB, Tabor NJ (2005) Middle Miocene paleoaltimetry of southern Tibet: Implications for the role of mantle thickening and delamination in the Himalayan orogen. Geology 33:181-184

Cyr A, Currie BS, Rowley DB (2005) Geochemical and stable isotopic evaluation of Fenghuoshan Group lacustrine carbonates, north-central Tibet:Implications for the paleoaltimetry of Late Eocene Tibetan Plateau. J Geol 113:517-533

Dansgaard W (1964) Stable isotopes in precipitation. Tellus 16:436-468

Dettman DL, Kohn MJ, Quade J, Ryerson FJ, Ojha TP, Hamidullah S (2001) Seasonal stable isotope evidence for a strong Asian monsoon throughout the past 10.7 m.y. Geology 29:31-34

Dewey JF, Shackleton R, Chang CF, Sun YY (1988) The tectonic evolution of Tibet. *In*: The Geological Evolution of Tibet. Phil Trans Roy Soc London 327:379-413

Dutton A, Wilkinson BH, Welker JM, Bowen GJ, Lohman KC (2005) Spatial distribution and seasonal variation in $^{18}O/^{16}O$ of modern precipitation and river water across the conterminous USA. Hydrol Process 19:4121-4146

England P, Housemann G (1986) Finite strain calculations of continental deformation 2. Comparison with the India-Asia collision zone. J Geophys Res 91:3664-3676

France-Lanord C, Sheppard SMF, Le Fort P (1988) Hydrogen and oxygen isotope variations in the High Himalaya peraluminus Manaslu leucogranite: Evidence for heterogeneous sedimentary source. Geochim Cosmochim Acta 52:513-526

Friedman I, Harris JM, Smith GI, Johnson CA (2002a) Stable isotope composition of waters in the Great Basin, United States 1. Air-mass trajectories. J Geophys Res 107:ACL 14-11 - ACL 14-14

Friedman I, Smith GI, Johnson CA, Moscati RJ (2002b) Stable isotope composition of waters in the Great Basin, United States 2. Modern Precipitation. J Geophys Res 107:ACL 15-11 - ACL 15-21

Gajurel AP, France-Lanord C, Huyghe P, Guilmette C, Gurung D (2006) C and O isotope compositions of modern fresh-water mollusc shells and river waters from the Himalaya and Ganga plain. Chem Geol 233:156-183

Garzione CN, Dettman DL, Quade J, DeCelles PG, Butler RF (2000a) High times on the Tibetan Plateau: Paleoelevation of the Thakkhola graben, Nepal. Geology 28:339-342

Garzione CN, Quade J, DeCelles PG, English NB (2000b) Predicting paleoelevation of Tibet and the Himalaya from $\delta^{18}O$ vs. altitude gradients in meteoric water across the Nepal Himalaya. Earth Planet Sci Lett 183:215-229

Garzione CN, Molnar P, Libarkin JC, MacFadden BJ (2006) Rapid late Miocene rise of the Bolivian Altiplano: Evidence for removal of mantle lithosphere. Earth Planet Sci Lett 241:543-556

Garzione CN, Molnar P, Libarkin JC, McFadden BJ (2007) Reply to Comment on "Rapid late Miocene rise of the Bolivian Altiplano:Evidence for removal of mantle lithosphere" by Garzione et al. (2006), Earth Planet Sci Lett 241 (2006) 543–556. Earth Planet Sci Lett 259:630-633

Ghosh P, Garzione CN, Eiler JM (2006) Rapid uplift of the Altiplano revealed through ^{13}C-^{18}O bonds in paleosol carbonates. Science 311:511-515

GLOBE Task Team and others (Hastings DA, Dunbar PK, Elphingstone GM, Bootz M, Murakami H, Maruyama H, Masaharu H, Holland P, Payne J, Bryant NA, Logan TL, Muller J-P, Schreier G, MacDonald JS) (1999) The Global Land One-kilometer Base Elevation (GLOBE) Digital Elevation Model, Version 1.0. National Oceanic and Atmospheric Administration, National Geophysical Data Center. *http://www.ngdc.noaa.gov/mgg/topo/globe.html*

Gonfiantini R, Roche MA, Olivry JC, Fontes JC, Zuppi GM (2001) The altitude effect on the isotopic composition of tropical rains. Chem Geol 181:147-167

Harrison TM, Copeland P, Hall S, Quade J, Burner S, Ojha TP, Kidd WSF (1993) Isotope preservation of Himalayan/Tibetan uplift denudation and climatic histories of two molasse deposits. J Geol 101:157-176

Horita J, Wesolowski DJ (1994) Liquid-vapor fractionation of oxygen and hydrogen isotoopes of water from freezing to the critical temperature. Geochim Cosmochim Acta 58:3425-3437

Horton TW, Chamberlain CP (2006) Stable isotopic evidence for Neogene surface downdrop in the central Basin and Range Province. Geol Soc Amer Bull 118:475-490

Huber M, Caballero R (2003) Eocene El Niño: Evidence for robust tropical dynamics in the "hothouse". Science 299:877-881

Hydro1k (2005) Hydro1k Asia, *http://edc.usgs.gov/products/elevation/gtopo30/hydro/asia.html*

IAEA-Monthly (2004) GNIP2001Monthly.xls, *http://isohis.iaea.org/userupdate\GNIPMonthly.xls*

IAEA-Yearly (2004) GNIP2001Yearly.xls, *http://isohis.iaea.org/userupdate\GNIPYearly2001.xls*

Kalnay E, Kanamitsu M, Kistler R, Collins W, Deaven D, Gandin L, Iredell M, Saha S, White G, Woollen J, Zhu Y, Chelliah M, Ebisuzaki W, Higgins W, Janowiak J, Mo K, Ropelewski C, Wang J, Leetmaa A, Reynolds R, Jenne R, Joseph D (1996) The NCEP/NCAR 40-year reanalysis project. Bull Amer Met Soc 77:437-471

Kent-Corson ML, Sherman LS, Mulch A, Chamberlain CP (2006) Cenozoic topographic and climatic response to changing tectonic boundary conditions in western North America. Earth Planet Sci Lett 252:453-466

Kohn MJ, Miselis JL, Fremd TJ (2002) Oxygen isotope evidence for progressive uplift of the Cascade Range, Oregon. Earth Planet Sci Lett 204:151-165

Kohn MJ, Dettman DL (2007) Paleoaltimetry from stable isotope compositions of fossils. Rev Mineral Geochem 66:119-154

Krisnamurthy RV, Bhattacharya SK (1991) Stable oxygen and hydrogen isotope ratios in shallow ground waters from India and a study of the role of evapotranspiration in the Indian monsoon. *In:* Stable Isotope Geochemistry: A tribute to Samuel Epstein. Taylor HP, O'Neil JR, Kaplan IR (eds), Geochemical Society, San Antonio, Texas, p 187-203

Majoube M (1971a) Fractionnement en 18O entre la glace et la vapeur d'eau. J Chim Phys 68:625-636

Majoube M (1971b) Fractionement en oxygen 18 et en deuterium entre l'eau et sa vapeur. J Chim Phys 68:1423-1436

Merlivat L, Nief G (1967) Isotopic fractionation of solid-vapor and liquid-vapor changes of state of water at temperatures below 0°C. Tellus 19:122-127

Molnar P, England P, Martinod J (1993) Mantle dynamics, uplift of the Tibetan Plateau, and the Indian monsoon. Rev Geophys 31:357-396

Mulch A, Graham SA, Chamberlain CP (2006) Hydrogen isotopes in Eocene river gravels and paleoelevataion of the Sierra Nevada. Science 313:87-89

Mulch A, Chamberlain CP (2007) Stable isotope paleoaltimetry in orogenic belts – the silicate record in surface and crustal geological archives. Rev Mineral Geochem 66:89-118

Poage MA, Chamberlain CP (2001) Empirical relationships between elevation and the stable isotope composition of precipitation and surface waters: Considerations for studies of paleoelevation change. Amer. Jour. Sci. 301:1-15

Putkonen JK (2004) Continuous snow and rain data at 500 to 4400 m altitude near Annapurna, Nepal, 1999–2001. Arctic Antarctic Alpine Res 36:244-248

Pysklywec RN, Beaumont C, Fullsack P (2000) Modeling the behaviour of the continental mantle lithosphere during plate convergence. Geology 28:655-658

Quade J, Garzione C, Eiler J (2007) Paleoelevation reconstruction using pedogenic carbonates. Rev Mineral Geochem 66:53-87

Ramesh R, Sarin MM (1995) Stable isotope study of the Ganga (Ganges) river system. J Hydrol 139:49-62

Roe GH, Montgomery DR, Hallet B (2002) Effects of orographic precipitation variations on the concavity of steady-state river profiles. Geology 30:143-146

Roe GH (2005) Orographic precipitation. Ann Rev Earth Planet Sci 33:645-671

Rowley DB, Pierrehumbert RT, Currie BS, Hosman A, Clayton RN, Hutardo J, Whipple K, Hodges K (1999) Stable isotope-based paleoaltimetry and the elevation history of the High himalaya since the late Miocene. Geol Soc Amer Abstracts with Prog #50213

Rowley DB, Pierrehumbert RT, Currie BS (2001) A new approach to stable isotope-based paleoaltimetry: implications for paleoaltimetry and paleohypsometry of the High Himalaya since the Late Miocene. Earth Planet Sci Lett 188:253-268

Rowley DB, Currie BS (2006) Palaeo-altimetry of the late Eocene to Miocene Lunpola basin, central Tibet. Nature 439:677-681

Rowley DB, Garzione CN (2007) Stable isotope-based paleoaltimetry. Ann Rev Earth Planet Sci 35:463-508

Royden LH, Burchfiel BC, King RW, Wang EC, Chen ZL, Shen F, Liu YP (1997) Surface deformation and lower crustal flow in eastern Tibet. Science 276:788-790

Rozanski K, Araguas-Araguas L, Gonfianti R (1993) Isotopic patterns in modern global precipitation. *In:* Climate Change in Continental Isotopic Records. American Geophysical Union, Washington, p 1-36

Zhu B, Kidd WSF, Rowley DB, Currie BS, Shafique N (2005) Age of initiation of the India-Asia collision in the East-Central Himalaya. J Geol 113:265-285

Reviews in Mineralogy & Geochemistry
Vol. 66, pp. 53-87, 2007
Copyright © Mineralogical Society of America

3

Paleoelevation Reconstruction using Pedogenic Carbonates

Jay Quade

Department of Geosciences
University of Arizona
Tucson, Arizona, 85721, U.S.A.
quadej@email.arizona.edu

Carmala Garzione

Department of Earth and Environmental Sciences
University of Rochester
Rochester, New York, 14627, U.S.A.
garzione@earth.rochester.edu

John Eiler

Division of Geological and Planetary Sciences
California Institute of Technology
Pasadena, California, 91125, U.S.A.
eiler@gps.caltech.edu

ABSTRACT

Paleoelevation reconstruction using stable isotopes, although a relatively new science, is making a significant contribution to our understanding of the recent growth of the world's major orogens. In this review we examine the use of both light stable isotopes of oxygen and the new "clumped-isotope" (Δ_{47}) carbonate thermometer in carbonates from soils. Globally, the oxygen isotopic composition ($\delta^{18}O$) of rainfall decreases on average by about 2.8 ‰/km of elevation gain. This effect of elevation will in turn be archived in the $\delta^{18}O$ value of soil carbonates, and paleoelevation can be reconstructed, provided (1) temperature of formation can be estimated, (2) the effects of evaporation are small, (3) the effects of climate change can be accounted for, and (4) the isotopic composition of the carbonate is not diagenetically altered. We review data from modern soils to evaluate some of these issues and find that evaporation commonly elevates $\delta^{18}O$ values of carbonates in deserts, an effect that would lead to underestimates of paleoelevation. Some assessment of paleoaridity, using qualitative indicators or carbon isotopes from soil carbonate, is therefore useful in evaluating the oxygen isotope-based estimates of paleoelevation. Sampling from deep (> 50 cm) in paleosols helps reduce the uncertainties arising from seasonal temperature fluctuations and from evaporation.

The new "clumped-isotope" (Δ_{47}) carbonate thermometer, expressed as Δ_{47}, offers an independent and potentially very powerful approach to paleoelevation reconstruction. In contrast to the use of $\delta^{18}O$ values, nothing need be known about the isotopic composition of water from which carbonate grew in order to estimate of temperature of carbonate formation from Δ_{47} values. Using assumed temperature lapse rates with elevation, paleoelevations can thereby be reconstructed.

Case studies from the Andes and Tibet show how these methods can be used alone or in combination to estimate paleoelevation. In both cases, the potential for diagenetic alteration

     DOI: 10.2138/rmg.2007.66.3

of primary carbonate values first has to be assessed. Clear examples of both preservation and alteration of primary isotopic values are available from deposits of varying ages and burial histories. Δ_{47} values constitute a relatively straightforward test, since any temperature in excess of reasonable surface temperatures points to diagenetic alteration. For $\delta^{18}O$ values, preservation of isotopic heterogeneity between different carbonate phases offers a check on diagenesis. Results of these case studies show that one area of south-central Tibet attained elevations comparable to today by the late Oligocene, whereas 2.7 ± 0.4 km of uplift occurred in the Bolivian Altiplano during the late Miocene.

INTRODUCTION

There is considerable debate over the timing and causes of uplift of the earth's great mountain ranges and plateaux, such as Tibet, the Himalaya, and the Andes (Fig. 1). Estimates of paleoelevation change provide key constraints for competing general models that link such large-scale orogenic events with lithospheric-scale geodynamic processes (Currie et al. 2005; Garzione et al. 2006; Rowley and Currie 2006; Kent-Corson et al. 2007). The use of the isotopic composition of carbonates, silicates, and oxides is at the forefront of these paleoelevation reconstructions. In this chapter we review use of one of several geologically abundant secondary carbonates—calcite formed in soils (or "pedogenic carbonate")—in paleoelevation reconstruction.

The H and O isotopic composition of rainfall (expressed as $\delta^{18}O_{mw}$ and δD_{mw} values in per mil, or ‰, relative to SMOW) varies primarily as a function of the degree of rainout (removal of water condensate) from a vapor mass. As rainout proceeds, ^2H (D) and ^{18}O are preferentially removed from the vapor mass, decreasing δD and $\delta^{18}O$ values of both the water vapor and rainfall derived from it. As a vapor mass ascends over mountains, adiabatic expansion causes cooling and condensation of water as so-called "orographic precipitation" (e.g., Roe 2005), producing some of the largest isotopic gradients in rainfall observed on Earth. A number of studies document large fractionations in the H and O isotopes in rainfall, snowfall, and associated surface water with increasing elevation (e.g., Ambach et al. 1968; Siegenthaler and Oeschger 1980; Rozanski and Sonntag 1982, Ramesh and Sarin 1995; Garzione et al. 2000a; Gonfiantini et al. 2001; Stern and Blisniuk 2002; Rowley and Garzione 2007; but see Dutton et al. 2005). The $\delta^{18}O$-elevation relationship has been calibrated both empirically (e.g., Garzione et al. 2000a, Poage and Chamberlain 2001) and theoretically using basic thermodynamic principles that govern isotopic fractionation of an ascending vapor mass (Rowley et al. 2001; Rowley and Garzione 2007). Globally, the $\delta^{18}O_{mw}$ value of rainfall on average decreases by 2.8 ‰/km (Poage and Chamberlain 2001), although the scatter around this average is very considerable (Blisniuk and Stern 2005), as we discuss for the cases of Tibet and the Andes in the next section of this review.

Studies using oxygen (and hydrogen) isotopes take advantage of this strong sensitivity of $\delta^{18}O_{mw}$ values to elevation change to reconstruct paleoelevation. Stated simply, the basic approach to paleoelevation reconstruction is this: the oxygen isotope composition of paleocarbonate buried in geologic sections can be used to estimate the oxygen isotope composition of soil water, which in turn is used as an estimate of $\delta^{18}O_{mw}$ values; and it is this estimate of $\delta^{18}O_{mw}$ values that is compared with empirical or theoretical isotopic lapse rates to estimate paleoelevation.

Two key complications in this approach to paleoelevation reconstruction are that carbonates, including pedogenic carbonate which forms in soils, record the oxygen isotopic composition of soil waters (which may differ from meteoric water) *and* of the temperature of carbonate formation, both of which vary with elevation. Early studies used the $\delta^{18}O$ value of pedogenic carbonates (or $\delta^{18}O_{sc}$, expressed in ‰ relative to PDB) to reconstruct elevation by assuming temperatures of formation that allowed calculation of the $\delta^{18}O$ value of soil water

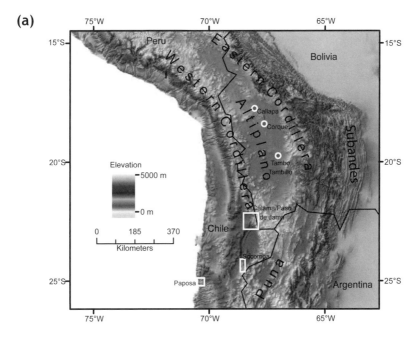

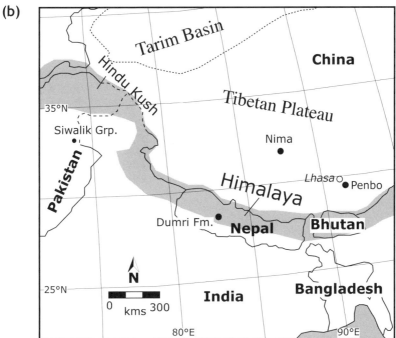

Figure 1. (a) Elevations of the Central Andean plateau between 15°S and 26°S, constructed with SRTM30 dataset. White circles show our late Oligocene to late Miocene sites discussed in the text. White boxes enclose study sites for modern soil carbonate in the Atacama Desert in Quade et al. (2007); (b) study locations in the Himalayan-Tibet orogen (Himalaya/Hindu Kush shaded).

(or $\delta^{18}O_{sw}$, expressed in ‰ relative to SMOW) (Garzione et al. 2000a,b; Rowley et al. 2001; Blisniuk et al. 2005; Currie et al. 2005; Rowley and Currie 2006). The $\delta^{18}O$ value of soil water can differ from $\delta^{18}O_{mw}$ values due to evaporative enrichment of ^{18}O in soil water (Cerling and Quade 1993). Moreover, the season of carbonate precipitation may also affect the isotopic composition of pedogenic carbonate, both in terms of seasonal variability in the temperature of carbonate formation and the isotopic composition of rainfall and soil water (e.g., Quade et al 1989a; Amundson et al. 1996; Liu et al. 1996; Stern et al. 1997).

These competing effects can be mitigated by the application of a new "clumped-isotope" carbonate thermometer that allows the determination of the temperature of carbonate formation (Ghosh et al. 2006a). This technique relies on the abundance of bonds between rare, heavy isotopes (i.e., ^{13}C-^{18}O) in the carbonate mineral lattice, the relative abundance of which increases at lower temperatures. Recent paleoelevation studies (Ghosh et al. 2006b) applied this carbonate thermometer to better constrain $\delta^{18}O_{sw}$ values from which paleoelevation was calculated using the traditional approach (Garzione et al. 2006).

We will begin this review by evaluating modern variability in $\delta^{18}O$ values of surface waters with elevation change across the Andean plateau (Fig. 1a) and the Himalayan-Tibet plateau (Fig. 1b). Global patterns of the $\delta^{18}O_{mw}$/elevation relationship are comprehensively reviewed in Poage and Chamberlain (2001) and Blisniuk and Stern (2005). We will then compare $\delta^{18}O_{mw}$ to $\delta^{18}O_{sc}$ results from the Mojave Desert, the Andes, and Tibet, among the few regions where coupled water and carbonate isotopic data from modern soils are available, providing an essential interpretive backdrop for the paleosol carbonate records. Finally, we present case studies from the Andes and Himalaya-Tibet as examples of the use of $\delta^{18}O$ and "clumped-isotope" values from pedogenic carbonates in paleoelevation reconstruction.

DEPENDENCE OF THE ISOTOPIC COMPOSITION
OF RAINFALL ON ELEVATION

$\delta^{18}O_{mw}$ values decrease with elevation gain on both sides of the Andes and Tibet (Fig. 2a). On the windward side, these changes reflect the expected effects of progressive distillation on air masses as they rise, cool, and rain out over an orographic barrier. Interestingly, $\delta^{18}O_{mw}$ values also decrease with elevation on the leeward sides of large mountain ranges, especially in warm climates (summarized in Blisniuk and Stern 2005), including the western flank of the central Andes, the Himalaya, and to some extent the Sierra Nevada. This is not predicted by Rayleigh distillation models of cloud mass rain out (Rowley and Garzione 2007). Some research suggests that in dry leeward settings, raindrops are evaporatively enriched in ^{18}O during descent, and a small fraction of the original raindrop reaches the ground (Beard and Pruppacher 1971; Stewart 1975). Thus, evaporative gains in raindrop ^{18}O apparently more than offset Rayleigh-type depletions in ^{18}O in the original cloud mass.

One striking example of this pattern is from the Andes, drawing mainly upon published data between 17-21°S from Gonfiantini et al. (2001) and our unpublished water data for the eastern side of the Andes, and from Fritz (1981) and Aravena et al. (1999) for the Pacific slope (Fig. 2a). For the eastern slope, $\delta^{18}O$ values decrease by ~2.6 ‰/km across an elevation gain of 2200 m. These results match well with $\delta^{18}O$ values of snow from Mt. Sajama (Hardy et al. 2003), and rain falling on Lake Titicaca (Cross et al. 2001). On the Pacific slope, samples collected from 22 stations over a four-year period show an even steeper isotopic lapse rate of −6.2 ‰/km. We will return to these data when we look at isotopic results from modern pedogenic carbonates in this region.

Stream-water sampling from Garzione et al. (2000a) along the Kali Gandaki up the south face of the Himalaya at about 78°E shows a very regular decrease in $\delta^{18}O_{mw}$ values (Fig. 2b). The best fit to these data is a second-order polynomial, with an average decrease of around 2.5

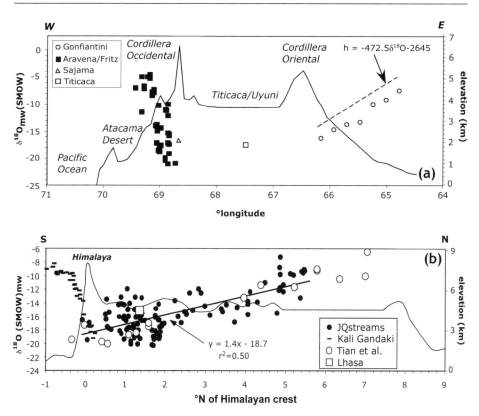

Figure 2. Topography and $\delta^{18}O_{mw}$ (SMOW) values for (a) the Andes, compiled from Aravena et al. (1999) and Fritz et al. (1981); Hardy et al. (2003) (Sajama); and Cross et al. (2001)(Titicaca); uncorrected Gonfiantini et al. (2001); dashed line and equation are corrected Gonfiantini data combined with our unpublished data (see also Fig. 14 and text); and (b) for Tibet, compiled from Garzione et al. (2000a) (Kali Gandaki); Tian et al. (2001); Quade unpublished data (JQ streams); and Araguás-Araguás et al. (1998) (Lhasa).

to 2.7 ‰/km. $\delta^{18}O_{mw}$ values decrease to below −20‰ for elevations above 5000 m, as sampled near high-elevation glaciers (Zhang et al. 2002; Tian et al. 2005) and from streams flowing from the highest elevations on both sides of the Himalaya (Tian et al. 2001). The more unexpected pattern is the increase in $\delta^{18}O_{mw}$ values north of the Himalayan crest, as observed by Tian et al. (2001) and Zhang et al. (2002). The sampling stations are widely dispersed, and we can add our own unpublished stream water samples from about 29-34°N. The combined data display an increase of ~1.5 ‰/° latitude. Virtually all the collection stations range from 3700 m to 5000 m. With our own stream-water samples, we show sampling elevations (Fig. 2b) unadjusted for the more relevant mean elevation and latitude of the catchment. But since we sampled mostly small catchments, we believe the correlation with Tian's and Zhang's data is not a coincidence.

The southwest Indian monsoon dominates rainfall south of the Himalayan front as well as stations immediately north of the crest, perhaps mixed from some contribution from the East Asian monsoon. The seasonal distribution of $\delta^{18}O_{mw}$ values for Lhasa, located 1.5° north of Himalayan ridgecrest, illustrates this. Over ninety per cent of the rain falls in the summer rainy season, and that rainfall displays much lower $\delta^{18}O_{mw}$ values than winter snow and rain (Araguás-Araguás et al. 1998). The strong ^{18}O and D depletions in summer rainfall can be modeled in terms of gradual, Rayleigh distillation of moisture (Rowley et al. 2001) that derives water vapor from the equatorial Indian Ocean or East China Sea, is drawn northward and westward over land in the summer, and rises, cools and rains out as it passes over the Himalaya and Tibet.

This simple picture does not explain the gradual increase in $\delta^{18}O_{mw}$ values north of the Himalayan crest. Mean elevation of sampling sites north of the crest lies almost entirely within 4500 ± 500 m, and even increases slightly northward, the opposite of the expected pattern where elevation is the main determinant of $\delta^{18}O_{mw}$ values. Likewise, the continentality effect—simple rain out of air masses with distance from coastlines—should cause $\delta^{18}O_{mw}$ values to decrease inland, not decrease.

The explanation provided by Tian et al. (2001) and supported by a variety of data, including seasonality of rainfall and deuterium excess, is that other, either local or northerly derived air masses penetrate southward during the winter months. The $\delta^{18}O_{mw}$ value of stream samples and amount-weighted rainfall samples increase northwards, reflecting a decreasing proportion of summer (monsoonal) rainfall in year-round weighted rainfall. The key point for paleoelevation reconstruction is that $\delta^{18}O_{mw}$ values change quite significantly across the Tibetan Plateau for climatic, not elevational reasons. The result is that lapse rates in the southern plateau (-2.6 ‰/ km) are almost twice those of the northern plateau (-1.5 ‰/km). Consequently, paleoelevation estimates depend on distance of sampling sites from the Himalayan range crest, and how climate change in the past, such as a strengthened or weakened Asian monsoon, might have increased or decreased isotopic lapse rates northward on the plateau.

MODERN PEDOGENIC CARBONATES

Pedogenic carbonate formation

The general equation for weathering (to the right) and calcite precipitation (to the left) in soils is:

$$CaCO_3(s) + CO_2(g) + H_2O = Ca^{++}(aq) + 2HCO_3^-(aq) \tag{1}$$

Here we use calcite as the parent material but in its place just as easily could have used Ca-Mg silicate minerals. Pedogenic carbonate formation (to the left) is driven by both soil water and CO_2 loss. Seasonality of carbonate formation is not known for certain, but in most settings it is likely concentrated in the summer half year when soils are thawed or warmer, plants are active and evapotranspiring, and evaporation is greatest.

In well-drained soil profiles, water for weathering reactions like (1) is supplied directly by local rainfall. Pedogenic carbonate thus enjoys a key advantage over other types of carbonate in paleoelevation reconstruction in that it forms from rainfall that fell on the site, and not runoff from higher elevations, as in the case of many riverine and lacustrine carbonates. For the oxygen system, molar water to rock ratios are extremely high in soils, and thus the $\delta^{18}O$ value of secondary weathering phases such as clays, iron oxides, and carbonates should not be influenced by parent material during weathering. Analogously, for the carbon system in most soils, the molar $C_{plant}/C_{parent\ material}$ is also very large, and hence plant CO_2 mixed with the atmospheric CO_2, and not carbon from local parent material such as limestone, will determine the $\delta^{13}C$ value of pedogenic carbonate. In dry climates, soils dewater largely by a combination of evapotranspiration (ET) and evaporation (E), with some percolation through to the local water table. The ratio of dewatering by these processes (F_E/F_{ET}) decreases with increasing soil depth and increasing rainfall. For oxygen isotopes in soils, evaporation is a fractionating process, enriching residual soil water in ^{18}O. By contrast, evapotranspiration is a non-fractionating process; water is drawn unfractionated into plant roots from the soil. Hence, in drier climates and shallower in soils, evaporative enrichment in ^{18}O has been widely observed (e.g., Allison et al. 1983; Hsieh et al. 1998). After correction for temperature (discussed later), this effect should also be expressed in secondary mineral phases, such as carbonates in soils.

In the next section we compare $\delta^{18}O_{sc}$ values to local $\delta^{18}O_{mw}$ values from rainfall collected sometime in the last few decades. To make this comparison, we use pedogenic carbonate samples that formed recently, preferably over the latter part of the Holocene, on the assumption that this is the carbonate most likely to represent isotopic equilibrium conditions between calcite and the sampled water. In the absence of radiometric dates from our samples, we use pedogenic carbonate morphology as a qualitative indicator of age, following the conventions of Gile et al. (1966) (see also Machette 1985), in which thin carbonate coatings on alluvial clasts represent Holocene-age cements, and continuous filling of the soil matrix by carbonates represents older (10^4-10^6 yrs) Quaternary cementation.

The oxygen isotopic composition of Holocene pedogenic carbonate

Salomans et al. (1978), Talma and Netterberg (1983) and Cerling (1984) were among the first to show that $\delta^{18}O_{mw}$ values are positively correlated with the $\delta^{18}O_{sc}$ values, a relationship later shown to hold true for a variety of meteoric cements (Hays and Grossman 1991). The details of this relationship are worth exploring in light of newer data and from the perspective of paleoelevation reconstruction. To do this we selected three data sets that encompass the broad range of rainfall (~20 to 1000 mm/yr) in which pedogenic carbonates develop. These include modern pedogenic carbonates from the relatively wet mid-western USA (Cerling and Quade 1993), the Mojave Desert (Quade et al. 1989a), and from the extremely arid Atacama Desert (Fig. 1a; Quade et al. 2007). We compare $\delta^{18}O_{sc}$ values from Holocene-age soils against predicted $\delta^{18}O_{sc}$ based on local mean annual temperature (MAT) + 2 °C (see defense of this formulation in the next section of the paper) and known $\delta^{18}O_{mw}$ values (Fig. 3). In higher rainfall areas, predicted and observed $\delta^{18}O_{sc}$ values are positively correlated, as observed by others previously (i.e., Salomans et al 1978; Cerling 1984). The departures of observed $\delta^{18}O_{sc}$ values from the 1:1 predicted line are larger where the climate is drier, consistent with evidence from modern soil water studies (e.g., Hsieh et al. 1998). Hence, pedogenic carbonates from the driest sites in the Mojave (MAP < 100 mm/yr) and in nearly all sites in the hyperarid Atacama Desert

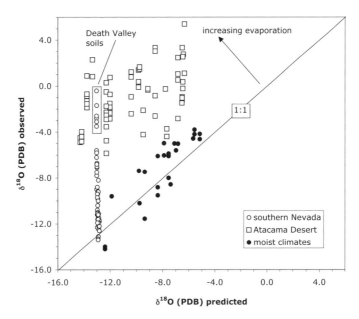

Figure 3. Observed $\delta^{18}O_{sc}$ (data from Quade et al. (1989a) and Cerling and Quade (1993), and Quade et al. (2007)) versus predicted $\delta^{18}O_{sc}$ values based on local mean annual temperature and $\delta^{18}O_{mw}$ values.

(2-114 mm/yr) show the strongest soil evaporation effects. Examination of $\delta^{18}O_{sc}$ values collected with depth in soil profiles is also instructive. In dry settings like the Mojave Desert, $\delta^{18}O_{sc}$ values increase toward the surface (Fig. 4), whereas from moister settings like Kansas, $\delta^{18}O_{sc}$ values show little variation with soil depth (Cerling and Quade 1993).

In the context of paleo-elevation reconstruction, the following conclusions can be drawn: (1) in moister climates, $\delta^{18}O_{sc}$ values should provide a close approximation of $\delta^{18}O_{mw}$ values after correction for temperature, (2) in drier settings, $\delta^{18}O_{sc}$ values will be influenced by evaporative enrichment, and hence in this instance (3) sampling laterally and deeper (≥ 50 cm) in soils should produce a better approximation of unevaporated $\delta^{18}O_{mw}$ values. Because of evaporation effects, $\delta^{18}O_{sc}$ values should be viewed to provide only maximum estimates of $\delta^{18}O_{mw}$ values and thus minimum estimates of paleo-elevation.

Soil temperature considerations

One must be able to constrain soil temperature (T) to account for the fractionation between calcite and water. In this regard, several factors favor the use of pedogenic carbonates. One is that the fractionation factor between calcite and water is T small (~ -0.22 to -0.24 ‰/°C in the range 0-30 °C) when compared, for example, to the steep average dependence of $\delta^{18}O_{mw}$ values on elevation of -2.8 ‰/km. Thus, $\delta^{18}O_{sc}$ values are widely used

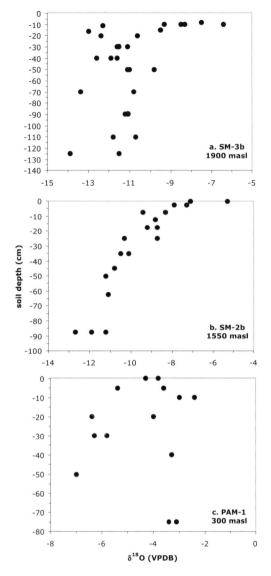

Figure 4. Depth profiles of $\delta^{18}O_{sc}$ (PDB) versus soil depth, from three sites in the Mojave Desert, (a) SM-3 (1990 masl), (b) SM-2b (1550 masl), and (c) PaM-1 (300 masl), all from

by geologists to constrain $\delta^{18}O_{mw}$ values, as in paleo-elevation reconstructions, but only used for paleo-temperature reconstructions in a narrow set of circumstances where paleo-$\delta^{18}O_{mw}$ values are relatively invariant. Having said this, the effects on $\delta^{18}O_{sc}$ values of decreasing T with elevation can partially to entirely offset the influence of decreases in $\delta^{18}O_{mw}$ values with elevation. Temperature lapse rates globally are about 5-6°C/km. This temperature lapse rate combined with a -2.8 ‰/km decrease in $\delta^{18}O_{mw}$ values would produce a ca. -1.5 ‰/km change with elevation gain in $\delta^{18}O_{sc}$ values.

One feature of soils that confers a significant advantage compared to other carbonates (such as lacustrine carbonate) is that soil T converges toward mean annual air T at depth. In contrast, shallow aquatic systems where many carbonates form can experience sharp diurnal and seasonal changes. Soil T varies with time (t) and exponentially as a function of soil depth (z) following (from Hillel 1982):

$$T(z,t) = T_{avg} + A_\circ \left[\sin(\omega t - z/d) \right] / e^{-z/d}$$

where T_{avg} is average air temperature, A_o is seasonal or diurnal maximum and minimum T amplitude at $z = 0$, ω is radial frequency, z is soil depth, d is the "damping" depth (or $1/e$ folding depth) characteristic of the soil:

$$d = \left(2\kappa / C_V \omega \right)^{1/2} \tag{2}$$

where κ is thermal conductivity and C_V is volumetric heat capacity. As regards surface-T fluctuations, the $1/e$ folding depth of diurnal fluctuations in the order of centimeters, is much shallower than the average depth of carbonate accumulation. This is due mainly to the high diurnal radial frequency (in which $\omega = 2\pi/86,400$ seconds), which in Equation (2) is inversely related to the damping depth. Seasonal temperature fluctuations have a much lower radial frequency, hence greater damping depth. Therefore, it is seasonal fluctuations in T, and their attenuation at deep soil depths (> -20 cm) where soil carbonate forms, that is the chief concern here.

Taking the hypothetical example of a Tibetan soil near Lhasa at 4100 m, we assume in the following simulation 0.0007 cal/cm sec, and 0.3 cal/cm³ °C for κ and C_V (from Hillel 1982), respectively. This yields $1/e$ folding lengths for seasonal temperature change of plateau soils of about 150 cm.

Average ground temperature differs by 1-3 °C from average air temperature due to excess ground warming during the summer months (Bartlett et al. 2006). This offset is correlated with mean radiation received at a site (1.21 K/100 W m⁻²), which is in turn is a function of vegetation cover, slope aspect, and latitude. For our calculations we added a constant 2 ± 1 °C to average air temperature to arrive at average ground temperature at depth.

Depth of pedogenic carbonate formation is typically between 50 and 150 cm on the Tibetan plateau. At these high elevations the soil is at or below freezing half of the year. Thus we can assume that pedogenic carbonate formation occurs during the summer half year, when soils thaw or are warmer, evaporation is greatest, and when plants are actively transpiring soil water. For our Tibetan soil site at 4100 m, monthly mean extremes in air temperature range between 11.6 and −6.4 °C, around a mean of about 2.6 °C. This yields a modeled seasonal T amplitude of about 12.7 °C at 50 cm to 6.7 °C at 150 cm (Fig. 5). For the summer half-year, this produces a range of 1.5 to 0.7‰ uncertainty in the estimation of paleo-$\delta^{18}O_{mw}$ from pedogenic carbonate sampled from 50 to 150 cm soil depth, respectively (Fig. 5b). Here we use the fractionation factor (α) between calcite and water of $1000 \ln \alpha_{calcite-water} = (18030/T) - 32.42$, where T is in Kelvin (Kim and O'Neil 1997). We can conclude that deeper sampling is better for reducing the error in paleo-elevation reconstruction arising from seasonal fluctuations in temperature. The contribution to the error can be no less than on the order ~0.8‰ for typical Tibetan soils; in other words, about 300 m in paleoelevation terms.

Evaluation of aridity

Aridity is of interest in paleoelevation reconstruction for two reasons: (1) in extreme aridity cases, the $\delta^{18}O_{sc}$ values are dominated by evaporation, and realistic estimates of paleoelevation are probably not obtainable, no matter the sampling density, and (2) aridity develops in rain shadows, and thus may provide qualitative evidence of orogenic blockage of moisture.

Evaluation of paleoaridity can (and should) be approached both qualitatively and quanti-

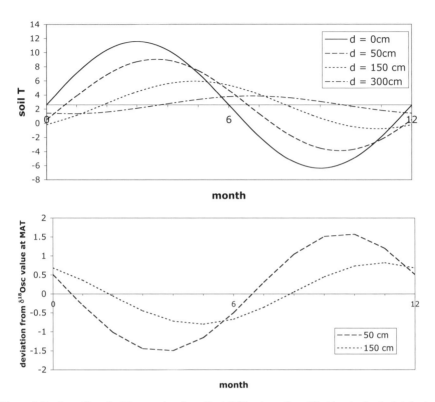

Figure 5. Depth profiles of soil temperature for a site at 4100 m in southern Tibet (see text) calculated using MAT from nearby Lhasa (3660 m) of 8°C and a local air-T lapse rate of 6 °C/km, a) showing modeled seasonal fluctuations in soil temperature at various depths, and (b) seasonal deviations from predicted $\delta^{18}O_{sc}$ values at MAT.

tatively. The qualitative approach involves both soil morphology and carbonate leaching depth. Rech et al (2006) provides one recent example from the central Atacama Desert of the utility of paleosols in establishing the limits on paleo-aridity (Fig. 1a). They showed that hyperarid (< 20 mm/yr) conditions have prevailed—based on the presence of buried nitrate and gypsic paleosols—in the central Atacama Desert since at least 12 Ma. In this setting where evaporation strongly distorts $\delta^{18}O_{sc}$ values, useful estimates of paleo-elevation are hard to obtain, as we will argue in a later section of the paper. Rech et al (2006) go on to describe older calcareous paleosols in the ~20 Ma range characterized by strong reddening, visible secondary clay accumulations, and vertic fracturing of the paleosol. These are all features uncharacteristic of calcareous paleosols of the current Atacama Desert, and a good match to modern soils now found in moister central Chile where mean annual rainfall (MAP) is > 200 mm/yr. These kinds of pedogenic carbonates are much better targets for paleoelevation reconstruction.

In addition to reddening and clay content (Birkeland 1984), depth of pedogenic carbonate leaching is a useful qualitative index of paleoaridity. In modern soils, leaching depth increases by 0.2 to 0.3 cm/cm rainfall (Jenny and Leonard 1934; Royer 1999; Wynn 2004). The scatter around this relationship is very large, due to a host of factors such as erosion of paleosol surfaces, texture of parent material, local drainage conditions, and so forth (Royer 1999). In our experience, depth of leaching is useful where multiple paleosols can be observed, vertically and along strike, in an attempt to average out some of the non-climatic influences listed above.

For a more quantitative approach to paleoaridity, carbon isotopes from pedogenic carbonate can provide evidence of soil respiration rates, which are closely related to plant cover at a site, and hence aridity. Respiration rates cannot be calculated from $\delta^{13}C_{sc}$ values without independent knowledge of $\delta^{13}C$ values of local plant cover, because both C_3 and C_4 plants compose modern vegetation. C_4 plants expanded across low-latitude and mid-latitude ecosystems between 8 and 4 Ma, prior to which time C_3 plants appear to dominate most ecosystems (Cerling et al. 1997; but see also Fox and Koch 2004). This ecologic simplification makes estimation of paleo-respiration rates from $\delta^{13}C_{sc}$ values possible prior to about 8 Ma, since the only two end members contributing to soil CO_2 are C_3 plants and the atmosphere (Fig. 6). Prior to the mid-Oligocene, pCO_2 appears to have been much higher (Pagani et al. 1999; Pearson and Palmer 2000; Pagani et al. 2005), once again complicating estimation of paleo-respiration rates. But for the period 30-8 Ma, paleo-respiration rates can be calculated from $\delta^{13}C_{sc}$ values alone, allowing us to trace the development of rain shadows in the lee of both the emerging Himalaya and Andes.

The basic principle, first modeled in Cerling (1984) and modified in Davidson (1995) and Quade et al. (2007), is that the $\delta^{13}C$ value of soil CO_2, and the pedogenic carbonate from which it forms, is determined by the local mixture of C_3 plant CO_2 and atmospheric CO_2. In general, the deeper in the soil profile or the higher the soil respiration rate, the greater the influence of plant-derived CO_2. Thus, for a given soil depth, $\delta^{13}C_{sc}$ values are determined by soil respiration rate. Temperature has a very modest effect on carbon isotope fractionation ($\sim -0.11\ ‰/°C$). As an example, $\delta^{13}C_{sc}$ values will increase from -4.6 to $-9.4‰$ as respiration rates increase from 0.5 to 8 mmoles/m²/hr at 9 °C (Fig. 6), where > 0.5 mmoles/m²/hr roughly translates into MAP > 200 mm/yr in the Mojave Desert. Later in the paper we will use this framework to interpret $\delta^{13}C_{sc}$ values from Tertiary-age carbonates in Himalaya/Tibet and the Andes.

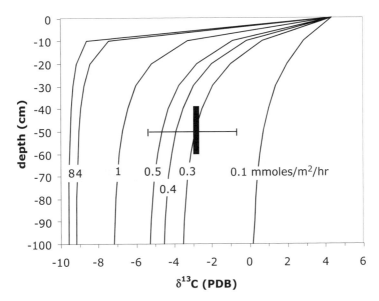

Figure 6. Changes in the $\delta^{13}C$ (PDB) value of soil carbonate at various respiration rates (indicated) in a pure C_3 world and at low pCO_2 (400 ppmV) from 30 to 8 Ma, as predicted by the one-dimensional diffusion model of Cerling (1984) and Quade et al. (2007), assuming $\delta^{13}C$ (PDB) of plants = $-25‰$, $\delta^{13}C$ (PDB) of the atmosphere = $-6.5‰$, an exponential form to CO_2 production with depth, and characteristic CO_2 production depth (k) of 32 cm. Black bar denotes range of sampled carbonate soil depths and average $\delta^{13}C_{sc}$ (PDB) values from late Oligocene paleosol carbonate samples from Nima, southern Tibet; horizontal error bar reflects uncertainties in soil porosity = 0.5 ± 0.1 (see Hillel 1982), $T = 9\pm3°C$, and $pCO_2 = 400 \pm 100$ ppmV (see Pagani et al. 2005).

Elevation variation in the $\delta^{18}O_{sc}$ value of pedogenic carbonate

It follows from our previous discussion of modern pedogenic carbonates that there should be a strong correlation between elevation of sites and local $\delta^{18}O_{mw}$, and hence $\delta^{18}O_{sc}$ values. For all the interest in paleoelevation reconstruction, there is surprisingly little ground-truthing of the elevation-$\delta^{18}O_{sc}$ relationship by direct sampling of modern carbonates. Most studies simply calculate theoretical $\delta^{18}O_{sc}$ values using some assumed $\delta^{18}O_{mw}$/elevation relationship, and local MAT. Here we explore the validity of these assumptions by looking at actual, not theoretical, $\delta^{18}O_{sc}$ values from modern soils from different elevations in southern Nevada, Tibet, and the Pacific slope of the Andes. Our conclusion is that $\delta^{18}O_{sc}$ values from the warm and arid Mojave and hyperarid Atacama are severely affected by evaporation, whereas evaporation effects are much reduced in carbonates from moist India/Nepal and dry but very cold Tibet.

$\delta^{18}O$ values of modern (= Holocene) pedogenic carbonate collected from 50 cm depth in Death Valley and up the east faces of the Spring and Grapevine Mountains of southern Nevada decreases steeply with elevation gain by about 5.5 ‰/km ($r^2 = 0.65$) (Quade et al. 1989a) (Fig. 7). By comparison, the $\delta^{18}O_{mw}$ values compiled from 32 stations located across the region decreases with elevation by a modest 1.3 ‰/km ($r^2 = 0.4$) (Friedman et al. 1992). If these $\delta^{18}O_{mw}$ values are representative of rain falling on soil sites, the $\delta^{18}O_{sc}$ values predicted to form in equilibrium with them (adjusting for temperatures at each site) would all be around −13‰ (Fig. 7), and virtually invariant with elevation. We only observe these very low values deep in soils at the wetter sites (MAP > 200 mm/yr). The residual between observed and predicted $\delta^{18}O_{sc}$ values is strongly correlated (Fig. 8; $r^2 = 0.76$) with mean annual rainfall. This shows that almost all of the steep gradient in $\delta^{18}O_{sc}$ values that we observe over a 3000 m elevation range

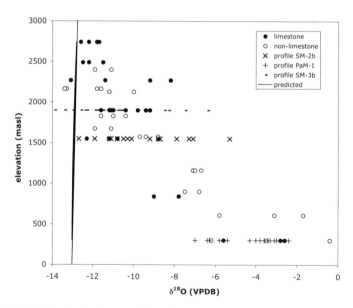

Figure 7. $\delta^{18}O_{sc}$ (PDB) of soil carbonate collected from ~50 cm soil depth versus elevation (masl) from Holocene-age soils in the Mojave Desert (from Quade et al. 1989a). Soil carbonate collected on both limestone and non-limestone parent materials show the same approximate decrease with elevation of −5.5 ‰/km. Soil profile values are the same as in Figure 4. The line denotes the $\delta^{18}O_{sc}$ (PDB) predicted from local $\delta^{18}O_{mw}$ values from 32 rainfall collection sites in the Mojave Desert ($\delta^{18}O_{mw}$ (SMOW) = −0.0013 × elevation (m) − 10.68; $r^2 = 0.39$; converted from δD values using $\delta D = 7.2\delta^{18}O + 7.8$) compiled from Friedman et al. (1992), using MAT + 2 °C at each site (T (°C) = −0.0079 × elevation (m) − 23.07; $r^2 = 0.93$; 8 climate stations).

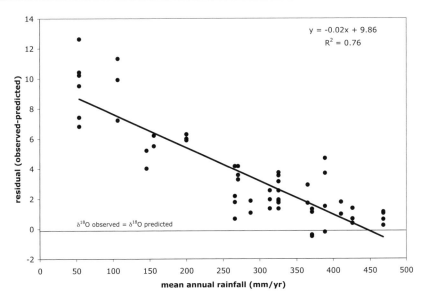

Figure 8. The residuals calculated from Figure 7 plotted against mean annual rainfall for each Mojave Desert site.

is produced by evaporation, and changes in $\delta^{18}O_{mw}$ values have little influence. At rainfall rates below 200 mm/yr, $\delta^{18}O_{sc}$ values from the same pedogenic carbonates are highly enriched in ^{18}O by evaporation (Fig. 8), and can yield large (> 2 km) underestimates of actual elevation.

We recently completed a transect up the Pacific slope of the Andes at 24°S (Fig. 1a; Fig. 9). The transect spans the central sector of the Atacama Desert, one of the driest regions in the world. $\delta^{18}O_{sc}$ values for samples > 50 cm depth from this transect vary between −5.9 and +5.4‰. Scatter of data within one profile or across several profiles in a narrow elevation range is very large, often > 5‰, even from samples deeper than 50 cm. In spite of this scatter, the most negative $\delta^{18}O_{sc}$ values decrease modestly with elevation, from as low as −2‰ on the coast to −5.9‰ at 3400 m, with an excursion toward higher values in the driest part of the transect between about 1000 and 2500 m, a region so dry that it is virtually plantless.

The $\delta^{18}O_{mw}$ values from Fritz (1981) and Aravena et al. (1999) can be used to predict $\delta^{18}O_{sc}$ values assuming MAT + 2 °C at soil depth, with an error of ≤ 1.5‰ at ≥ 50 cm, as calculated in a previous section. In the coastal Atacama, fog drip outstrips rainfall from the rare Pacific storm (Larrain et al. 2002), and hence we compare $\delta^{18}O_{mw}$ values of fog (Aravena et al. 1989) and of local rainfall as assumed parent waters in the coastal zone. We find that observed $\delta^{18}O_{sc}$ values always exceed predicted $\delta^{18}O_{mw}$ values (Fig. 9). Predicted $\delta^{18}O_{sc}$ values along the transect between sea level and 4000 m are between about −3 and −15‰, whereas the *most negative* observed $\delta^{18}O_{sc}$ values range between −3 and −5.9‰. Clearly, the prognosis for meaningful paleoelevation reconstruction in such arid conditions is poor. For example, even with our very large sample size of >60 from between 3000 and 4000 m, we would underestimate paleoelevation using only the most negative values by at least 1 km. This is why it is vital to make some assessment of paleoaridity independent of the conventional $\delta^{18}O_{sc}$ measurements.

The second data set that we briefly consider comes from selected locations along a north-south transect across the Himalaya and Tibet (Fig. 10). This is work in progress, but enough information is already in place to make some useful observations. As with other data sets, we focused on sampling deeper (> 50 cm) carbonate from younger profiles, although in the case of

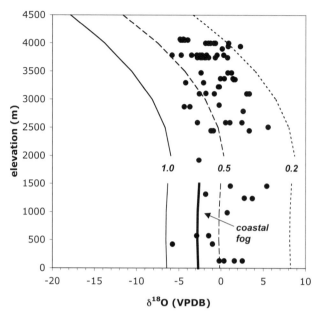

Figure 9. $\delta^{18}O_{sc}$ (PDB) values of Holocene soils versus elevation in the Atacama Desert, between 22.5 and 25°S. The thin lines (solid, dashed, dotted) denote modeled fraction (f) of soil water remaining after evaporation, using local site MAT and $\delta^{18}O_{mw}$ (SMOW) values for rainwater (from Aravena et al. 1999 and Fritz 1981); and the thick line for f = 1.0 for coastal fog from Aravena et al. (1989).

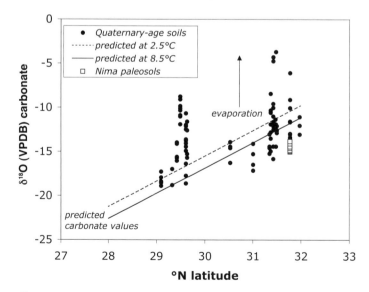

Figure 10. $\delta^{18}O_{sc}$ (PDB) values from Quaternary-age soils across the Himalayan-Tibetan orogen, and "corrected" values (see text) from late Oligocene paleosols at Nima. Lines denote $\delta^{18}O_{sc}$ (PDB) values predicted at a range of soil temperatures (indicated), and using the regression equation of unpublished water data across the Tibetan plateau shown in Figure 2b.

Tibet we also obtained carbonates from older geomorphic surfaces, in order to develop some perspective on the isotopic effects of climate variability during the Quaternary.

In general, $\delta^{18}O_{sc}$ values follow the expected pattern based on changes in modern $\delta^{18}O_{mw}$ values across the Himalayan-Tibetan orogen (Fig. 10). To make this comparison we drew on our published and unpublished data on Holocene-age carbonates from three areas: modern Pakistan, southern Tibet, and a few scattered sites in NE Tibet. In Pakistan, twelve analyses of modern pedogenic carbonates from three profiles yield an average value of −6.3‰ (Quade and Cerling 1995). This value is very similar to the long-term average for the Quaternary-age paleosols from Pakistan (−6.4‰) and from central Nepal (−7.0) (Quade et al. 1995b). The most negative modern $\delta^{18}O_{sc}$ values of −7.7‰ in Pakistan, and −8.5‰ for the late Neogene Siwalik paleosols are also very similar, and are close to that predicted (−8.7‰) to form from weighted modern $\delta^{18}O_{mw}$ values from New Delhi (Araguás-Araguás et al. 1998). So, based on this admittedly restricted sample of carbonates and rainfall, the system appears to be well-behaved: we can accurately reconstruct paleo-elevation for these sites during the late Neogene, which have always been at low elevation (< 500 m), using the most negative $\delta^{18}O_{sc}$ values, and the scatter of $\delta^{18}O_{sc}$ values is relatively low (< 3‰), indicating only modest evaporative modification.

Turning briefly to southern Tibet, we have preliminary data from sixteen surface soils collected between 29 and 32°N, 79 to 92°E, and 3800-5400 m. Pedogenic carbonate from these surface soils displays a range of morphologies from thin gravel coatings to pervasive matrix cements, suggestive in some cases of ages >Holocene, unlike our Mojave and Atacama soils, which are all very young (Holocene). The Tibetan samples yield a wide range of results, from −18 to +3‰ (Fig. 10). For comparison, we can predict $\delta^{18}O_{sc}$ values using $\delta^{18}O_{mw}$ values from Figure 2b and assuming local the range of MAT + 2 °C encompassing these sites. As with the Mojave and Atacama Desert samples, many observed $\delta^{18}O_{sc}$ values exceed predicted values, again because of evaporation. Interestingly, in many cases the observed $\delta^{18}O_{sc}$ values are 1 to 5‰ more negative than expected $\delta^{18}O_{sc}$ values. Detailed consideration of this interesting but incomplete data set is beyond the scope of this review, but we believe that it may illustrate the effects of perhaps the largest uncertainty in all paleo-elevation analysis using $\delta^{18}O_{sc}$ values, that of changing climate. We suggest that modern meteoric waters probably have higher $\delta^{18}O$ values than average longer-term waters as archived by Quaternary-age pedogenic carbonates. This may be due to cyclical changes during the Quaternary in the average penetration northward of Asian monsoon rainfall into southern Tibet. If true, this would lead to overestimates of paleoelevation using modern $\delta^{18}O_{mw}$-elevation relationships. Our analyses illustrate the key advantage of sampling Quaternary pedogenic carbonates as opposed to just modern meteoric waters in calibration studies. Carbonates provide more time-depth and will tend to average out the swings in Quaternary climate.

A few samples of pedogenic carbonate from 3000-3500 m in northeast Tibet fall between −11 and −4‰ in $\delta^{18}O_{sc}$ (PDB), more positive than values in southern Tibet at comparable elevations, and consistent with overall regional trends of increasing $\delta^{18}O_{mw}$ values with latitude. $\delta^{18}O$ values of surface waters from this area and approximate elevation (Garzione et al. 2004) range from −11 to −8‰, consistent with the higher observed $\delta^{18}O_{sc}$ values.

PALEO-ELEVATION ESTIMATES USING
CARBONATE "CLUMPED ISOTOPE" PALEOTHERMOMETRY

Up to this point our review has focused on reconstruction of paleoelevation using $\delta^{18}O_{sc}$ values. This approach has the advantage that suitable samples are common constituents of paleosols, the necessary analytical methods are widely available and straightforward, and the various complicating factors, such as soil-water evaporation and post-depositional diagenesis, have been studied in samples with relatively well-understood histories. Nevertheless, the validity of this "conventional" method ultimately rests on the ability to: (1) account for the effect of temperature on the isotopic fractionation between soil water and pedogenic carbonate; and (2) distinguish orographic effects on the isotopic composition of meteoric water from other possible influences, such as seasonality, and climate and latitude change.

Ghosh et al. (2006b) present an innovation to paleoelevation reconstruction using pedogenic carbonate that potentially addresses these issues. The Ghosh et al. approach uses carbonate "clumped isotope" thermometry (Ghosh et al. 2006a; Schauble et al. 2006) to impose an independent constraint on the temperatures of pedogenic carbonate growth. This information permits a more robust estimate of the $\delta^{18}O_{mw}$ from which carbonate grew and provides an independent constraint on paleoelevation by comparison with the altitudinal gradient in surface temperature.

Principles, methods and instrumentation

Carbonate clumped isotope thermometry is based on the temperature dependence of the abundances of bonds between ^{13}C and ^{18}O in carbonate minerals. This temperature sensitivity stems from the fact that isotope exchange reactions such as:

$$Ca^{13}C^{16}O_3 + Ca^{12}C^{18}O^{16}O_2 = Ca^{13}C^{18}O^{16}O_2 + Ca^{12}C^{16}O_3 \quad (3)$$

are driven to the right with decreasing temperature. That is, ordering, or "clumping," of heavy isotopes into bonds with each other is favored at low temperatures.

We are aware of no analytical method for directly measuring the abundances of reactant and product species in equation (3) with precision sufficient for useful geothermometry. Therefore, the carbonate clumped isotope thermometer uses the abundances of analogous species in CO_2 produced by phosphoric acid digestion of carbonate (i.e., $^{13}C^{16}O_2$, $^{12}C^{18}O^{16}O$, $^{13}C^{18}O^{16}O$ and $^{12}C^{16}O_2$). These measurements are made using a gas-source isotope ratio mass spectrometer that has had its collection array modified to permit simultaneous collection of all cardinal masses 44 through 49 corresponding to CO_2 isotopologues.

Ion corrections, standardization and nomenclature for analyses of $^{13}C^{16}O_2$, $^{12}C^{18}O^{16}O$, and $^{12}C^{16}O_2$ follow established conventions (Santrock et al. 1985; Allison et al. 1995; Gonfiantini et al. 1995). $^{13}C^{18}O^{16}O$ is a special case in several respects (Eiler and Schauble 2004; Wang et al. 2004; Affek and Eiler 2006): most of the variations in its abundance in natural CO_2 (including that produced by acid digestion of naturally occurring carbonates) arise from variations in bulk isotopic composition. In other words, if a population of CO_2 molecules contains an unusually large amount of ^{13}C and/or ^{18}O, then it will contain an unusually large number of $^{13}C^{18}O^{16}O$ molecules simply because the probability of a molecule incorporating both rare isotopes is relatively high. This cause of variations in $^{13}C^{18}O^{16}O$ abundance is relatively uninteresting because it tells nothing beyond what is known from measuring $\delta^{13}C$ and $\delta^{18}O$. Therefore, the nomenclature for reporting analyses of $^{13}C^{18}O^{16}O$ considers its enrichment or depletion relative to the amount that would be present in the analyzed population of molecules if that population had a "stochastic distribution" of isotopes among all possible isotopologues; that is, if the probability that a molecule contains a given isotope in a given location within the molecule were entirely random. This enrichment or depletion is reported using values of "Δ_{47}," which are defined as the difference, in per mil, of the ratio of mass 47 to mass 44 isotopologues in a sample from that ratio expected for a stochastic distribution of isotopes in that sample. Note that the population of mass 47 isotopologues includes contributions from $^{12}C^{17}O^{18}O$ and $^{13}C^{17}O_2$, which cannot be mass discriminated from $^{13}C^{18}O^{16}O$ using existing gas source isotope ratio mass spectrometers. These species must be accounted for when calculating the expected stochastic distribution and when interpreting observed values of Δ_{47}, but are not a significant complicating factor for most materials of interest to the present discussion. Standardization of Δ_{47} values requires comparison of samples with materials that are known to have a stochastic distribution. This is generally accomplished by comparison with gases that have been internally equilibrated by heating to 1000 °C.

The relationship between growth temperature of a carbonate mineral and the Δ_{47} value of CO_2 produced by acid digestion of that carbonate is known for a variety of natural and synthetic materials (including recent pedogenic carbonate) (Fig. 11). The temperature dependence of -0.005% per °C is clear but subtle. Moreover, $^{13}C^{18}O^{16}O$ is an exceedingly rare isotopic species, only ~45 ppm of most natural CO_2. These factors give rise to the two principal technical limitations to carbonate clumped isotope thermometry: (1) exceptionally long analyses of ~one hour and large samples of ~5 mg are required to generate measurements of Δ_{47} having precision good enough for many problems in paleothermometry; and, (2) samples must be thoroughly cleaned by cryogenic, gas chromatographic and other procedures prior to analysis (Eiler and Schauble 2004; Affek and Eiler 2006; Ghosh et al. 2006a). The primary effect of these limitations is to slow the rate at which usefully precise measurements can be made. Previous experience suggests that the highest precision that can be consistently achieved is ~ 0.005 to 0.010 ‰ (Came et al. 2007), and involves such a high level of replicate measurements and standardization that only two or three unknowns per day can be analyzed.

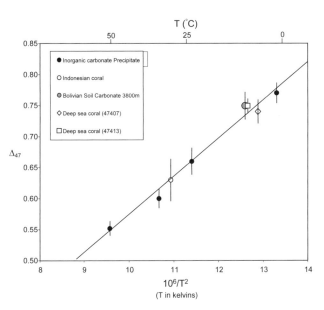

Figure 11. Data documenting the calibration of the carbonate clumped-isotope thermometer for inorganic calcite grown in the laboratory (filled circles) and aragonitic corals grown in nature at known temperatures (an example of one of several biogenic materials we have also calibrated; unfilled symbols). The large, gray circle shows the result of analyses of a modern soil carbonate collected from the Bolivian Altiplano plateau. The horizontal position of this data point is based on the mean annual surface temperature near the site of collection between 2004 and the present.

Advantages and disadvantages

Carbonate clumped-isotope thermometry brings several significant capabilities to the problem of paleoelevation reconstruction. Most importantly, the thermometer is based on the thermodynamics of a homogeneous reaction, involving a reaction among components of a single phase. Therefore, the temperature of carbonate growth is defined rigorously by analysis of the isotopic constituents of carbonate alone, or, more precisely, CO_2 produced from that carbonate. Nothing need be known about the isotopic composition of water from which carbonate grew to determine the temperature of carbonate growth.

Secondly, the material analyzed to determine growth temperature is the same as that used to define the $\delta^{18}O_{sc}$ value (in fact, the two measurements are made simultaneously on the same aliquot of CO_2). Because one knows the growth temperature from Δ_{47} measurements and $\delta^{18}O_{sc}$ values through conventional measurements, one can calculate the $\delta^{18}O_{sw}$ value from which it grew based on the known temperature dependence of the carbonate-water oxygen isotope fractionation (e.g., Kim and O'Neil 1997). Thus, clumped isotope thermometry, combined with conventional stable isotope geochemistry, provides two complementary bases for estimating paleoelevation: comparison of the record of carbonate growth temperatures with inferred altitudinal gradients in surface temperature, and comparison of calculated values of the $\delta^{18}O_{sw}$ with inferred altitudinal gradients in the $\delta^{18}O_{mw}$ values.

Finally, the correlation of calculated $\delta^{18}O_{sw}$ values with carbonate growth temperature imposes one more potentially useful constraint: the slope of this trend, $\partial \delta^{18}O_{water}/\partial T$, for an altitudinal transect is, in many cases, a known or predictable quantity that differs from slopes resulting from variations in season, latitude or climate (Ghosh et al. 2006b; see also Figure 15 of this chapter). Therefore, knowledge of this slope for a suite of related pedogenic carbonates can potentially distinguish between variations in surface temperature and/or the $\delta^{18}O_{mw}$ value caused by changes in elevation versus variations caused by other factors. In practice, this constraint will be more useful in some cases than in others because of natural variations in $\partial \delta^{18}O_{water}/\partial T$ slopes for processes of interest, and because some suites of pedogenic carbonates will be influenced by more than one factor. Nevertheless, it provides another basis for improving the confidence and precision of paleoelevation estimates.

Approaches to paleoelevation reconstruction based on carbonate clumped isotope thermometry also carry with them several disadvantages, most importantly, that the analytical techniques involved in a clumped isotope paleotemperature estimate (Eiler and Schauble 2004; Affek and Eiler 2006; Ghosh et al. 2006b) are slow, technically difficult, and presently made in only one laboratory. It seems likely that these measurements will become more routine as they are adopted in other laboratories and improved through various possible innovations (such as automation). Nevertheless, the supply of data is a limitation at present. Secondly, the precision of data is poor for some problems. This too appears likely to improve through time (e.g., Came et al. 2007) report carbonate clumped isotope measurements with precisions corresponding to errors of only $\pm 0.9\ °C$), but will remain a significant issue for the foreseeable future due to the subtle temperature sensitivity of isotopic clumping in Reaction (1). Finally, because carbonate clumped isotope thermometry is a relatively new technique, the potential exists that there are unrecognized systematic errors in its application to pedogenic carbonates, such as unexpected fractionations during pedogenic carbonate growth, or poor preservation.

DIAGENESIS

Any attempt to reconstruct elevation using geochemical proxies must consider the possibility that diagenesis and burial metamorphism have overprinted the carbonate record. The $\delta^{18}O$ values of all carbonates are quite susceptible to diagenetic alteration, as witnessed by the extraordinary difficulty in reconstructing $\delta^{18}O$ values of sea-water through deep geologic time using carbonates, as contrasted with, for example, $\delta^{13}C$ values and $^{87}Sr/^{86}Sr$ ratios (e.g., Veizer et al. 1999). Carbonates buried to depths of several km or less can be overprinted through dissolution and re-precipitation reactions with ground and formation waters. The best approach to evaluating the potential effects of diagenesis depends on the material under study. For example, studies that use non-marine shells can take advantage of the fact that unaltered shell retains its primary structure (Veizer et al. 1999) and has not been converted to calcite from aragonite. In the case of soil nodules, recrystallization to spar is a clear warning sign, since the primary texture of pedogenic carbonate is mostly micritic to microsparitic (Chadwick et al.

1988; Deutz et al. 2002; Garzione et al. 2004). These replacement textures can be identified with the naked eye, optical microscope or cathodoluminiscope, and avoided by sampling with a slow-speed drill or similar tool, as in both our Andean and Tibet studies discussed below. This strategy appears to be sufficient for recovering records of primary variation in $\delta^{18}O_{sc}$ and $\delta^{13}C_{sc}$ values (e.g., Koch et al. 1995). However, there is no means of knowing, *a priori*, whether this approach will avoid overprinting of the ^{13}C-^{18}O ordering that governs the carbonate clumped isotope thermometer. Hence, in the next section on the Andes we discuss other ways to evaluate the fidelity of the clumped isotope thermometer.

Another approach to diagenesis discussed in the final section of this paper is to look for heterogeneity in $\delta^{18}O_{sc}$ values of adjacent calcitic phases such as interbedded marine, detrital (Quade and Cerling 1995), lacustrine (e.g., Cyr et al. 2005; DeCelles et al. 2007a), and paleosol carbonates. The concept here is that diagenesis should tend to homogenize $\delta^{18}O$ values that in a primary state may have had very different values.

PALEOSOL RECORDS OF PALEOELEVATION CHANGE: CASE STUDIES

We now turn to published and unpublished results of case studies from the Andes and Asia as illustrations of how $\delta^{18}O_{sc}$ and Δ_{47} measurements from paleosol carbonates can be used in paleoelevation reconstruction. The key points that we address include: (1) accounting for the role of climate change in reconstructing paleoelevation; (2) testing for diagenetic alteration of primary $\delta^{18}O_{cc}$ values due to burial in deep sedimentary basins; (3) evaluating paleoaridity, and therefore the extent to which evaporation has increased the $\delta^{18}O_{mw}$ values of paleowaters, producing underestimates of paleoelevation, (4) comparing pedogenic carbonate records to other sedimentary carbonates to determine the fidelity of various elevation proxies, and (5) the power of combining conventional $\delta^{18}O_{mw}$ and Δ_{47} measurements in paleoelevation reconstruction.

Paleoelevation reconstruction of the Bolivian Altiplano

We can combine both $\delta^{18}O$ and Δ_{47} measurements of pedogenic carbonate from the Miocene deposits of Bolivia to trace uplift of the Bolivian Altiplano through time. The northern Altiplano is an attractive target for both approaches to paleoelevation reconstruction for several reasons. It preserves carbonate-bearing sediments deposited over much of its uplift history; the modern altitudinal gradients in temperature and the $\delta^{18}O_{mw}$ values are known (Gonfiantini et al. 2001), providing a locally calibrated basis for reconstructing past elevation; modern rainfall rates are > 250 mm/yr, providing adequately wet conditions to reconstruct meteoric water compositions; previous studies of the mountain belt history and geomorphology of the Altiplano provide observations that can be used to test and elaborate on one's results; and, the amplitude of elevation change is large (several km) and thus the analytical precision of clumped isotope measurements is not a severe limitation.

The Altiplano basin is a broad internally drained basin that occupies the central Andean plateau between 15°S and 25°S. At an average elevation of ~3800 m, the Altiplano is situated between the Eastern and Western cordilleras that reach peak elevations in excess of 6 km (Figs. 1a, 2a). The Western Cordillera is the modern magmatic arc of the Andes, and the Eastern Cordillera is a fold-thrust belt made up of deformed Paleozoic metasedimentary rocks. The Altiplano basin has been internally drained since at least late Oligocene time, evidenced by both westward paleoflow and derivation of sedimentary sources from the Eastern Cordillera (Horton et al. 2002; DeCelles and Horton 2003). The late Miocene stratigraphic sections near Callapa are ~3500 m thick and are exposed in the eastern limb of the Corque synclinorium. Three measured sections include fluvial and floodplain deposits in the lower 1200 m and the upper 800 m and widespread lacustrine deposits (laterally continuous for more than 100 km along strike) in the middle part of the section (Garzione et al. 2006). Age constraints on these rocks come from $^{40}Ar/^{39}Ar$ dates on tuffs within our measured section (Marshall et al. 1992) and magnetostratigraphy (Roperch et al. 1999; Garzione et al. 2006).

Fluvial-floodplain and lacustrine sediments contain authigenic carbonates for which $\delta^{18}O_{sc}$ values and $\delta^{13}C_{sc}$ values were obtained (Garzione et al. 2006). The floodplain deposits contain both paleosol carbonate nodules and palustrine carbonates. Paleosols are massive and red to reddish brown. Discrete carbonate (Bk) horizons include rare carbonate rhizoliths and occur below the upper part of the B horizon that has been leached of carbonate. Paleosol carbonate nodules, 0.5 to 3 cm in diameter, were sampled between ~20 and 80 cm below the top of the paleosol where their depth within the soil profile could be measured. In many instances, it was not possible to determine the top of the soil profile because, in general, floodplain lithofacies show extensive oxidation and pedogenesis that makes it difficult to identify the top of individual soil profiles. Palustrine carbonates represent marsh or shallow pond deposits in the floodplain adjacent to fluvial channels. These laminated, mud-rich micrites presumably precipitated seasonally when evaporation rates and productivity were higher. Within the lacustrine interval in the middle part of the section, carbonates are rare and include laminated, very thinly bedded, micritic limestone that contains vertical worm burrows and laminated, thinly to thickly bedded, calcareous mudstone. In thin section, paleosol, palustrine, and lacustrine carbonates lack sparry calcite, suggesting that they have not undergone extensive, late-stage diagenesis. This inference is supported by Δ_{47} paleothermometry data that indicate that pedogenic carbonates do not show a systematic increase in formation temperature with burial depth (Eiler et al. 2006; Fig. 12).

Garzione et al. (2006) excluded lacustrine carbonates from their paleoelevation analysis while including palustrine and pedogenic carbonates. Both modern and ancient lake studies

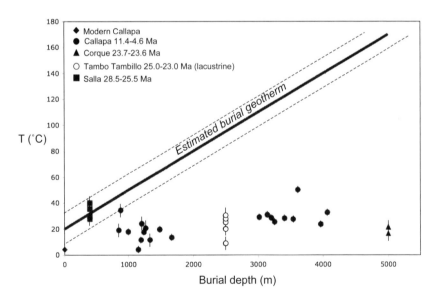

Figure 12. Apparent growth temperatures for various Altiplano carbonates based on clumped isotope thermometry, plotted as a function of estimated maximum burial depth. Symbols discriminate among soil carbonates from sections near Callapa, Corque and Salla and lacustrine carbonates from near Tambo Tambillo, as indicated by the legend. The heavy solid line indicates an estimated burial geotherm, assuming a surface temperature of 20 °C and a gradient of 30 °C per km. The dashed lines define a ±10° offset from this trend, which we consider a reasonable estimate of its uncertainty. Carbonates deposited within the last 28.5 Ma and buried to 5000 meters or less exhibit no systematic relationship between apparent temperature and burial depth, and show no evidence for pervasive resetting of deeply buried samples. Error bars are ±1σ (when not visible, these are approximately the size of the plotted symbol).

show that closed-basin lakes, such as the paleolake that occupied the northern Altiplano in late Miocene time, experienced moderate to extreme evaporative enrichments in the ^{18}O of water and carbonate that precipitated from it (e.g., Talbot 1990). Studies in the Altiplano region support the inference that closed-lake waters in the Altiplano and Eastern Cordillera have higher values of δ^{18}O than local rainfall (Wolfe et al. 2001). In addition, Altiplano lacustrine deposits contain abundant gypsum suggestive of extreme evaporation that would systematically bias carbonates toward lower estimates of paleoelevation. Comparing lacustrine deposits to paleosol carbonates above and below the lacustrine interval shows that $\delta^{18}O_{mw}$ values of lacustrine carbonates are on average 2.8‰ to 3.8‰ higher than pedogenic carbonates (Table 1), which reinforces the notion that they record the effects of evaporative enrichment in ^{18}O. Palustrine carbonates form in settings comparable to open-basin lakes through which water flows and where evaporative enrichment is much less than in closed basins. These depositional environments are in communication with river waters, lack interbedded gypsum, and therefore should not be as evaporitic. Despite lack of evidence for extreme evaporation, these carbonates generally record δ^{18}O values that are ~1‰ higher than $\delta^{18}O_{sc}$ values of similar age (Table 1). The consistently lower $\delta^{18}O_{sc}$ values indicate that paleosol carbonates are probably the most faithful recorders of the isotopic composition of meteoric water and therefore provide the best estimates of paleoelevation.

In modern soils globally, $\delta^{13}C_{sc}$ and $\delta^{18}O_{sc}$ values covary, with higher δ values reflecting more arid conditions and lower respiration rates (Cerling 1984; Quade et al. 1989a; Deutz et al 2002). In contrast, within the Altiplano basin, $\delta^{13}C_{sc}$ values increase at 7.6 Ma, whereas $\delta^{18}O_{sc}$ values decrease (Figs. 13a and b). A plot of δ^{18}O values versus δ^{13}C values (Fig. 13c) (excluding one outlier) show an inverse relationship between δ^{18}O and δ^{13}C values ($R^2 = 0.37$). The timing of the increase in δ^{13}C values could be interpreted to reflect lower soil respiration rates (e.g., Cerling and Quade 1993) or an increasing proportion of C_4 grasses in the Altiplano at this time, coincident with the global expansion of C_4 grasses (Cerling et al. 1997; Latorre et

Table 1. Oxygen and carbon isotope data from stratigraphic sections sampled in the eastern limb of the Corque syncline (from Garzione et al. 2006).

Sample type	δ^{18}O(VPDB) (mean ‰ ± 1σ)	δ^{13}C(VPDB) (mean ‰ ± 1σ)	# of samples
11.5 to 10.3 Ma			
paleosols	−11.8 ± 0.9	−9.0 ± 1.0	14
palustrine	−10.9 ± 1.5	−8.2 ± 0.8	9
10.1 to 9.1 Ma			
lacustrine	−9.0 ± 1.3	−8.8 ± 3.2	21
7.6 to 6.8 Ma			
paleosol	−12.8 ± 0.9	−4.3 ± 2.1	2
palustrine	−11.3 ± 1.4	−8.4 ± 1.6	12
6.8 to 5.8 Ma			
paleosol	−14.7 ± 0.7	−6.1 ± 1.0	4
s.s. cement	−14.8 ± 0.4	−8.6 ± 1.2	5

Standard deviation (1σ) of the mean of each sample set is reported. VPDB - Vienna Peedee belemnite. Two pedogenic carbonate data points have been excluded, one that shows much higher temperatures relative to other pedogenic nodules of the same age (based on Δ_{47} measurement) and one that has a much higher δ^{18}O relative to other pedogenic nodules of the same age.

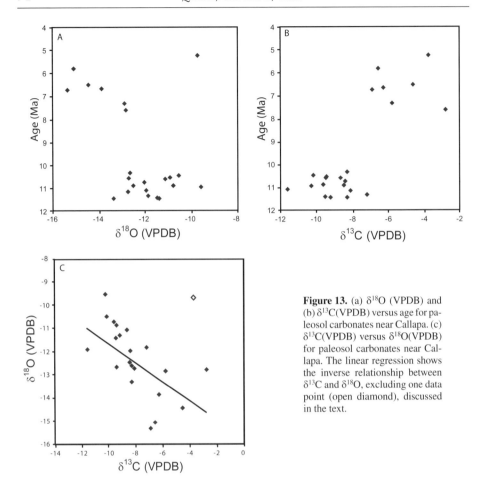

Figure 13. (a) $\delta^{18}O$ (VPDB) and (b) $\delta^{13}C$(VPDB) versus age for paleosol carbonates near Callapa. (c) $\delta^{13}C$(VPDB) versus $\delta^{18}O$(VPDB) for paleosol carbonates near Callapa. The linear regression shows the inverse relationship between $\delta^{13}C$ and $\delta^{18}O$, excluding one data point (open diamond), discussed in the text.

al. 1997). Analysis of both modern and fossil tooth enamel of large grazing mammals in the Altiplano shows no evidence for significant C_4 presence in the region today or at any time in the past 8 Myr (Bershaw et al. 2006). Therefore, an increase in the proportion of C_4 grasses is an unlikely cause for the observed increase in $\delta^{13}C_{sc}$ values. We suggest that this increase may be the result of lower respiration rates due to increasingly arid conditions in the Altiplano.

Increasing aridity in Altiplano basin is supported by several lines of evidence from the sedimentary record. First, both the thickness of fluvial channel deposits and their lateral continuity decrease up-section. In the oldest part of the section, channel sandstone bodies up to 15 m thick continue laterally for 100s of meters. Internally, these sand bodies consist of amalgamated lenticular beds that show low-angle cross stratification, interpreted to reflected lateral accretion on migrating longitudinal bars. Individual bar deposits are up to 2 m thick, reflecting the maximum channel depth during high discharge events. Above the lacustrine interval, channel sandstone bodies are up to 5 m thick and continue laterally for tens of meters. Individual beds often contain trough cross-stratification and are <1 m thick, indicating shallower channel depths compared to older deposits. Both shallower channel depths and a decrease in the lateral extent and overall thickness of sandstone channel deposits suggest lower discharge in the fluvial systems above the lacustrine interval. The second line of evidence comes from the depth of pedogenic carbonate formation. Below the lacustrine interval, pedogenic carbonates were observed to have formed below 30 cm in the paleosol profile, whereas pedogenic carbonates formed at depths as shallow as 16 cm in paleosols above the lacustrine interval. The observation in Holocene soils that depth to the Bk horizon increases with increasing mean annual precipitation (Royer 1999) suggests that the youngest part of the soil record reflects the most arid conditions. The combined observations of increased aridity in late Miocene time point to lower soil respiration rates as the cause of the increase in $\delta^{13}C$ values up section.

We revise and update paleoelevation estimates for the northern Altiplano here and in the following section on "clumped isotope thermometry." Realizing that Gonfiantini et al. (2001) mislabeled data in their Table 5 as weighted means, when they are in fact unweighted means, we estimate elevation as in Garzione et al. (2006) using the weighted mean values for 3 years of rainfall data reported in Table 6 of Gonfiantini et al. (2001). Seven sites for which rainfall amount was also reported can be used for the regression. The linear regression to these data is:

$$h = -472.5\delta^{18}O_{mw} - 2645 \qquad (4)$$

where h = elevation in meters, and with an $r^2 = 0.95$ (Fig. 2a, 14). We choose a linear regression because there are not enough data points in the weighted mean data set to evaluate whether a polynomial provides a better fit to the data. Surface water data collected over 2004 and 2005 years from small tributaries along the Coroico River, where rainfall samples were collected, agree with the weighted mean values observed in the sparse rainfall data set and show a similar isotopic gradient to Equation (4), which indicates that the isotopic gradient observed in meteoric water is reflected in surface water. Use of Equation (4) produces elevation estimates of: 400 to 2200 m in carbonates deposited before 10.3 Ma, 2000 to 3800 m in carbonates deposited between 7.4 Ma and 6.8 Ma, and 4000-4700 m in carbonates deposited after 6.8 Ma. This reanalysis of the data based on weighted mean precipitation suggests essentially the

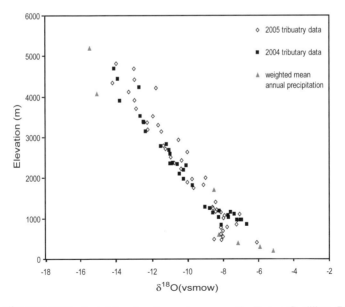

Figure 14. $\delta^{18}O$ (VSMOW) of rainfall and surface waters across the Eastern Cordillera. Rainfall data represent the weighted mean isotopic composition (1983-1985) from Gonfiantini et al (2001). Tributaries to the Coroico river were sampled in late May 2004 and early May 2005 and are plotted relative to the sampling elevation. A linear regression to the rainfall data (gray triangles) defines Equation (4) in the text.

same amount of surface uplift (2.5 to 3.6 km) as our original estimates of 2.5 to 3.5 km. It also brings the calculated elevations for the oldest part of the section into a more reasonable range, with all values above 400 m, as opposed to the negative elevations calculated for the oldest part of the record in Garzione et al. (2006). Given that paleoelevation estimates are prone to systematic biases that tend to underestimate elevations, we suggest using the difference between the highest elevation estimates prior to 10.3 Ma and after 6.8 Ma of 2.5 km as a conservative estimate for the amount of late Miocene surface uplift, with an uncertainty of ±1 km (see Rowley and Garzione 2007 for discussion of uncertainties)

Ideally, to apply stable isotopes to paleoelevation reconstruction, low-elevation records of regional climate are required to constrain variations in climate that may influence the isotopic record. For example, the Siwalik foreland basin deposits south of the Himalaya provide low-elevation constraints on climate and the isotopic composition of meteoric water used to estimate elevations in southern Tibet (Garzione et al. 2000a; Rowley at al. 2001; Currie et al. 2005). The accuracy of the current estimates of paleoelevation of the Altiplano is limited by lack of data from simultaneous low elevation records in the Subandean foreland (Rowley and Garzione 2007). Several potentially confounding variables that are currently unconstrained are the proximity and isotopic composition of sources of water vapor, secular changes in terrestrial temperature associated with late Cenozoic cooling, and the effects of climate change on the isotopic lapse rate. Because of extensive evapotranspiration over the Amazon basin, modern water vapor delivered to the eastern flank of the Andes is isotopically similar to water vapor at the Atlantic coast (Vuille et al. 2003). However, the continental isotopic gradient may have changed through time as a result of changes in plant cover and/or changes in the extent and nature of inland water bodies in the Amazon basin. Recent studies document a late Miocene marine incursion in the northern Subandean foreland. Although these marine rocks lack precise age constraints, Hernández et al. (2005) show that at least one marine unit was slightly younger than 7.7±0.3 Ma. Hernández et al. (2005), Uba et al. (2006), and Hoorn (2006) show that marine conditions existed briefly, while lacustrine conditions prevailed for longer time periods. A lacustrine origin is supported by stable isotopic evidence that indicates a predominantly Andean freshwater and cratonic freshwater origin for the water isotopic composition recorded in microfossils deposited between 18°S and 19°S (Hulka 2005). The changing paleogeography of the Andean foreland points to the need to better establish foreland records of both climate and paleogeography to more accurately constrain paleoelevation of the Altiplano in late Miocene time.

Results from clumped isotope thermometry

Burial effects. Eiler et al. (2006) examined the effects of post-depositional alteration on the carbonate clumped isotope thermometer in soil and lacustrine carbonates from the northern Altiplano, spanning a range in age from 0 to 28.5 Ma and maximum burial depths from ~0 to ~5 km. Age and burial depth estimates for these samples are based on data presented in MacFadden et al. (1985), Kennan et al. (1995), Kay et al. (1998), Garzione et al. (2006), and Ghosh et al. (2006b) and on unpublished magnetostratigraphic, U-Pb geochronologic, and stratigraphic data summarized in Eiler et al. (2006). The results (Fig. 12) demonstrate that the temperatures recorded by carbonate clumped isotope thermometry are uncorrelated with burial depth, generally (i.e., with one exception) indistinguishable from earth-surface temperatures, and preserve temperatures as low as ca. 20 °C in pedogenic carbonates as old as 23.6-23.7 Ma and buried as deep as 5 km. No systematic differences in apparent temperature were observed between soil and lacustrine carbonates, suggesting that the thermometer is insensitive to variations in textural characteristics such as grain size and porosity that might influence susceptibility to overprinting. These results suggest that burial diagenesis has no systematic effect on the temperatures recorded by the carbonate clumped isotope thermometer in micritic portions of pedogenic carbonates over timescales of tens of millions of years and up to burial temperatures approaching 200 °C. Nevertheless, there are reasons to remain cautious when applying this approach to new suites, because one of the samples originally examined by Ghosh et al. (2006b) recorded a temperature of 50 °C—higher than any plausible depositional temperature. This result is not representative of the record as a whole, but clearly indicates the possibility for overprinting during burial. Moreover, the samples in Figure 12 collected near Salla are shallowly buried but record temperatures that could lie on a burial geotherm. Eiler et al. (2006) speculate that hydrothermal activity associated with nearby ~15-25 Ma magmatism might have partially or completely overprinted these samples.

Paleoelevations of pedogenic carbonate from the Corque syncline. Only the 11.4 to 4.6 Ma paleosols exposed in the Callapa section (Figs. 1a, 12) of the Corque syncline have been studied in enough detail to justify a detailed reconstruction of elevation history. Ghosh et al. (2006b) present such a reconstruction, which we now review. Our discussion of these data differs in some respects from that presented in Ghosh et al. (2006b) because we have re-evaluated the constraints on long-term climate change (which influences the low-elevation "base line" of a paleoaltitude estimate) and re-calculated the elevation trend to $\delta^{18}O_{mw}$ values across the central Andes today. Figure 12 includes data for Callapa pedogenic carbonates from two sources; those from Ghosh et al. (2006b) were analyzed with a relatively high degree of replication and have uncertainties in Δ_{47} equivalent to ±2 to 3 °C, whereas data from Eiler et al. (2006) were not as thoroughly replicated and have uncertainties in apparent temperature averaging ±5 °C. This latter data set is sufficiently precise to test for burial metamorphic overprinting (Fig. 12), but provides inferior constraints on paleoelevation. We therefore focus our discussion on the data from Ghosh et al. (2006b).

Ghosh et al. (2006b) found that pedogenic carbonates from Corque syncline paleosols deposited between 11.4 and 10.3 Ma record carbonate precipitation temperatures of 28.4 ± 2.6 °C (standard error, or s.e., of ±0.9 °C); those grown between 7.6 and 7.3 Ma record temperatures of 17.7 ± 3.1 °C (s.e. of ±2.2 °C); and those grown between 6.7 and 5.8 Ma record temperatures of 12.6 ± 5.6 °C (s.e. of ±2.8 °C). This range in apparent temperature through time is broadly similar to the modern temperature variation with elevation between the east flanks of the Andes and the Altiplano, suggesting that the Altiplano may have risen from low elevation (ca. 1 km or less) to its modern elevation (ca. 4 km) between 10.3 and 6.7 Ma. Note that apparent temperatures for pedogenic carbonates grown between 11.4 and 10.3 Ma are at the high end of modern temperature variations at low elevations on the flanks of the Andes. This could reflect a warmer Miocene climate and/or preferential growth of paleosol carbonates during the austral summer. If we neglect this complicating factor, these data suggest cooling of 15.7 ± 2.9 °C (±1 se). Given the modern gradient in temperature with elevation in the Andes today (4.66 °C/km; Gonfiantini et al. 2001 and data compiled at *www.climate-zone.com*), this change is consistent with uplift of 3.4 ± 0.6 m, for an average uplift rate of 0.94 ± 0.17 mm/ yr between 10.3 and 6.7 Ma. Differences in climate and latitude can account for anywhere between 1.3 °C (Savin et al. 1975; Smith et al. 1981, Zachos et al. 2001) and 2.3 °C (Berner and Kothavala 2001) of the observed temperature change. If we correct for this systematic error, then the amount of cooling due to elevation change could be a minimum of 13.4 ± 2.9 °C (assuming a 2.3 °C change in low-elevation mean annual temperature, all between 10.3 and 6.7 Ma), implying elevation gain of 2.9 ± 0.6 km and an uplift rate of 0.80 ± 0.17 mm/yr.

The contrast in $\delta^{18}O$ (SMOW) values of waters in equilibrium with 11.4 to 10.3 Ma pedogenic carbonates (average −8.6 ± 0.4‰, ± 1 s.e.) versus those in equilibrium with 6.7 and 5.8 Ma pedogenic carbonates (average −14.6 ± 0.6‰, ± 1 s.e.) is 6.0 ± 0.7‰, (± 1 s.e. in the difference of the means). The modern gradient in the annual weighted average $\delta^{18}O_{mw}$ values on the slopes of the Andes (equation 4) implies an elevation gain of 2.8 ± 0.3 km (from 1.4 ± 0.2 to 4.2 ± 0.3 km) between 10.3 and 6.7 Ma, or an uplift rate of 0.78 ± 0.08 mm/yr.

Figure 15 plots the growth temperatures of Corque syncline pedogenic carbonates versus the $\delta^{18}O$ values of waters from which they grew, and compares the trend in those data to the

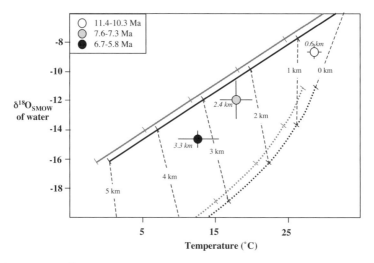

Figure 15. Plot of the $\delta^{18}O_{SMOW}$ value of water in equilibrium with soil carbonate nodules vs. the growth temperatures of those nodules. Circles are averages for the 11.4-10.3 Ma, 7.6-7.3 Ma and 6.7-5.8 Ma age groups, as indicated in the legend. Error bars are ±1 standard error of the population. Gray lines show the mean annual trend (solid) and trend of Jan/Feb extreme (dotted) for the modern relationships between surface temperature and $\delta^{18}O_{SMOW}$ of meteoric water. These trends are contoured for altitude in 1-km increments. The similar heavy and dotted black trends plot the expected location of the mean annual and Jan/Feb extremes in the mid-Miocene, based on inferred changes in the latitude of Bolivia, low-latitude climate change and secular variation in the $\delta^{18}O$ of sea water (Savin et al. 1975; Smith et al. 1981). Fine dashed lines connecting the mid-Miocene mean annual trend and Jan/Feb extreme trend show the slopes of seasonal variations in T and $\delta^{18}O$ of water at a fixed altitude. Paleoaltitudes of age-group averages are estimated by their intersections with this set of altitude contours, as indicated by the italicized text.

modern trends in surface temperatures and meteoric water $\delta^{18}O$ values across a range of elevations in the central Andes. These modern trends are shown both for the annual weighted mean (solid grey line) and Jan/Feb seasonal extreme (dotted gray line), and have been corrected for differences between the Miocene and today in climate, latitude of the Altiplano and the $\delta^{18}O$ value of the ocean (see the caption to Figure 15 for details). Fine, dashed black contour lines of constant elevation link the mean annual trend with the Jan/Feb extreme for the inferred Miocene trends. This figure is useful in two respects: (1) the fact that the trend defined by data for Miocene pedogenic carbonates parallels the trend for an elevation transect re-enforces the interpretation that variations in pedogenic carbonate growth temperature in the northern Altiplano and $\delta^{18}O$ primarily reflect elevation change rather than other factors (e.g., climate change; latitude change; changing seasonality of pedogenic carbonate growth); and (2) we can use a sample's position in relation to the iso-elevation contours in Figure 15 to estimate its paleoelevation, thereby circumventing systematic errors that could occur if pedogenic carbonates had any variations in the season of growth. This is important because the plot of points in Figure 15 suggests that the season of formation for the pedogenic carbonates examined in this study are biased toward warm, low-$\delta^{18}O$ conditions that prevail in the Bolivian austral summer. Our results suggest that the Altiplano rose 2.7 ± 0.4 km (from 0.6 ± 0.2 km to 3.3 ±0.4 km) between 10.3 Ma and 6.7, implying an average uplift rate of 0.73 ± 0.12 mm/yr.

Paleoatimetry of southern Tibet and the Himalaya

Burial effects. We have been involved in several studies in the Himalaya and Tibet that illustrate the need to carefully evaluate the potential effects of burial diagenesis on $\delta^{18}O_{sc}$ values before using them to reconstruct paleoelevation. The first comes from Penbo in southern Tibet

(Fig. 1b) where Leier (2005) analyzed paleosol carbonates in close stratigraphic proximity (< 200 m) to marine carbonates from the late Cretaceous Takena Formation. This association shows that the Takena paleosols formed near sea-level. The nodules from the paleosols are a mix of micrite with local sparry zones, both of which we analyzed. Both textures yielded $\delta^{18}O_{sc}$ values -13 to $-16‰$ (Fig. 16). These are implausibly low for a low-latitude coastal site and reflect isotopic exchange, possibly at elevated temperatures, with meteoric waters with very low $\delta^{18}O_{mw}$ values, such as those that occur in the region today. To test this, Leier (2005) analyzed stratigraphically adjacent marine limestone, including spars, micrite, and fossils fragments. Again, all phases—from matrix micrite to recrystallized fossils—yielded much lower $\delta^{18}O$ values (-13 to $-16‰$; Leier 2005; Fig. 16) than expected (0 to $-4‰$) for late Cretaceous marine carbonates (Veizer et al. 1999). This example clearly illustrates the effects of isotopic homogenization due to burial diagenesis of originally heterogeneous calcite phases.

The second example is unpublished data from paleosol carbonates of the Oligocene-age Dumri Formation in central Nepal (Fig. 1b). The Dumri Formation underlies the better known Siwalik Group, and represents the early stages (Oligo-Miocene) of Himalayan foreland deposition. Non-paleosol carbonates are unavailable from our study sections to check for isotopic homogenization expected from diagenesis. Petrographic evidence shows local but not pervasive recrystallization of paleosol nodules. Most $\delta^{18}O_{sc}$ values, however, are $< -17‰$, implausibly low given that the paleosols must have formed near sea-level along the ancestral Ganges River and the foreland reaches of its paleo-tributaries. The Dumri Formation at this location experienced deep (> 8 km) tectonic burial under a thick thrust wedge of Greater Himalayan nappe rocks (DeCelles et al. 1998), so resetting of $\delta^{18}O_{sc}$ values at elevated temperatures is unsurprising.

A final example comes from the Nima basin located ~450 km northwest of Lhasa at about 4500 m, in the southern part of the Bangong suture zone, which separates the Qiangtang and Lhasa terranes in central Tibet (Fig. 1b). The southern Nima basin contains more than 4 km of Tertiary alluvial, lacustrine, and lacustrine fan-delta deposits that accumulated next

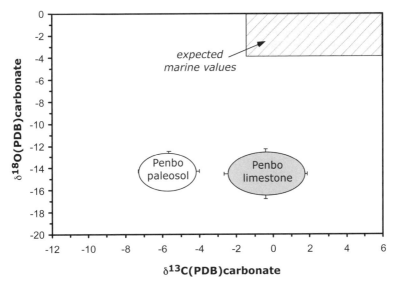

Figure 16. $\delta^{18}O_{sc}$ (PDB) versus $\delta^{13}C_{sc}$ (PDB) values from marine limestone and paleosol carbonate in the late Cretaceous Takena Formation at the Penbo locality, southern Tibet (Leier 2005). Range of expected marine isotopic values for the late Cretaceous from Veizer et al. (1999).

to growing thrust-faulted ranges (DeCelles et al. 2007b). Lacustrine marl beds and well-developed calcareous paleosols are common in the informally designated Nima Redbed unit which is exposed in a roughly 50 km-long outcrop belt. For age control, ^{40}Ar/^{39}Ar dates from six reworked tuffs in the Nima Redbed unit securely place the age of the analyzed deposits between 25 and 26 Ma (late Oligocene) (DeCelles et al. 2007b).

We devised an "isotopic conglomerate test" in order to assess the potential for diagenetic resetting of the isotopic paleoelevation signal (DeCelles et al. 2007a). The Nima Redbeds contain numerous conglomerates interbedded with the marls and paleosols that we sampled for paleoelevation reconstruction. Among other rock types, these conglomerates contain abundant marine limestone clasts derived from the Lower Cretaceous Langshan Formation cropping out around the basin. Where paleochannels carrying these gravels scour into contemporaneous marls and paleosols, reworked paleosol carbonate nodules are also present in channel lags. The limestone clasts are composed of both micrite and sparite and commonly contain obviously recrystallized marine fossils. In contrast, the Tertiary lacustrine marls and nodular paleosol carbonates are dense, well-indurated micrite containing well-preserved gastropod, ostracod, and *Chara* fossils.

Our hypothesis is that if resetting by diagenesis has occurred in the Nima basin, carbonates of all types should exhibit uniformly very low $\delta^{18}O_{sc}$ values ($< -6‰$), owing to diagenetic interaction at higher temperatures with ^{18}O-depleted meteoric water that is characteristic of recharge in the region today. Critical here is that clasts of marine limestone should not retain their primary marine $\delta^{18}O_{sc}$ values of $> -5‰$ (Veizer et al. 1999), as observed in the example from the upper Cretaceous Takena Formation.

Analysis of the reworked Cretaceous marine limestone pebbles and cobbles yielded several $\delta^{18}O_{sc}$ values $> -3‰$ (Fig. 17), in the range expected for unaltered Cretaceous marine carbonates. This indicates that some of the limestone pebbles are not diagenetically reset despite a long history of uplift, erosion, and burial over a time period of approximately 80 million years. Other Cretaceous-age limestone pebbles yielded $\delta^{18}O_{sc}$ values ranging between $-5‰$ and $-15‰$, indicating that they have been diagenetically reset by interaction with meteoric waters (Fig. 17). In sharp contrast, paleosol carbonate nodules—both reworked and *in situ*—in the Nima basin yielded tightly clustered $\delta^{13}C_{sc}$ values of $-3.5 \pm 0.6‰$ (range -2.3 to $-4.6‰$; n = 31) and $\delta^{18}O_{sc}$ values of $-17.0 \pm 0.3‰$ (range -16.3 to $-17.5‰$). The large spread in $\delta^{18}O$ values of the marine limestone clasts, as contrasted with the narrow range of values from *in situ* and reworked Nima paleosol carbonate (Fig. 17), indicates that resetting of the limestone must have occurred prior to deposition in the Nima basin, during previous burial of the limestone. Thus, we interpret paleosol carbonate in the Nima basin as preserving the original isotopic composition of meteoric water during the late Oligocene.

Results from marl and fossil Chara and ostracodes in section with the paleosols strongly reinforce the view that the paleosol carbonates are preserving primary values. Marls range in $\delta^{18}O$ values between -16.5 to $-8.9‰$, whereas those from fossil *Chara* and ostracodes fall between -18.1 to $-4.7‰$. The most negative values are very similar to those from pedogenic carbonates. The large spread in $\delta^{18}O$ values is consistent with periodic strong evaporation in the Nima paleolake. The most positive $\delta^{18}O$ values from fossils and marl are very similar to carbonate forming in modern Tibetan lakes (Quade, unpublished data).

Paleoaridity. As already discussed, we can use $\delta^{13}C_{sc}$ values from paleosols as potential archives of paleoaridity. Unlike $\delta^{18}O_{sc}$ values, $\delta^{13}C_{sc}$ values in carbonates are far less prone to diagenetic resetting because of the very low carbon (versus high oxygen) content of natural waters. Results from the upper Cretaceous Takena Formation cannot be used to reconstruct paleoaridity, because of much higher pCO$_2$ at that time (Ekart et al. 1999). However, results from the late Oligocene to early Miocene Dumri Formation and Nima redbeds are admissible.

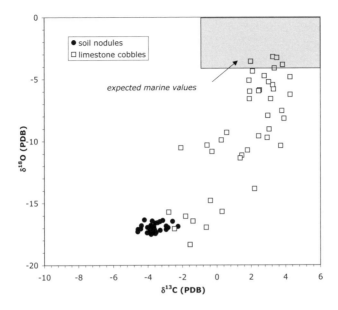

Figure 17. $\delta^{18}O_{sc}$ (PDB) versus $\delta^{13}C_{sc}$ (PDB) values from marine limestone cobbles (variable ages) and paleosol carbonate in the late Oligocene-age Nima Redbeds, southern Tibet (DeCelles et al. 2007a). Range of expected marine isotopic values for the late Cretaceous from Veizer et al. (1999).

$\delta^{13}C_{sc}$ values from the Dumri Formation average $-10.3 \pm 0.6‰$. These very low values are similar to those obtained from the overlying Siwalik Formation in paleosols > 8 Ma. They imply high respiration rates of 4-8 mmoles/m²/hr, assuming $\delta^{13}C$ values of local plants between -24 and $-25‰$. This in turn indicates moist climatic conditions, consistent with paleobotanic evidence for the period (Lakhanpal 1970).

By contrast, $\delta^{13}C_{sc}$ values from Nima paleosols in central Tibet (Fig. 1b) formed at about the same time as those in the Dumri Formation, but on the other side of the Himalaya, and are much higher ($-3.5 \pm 0.6‰$). These results are very close to the lowest $\delta^{13}C_{sc}$ values from modern pedogenic carbonates from immediately around Nima, which range from -3.4 to $+7.7‰$. The relatively high $\delta^{13}C_{sc}$ values for modern carbonates indicate mixing of atmospheric CO_2 with plant-derived CO_2 owing to low soil respiration rates, a result of arid local climate of $\sim$200-250 mm/yr. Similarly, the high $\delta^{13}C_{sc}$ values from paleosol carbonates indicate very low soil respiration rates at 25-26 Ma (Fig. 6). Ancient soil respiration rates of 0.15 to 0.7 mmoles/ m²/hr can be calculated from the one-dimensional soil diffusion model of Cerling (1984) and modified in (Quade et al. 2007), assuming model conditions given in the caption to Figure 6. These respiration rates are very low and typical of sparsely vegetated, low-elevation settings in the Mojave (Quade et al. 1989a). Aridity in modern Tibet is produced by orographic blockage of moisture arriving from outside the plateau, as well as by orographically induced stationary waves that tend to fix dry, descending air over the plateau during most of the year (Broccoli and Manabe 1992). The high $\delta^{13}C_{sc}$ values from Nima basin paleosols indicate that orographic barriers to moisture also existed during the late Oligocene, most likely the Himalaya, which was actively growing during Oligocene time and earlier.

Paleoelevation. The $\delta^{18}O_{sc}$ values of modern pedogenic carbonate from around Nima (12 analyses, 3 profiles) ranges from $-13.8‰$ to $-6.1‰$. $\delta^{18}O_{mw}$ values of meteoric water samples from springs and small creeks in the Nima area at modern elevations of 4500-5000 m range from $-12.6‰$ to $-16.2‰$. Using local mean annual temperature of $3°C$, $\delta^{18}O$ values of pedogenic carbonate forming in isotopic equilibrium with these waters should be -11.7 to $-13.3‰$. Thus, observed $(-13.8$ to $-6.1‰)$ and predicted $(-11.7$ to $-13.3‰)$ $\delta^{18}O_{sc}$ values overlap, but modern carbonates exhibit much more positive values (Fig. 10). These high values are consistent with the arid setting of Nima basin (annual rainfall is ~200 mm/yr).

The $\delta^{18}O_{sc}$ values of $-17.0 \pm 0.3‰$ from paleosol nodules are much lower than $\delta^{18}O_{sc}$ values of modern pedogenic carbonate from around Nima of $-13.8‰$ to $-6.1‰$ (Fig. 10). On the face of it this would appear to imply higher paleoelevations in the past compared to the present elevation of Nima at ~4500 m. However, this simple comparison is unrealistic because we know that corrections for decreased ice volume, a warmer Earth, lower latitude at 26 Ma, and likely changes in Asian monsoon intensity must be considered. Each of these factors can be considered separately. Or, as in the case of several previous studies (Garzione et al. 2000a; Rowley et al. 2001; Mulch et al. 2004; Currie et al. 2006; Rowley and Currie 2006), changes in $\delta^{18}O_{sc}$ values from a nearby low-elevation site can be used to make the correction. The second approach has great advantages, since it both "anchors" the temperature-$\delta^{18}O_{sc}$ relationship of the paleoclimate system delivering the moisture to Tibet at some known elevation, in this case near sea level, and it subsumes the effects on $\delta^{18}O_{sc}$ values of changes latitude, temperature, global ice volume, and regional climate through time. Critical here is that, today at least, northern India and southern Tibet share a common climate dominated by the SW Indian monsoon.

As with previous studies (Garzione et al. 2000a; Rowley et al. 2001), we can use the data from the Siwalik Group (Quade and Cerling 1995) in the Himalayan foreland of Nepal and Pakistan as our low-elevation reference site. These data only extend back to 17 Ma, since our attempts to extend the lowland isotopic record to the Oligocene failed, as already described. These records show increases in $\delta^{18}O_{sc}$ values in the late Miocene of about $2.5‰$, comparing the most negative values (not the means) of the pre-late Miocene pedogenic carbonates (Quade and Cerling 1995). We have interpreted this late Miocene increase in $\delta^{18}O_{sc}$ values as related to a weakening of the Asian monsoon (Dettman et al 2001), the opposite of earlier-held views (Quade et al. 1989b; Kroon et al. 1991; Ruddiman et al. 1997). Causes aside, this $~2.5‰$ "correction" can be added to $\delta^{18}O_{sc}$ values from paleosol carbonate $(-17‰)$ at Nima to yield about $-14.5‰$. This value is slightly lower than the lowest (least evaporated) modern pedogenic carbonate value from Nima of around $-13.8‰$. On this basis we conclude that $\delta^{18}O_{sc}$ values of paleosol carbonate—after correction—look very similar to modern values. This would imply little change in elevation of the site since the late Oligocene.

Central to our reconstruction is that the $\delta^{18}O_{sc}$ record from Pakistan back to 17 Ma can be applied to late Oligocene Nima, and the shift in the y-intercept of the system (Fig. 2b) of $~2.5‰$ encapsulates the changes in global temperature, latitude, sea-water $\delta^{18}O$ values, and monsoon strength over the past 26 Ma. An additional (and related) assumption is that the slope of the modern $\delta^{18}O_{sc}$ value-elevation relationship for southern Tibet has not appreciably changed back through time.

The last assumption is an important one, because just viewing climatic patterns in Tibet today, the slope of the $\delta^{18}O_{sc}$ value-elevation relationship in Tibet is highly dependent on latitude, and should vary according to the strength of the Asian monsoon. For example, a weaker monsoon at 26 Ma should reduce the slope of the $\delta^{18}O_{sc}$ value-elevation relationship in southern Tibet—that is, northerly or local air masses would have had a stronger influence further south in the late Oligocene. This would have the effect of increasing our estimates of paleoelevation compared to today. Alternatively, we might argue that the Asian monsoon was stronger at 26 Ma, resulting in a locally steeper isotopic lapse rates with elevation and overestimated paleoelevation. However, this reasoning is contradictory by invoking a stronger Asian monsoon at 26 Ma in the presence of a lower plateau.

This brings us to our current "best guess" about the Nima record, that paleoelevation in the Oligocene at Nima was at least as high as today, and no higher than today only if the modern Asian monsoon—or something that coincidentally looks a lot like it in oxygen isotope terms—was already in place at 26 Ma. We would stress, however, that Nima is only one of a handfull of well-dated Oligo-Miocene sites in Tibet (others examples are discussed in Garzione et al. 2000b, Currie et al. 2005, and Rowley and Currie 2006), distributed over a very broad and geologically diverse region. The challenge ahead is to see if our results from Nima and other sites in southern Tibet can be replicated, and to eventually reconstruct the uplift history of the rest of the plateau.

CONCLUDING REMARKS

Paleoelevation reconstruction is a relatively new science. Of the several approaches under development, the use of oxygen isotopes in pedogenic carbonates enjoys several advantages. As to soils themselves, pedogenic carbonate is relatively common in basin deposits of many orogens, especially on the drier leeside. Futhermore, soils form in the presence of local rainfall, and hence are not influenced by the $\delta^{18}O$ value of run-off from higher elevations in the way that, for example, riverine and lacustrine carbonates could be.

The $\delta^{18}O$ analysis of pedogenic carbonate has the advantage of being routine but the major disadvantage of being a function of both temperature and of the $\delta^{18}O$ value of rainfall variably modified by evaporation. Paleoelevation can be seriously underestimated due to the effects of evaporation. This effect can only be minimized by sampling deep in paleosol profiles and is probably an insurmountable problem in very dry settings (< 200 mm/yr) like the Atacama and low-elevation Mojave Desert. For this reason, Δ_{47} analyses hold much promise because they are only temperature dependent. Here, a final important advantage of soils comes into play: soil T collapses to slightly above (+1 to +3 °C) mean annual T at soil depth. Other common settings for carbonate formation such as oceans, lakes, and rivers can experience very large temperature swings diurnally or seasonally, and do not converge on mean annual temperature with depth. The combination of $\delta^{18}O$ and Δ_{47} analysis is especially powerful since it provides two largely independent estimates of paleoelevation. Δ_{47} measurements of pedogenic carbonate allow mean annual temperature to be reconstructed, and paleoelevations to be estimated using temperature lapse rates. These same paleotemperature estimates allow $\delta^{18}O_{mw}$ values to be calculated from $\delta^{18}O_{cc}$ values, providing a second estimate of paleoelevation, assuming a local isotopic lapse rate.

Diagenetic resetting poses a serious problem for future paleoelevation reconstructions using carbonate, especially as we delve into the older and deeper geologic record where resetting of $\delta^{18}O$ values is well documented in many situations. Our Tibetan and Bolivian studies offer two approaches to the problem, the Bolivian case where $\delta^{18}O$ and Δ_{47} values provide independent checks of each other, and the Tibetan case by analysis of co-occurring marine limestones.

In the longer view, the $\delta^{18}O$ and Δ_{47} combination entails the fewest assumptions and holds the greatest promise for paleoelevation reconstruction. However, this potential will only be fully realized when Δ_{47} analysis becomes more routine and can be calibrated and tested in very young pedogenic carbonates formed in soils with well-constrained temperature histories.

ACKNOWLEDGMENTS

JQ thanks Paul Kapp and Pete DeCelles for involving him in their Tibetan research, and discussions and data from Dave Dettman and Andrew Leier, and acknowledges support from

NSF-EAR-Tectonics 0438115. CG thanks David Rowley and Terry Jordan for insightful discussions, and acknowledges support from NSF-EAR-Tectonics 0230232.

REFERENCES

Affek HP, Eiler JM (2006) Abundance of mass-47 CO_2 in urban air, car exhaust and human breath. Geochim Cosmochim Acta 70:1-12

Allison GB, Barnes CJ, Hughes MW (1983) The distribution of deuterium and ^{18}O in dry soils 2. Experimental. J Hydrol 64:377-397

Allison CE, Francey RJ, Meijer HAJ (1995) Recommendations for the reporting of stable isotope measurements of carbon and oxygen in CO_2 gas. *In:* IAEA-TECDOC-825, Reference and intercomparison materials for stable isotopes of light elements. IAEA, Vienna. p 155-162

Ambach W, Dansgaard W, Eisner H, Mollner J (1968) The altitude effect on the isotopic composition of precipitation and glacier ice in the Alps. Tellus 20:595-600

Amundson R, Chadwick O, Kendall C, Wang Y, DeNiro M (1996) Isotopic evidence for shifts in atmospheric circulation patterns during the late Quaternary in mid-North America. Geology 24:23-26

Araguás-Araguás L, Froelich K, Rozanski K (1998) Stable isotope composition of precipitation over southeast Asia. J Geophys Res 103:28,721-28,742

Aravena R, Suzuki O, Pollastri A (1989) Coastal fog and its relation to groundwater in IV region of northern Chile. Chem Geol (Isotope Geos Sec) 79: 83-91

Aravena R, Suzuki O, Pena H, Pollastri A, Fuenzalida H, Grilli A (1999) Isotopic composition and origin of precipitation in northern Chile. Appl Geochem 14:411-422

Bartlett MG, Chapman DS, Harris RN (2006) A decade of ground-air temperature tracking at Emigrant Pass Observatory, Utah. J Clim 19:3722-3731

Beard KV, Pruppacher HR (1971) A wind tunnel investigation of the rate of evaporation of small water drops falling at terminal velocity in air. J Atmos Sci 28:1455-1464

Berner RA, Kothavala Z (2001) Geocarb III: a revised model of atmospheric CO_2 over Phanerozoic time. Am J Sci 301:182-204

Bershaw J, Garzione CN, Higgins P, MacFadden BJ, Anaya F, Alveringa H (2006) The isotopic composition of mammal teeth across South America: A proxy for paleoclimate and paleoelevation of the Altiplano: EOS, Transactions, American Geophysical Union 87:Abstract #T33C-0519

Birkeland PW (1984) Soils and Geomorphology. Oxford University Press, New York

Blisniuk PM, Stern LA (2005) Stable isotope paleoaltimetry---a critical review. Am J Sci 305:1033-1074

Blisniuk PM, Stern LA, Chamberlain CP, Idleman B, Zeitler PK (2005) Climatic and ecologic changes during Miocene surface uplift in the southern Patagonian Andes. Earth Planet Sci Lett 230:125-142

Broccoli AJ, Manabe S (1992) The effects of orography on midlatitude Northern Hemisphere dry climates. J Climate 5:1181-1201

Came R, Veizer J, Eiler JM (2007) Surface temperatures of the Paleozoic ocean based on clumped isotope thermometry. Nature (in press)

Cerling TE (1984) The stable isotopic composition of modern soil carbonate and its relation to climate. Earth Planet Sci Lett 71:229-240

Cerling TE, Quade J (1993) Stable carbon and oxygen isotopes in soil carbonates. *In:* Continental Indicators of Climate. Swart P, McKenzie JA, Lohman KC (eds) Proceedings of Chapman Conference, Jackson Hole, Wyoming, American Geophysical Union Monograph 78:217-231

Cerling TE, Harris JM, MacFadden BJ, Ehleringer JR, Leakey MG, Quade J, Eisenman V (1997) Global vegetation change through the Miocene/Pliocene boundary. Nature 389:153-157

Chadwick OA, Sowers JM, Amundson RA (1988) Influence of climate on the size and shape of pedogenic crystals. Soil Sci Soc Am J 53:211-219

Cross SL, Seltzer GO, Fritz SC, Dunbar RB (2001) Late Quaternary climate change and hydrology of tropical South America inferred from an isotopic and chemical model of Lake Titicaca, Bolivia and Peru. Quat Res 56:1-9

Currie BS, Rowley DB, Tabor NJ (2005) Middle Miocene paleoaltimetry of southern Tibet: implications for the role of mantle thickening and delamination in the Himalayan orogen. Geology 33(3):181-184

Cyr AJ, Currie BS, Rowley DB (2005) Geochemical evaluation of Fenghuoshan Group Lacustrine carbonates, north-central Tibet: implications for paleoaltimetry of the Eocene Tibetan Plateau. J Geol 113:517-533

Davidson GR (1995) The stable isotopic composition and measurement of carbon in soil CO_2. Geochim Cosmochim Acta 59(12):2485-2489

DeCelles PG, Horton BK (2003) Early to middle Tertiary foreland basin development and the history of Andean crustal shortening in Bolivia. Geol Soc Am Bull 115(1):58-77

DeCelles PG, Gehrels G, Quade J, Ojha TP (1998) Eocene-early Miocene foreland basin development and the history of Himalayan thrusting, western and central Nepal. Tectonics 17(5):741-765

DeCelles PG, Quade J, Kapp P, Fan M, Dettman DL, Ding L (2007a) High and dry in central Tibet during the late Oligocene. Ear Planet Sci Lett 253:389-401

DeCelles PG, Kapp P, Ding L, Gehrels GL (2007b) Late Cretaceous to middle Tertiary basin evolution in the central Tibetan Plateau: Changing environments in response to tectonic partitioning, aridification, and regional elevation gain. Geol Soc Am Bull 119(5-6):654-680

Dettman DL, Kohn MJ, Quade J, Ryerson FJ, Ojha TP, Hamidullah S (2001) Seasonal stable isotope evidence for a strong Asian monsoon throughout the past 10.7 Ma. Geology 29:31-34

Deutz P, Montanez IP, Moger HC (2002) Morphology and stable and radiogenic isotope composition of pedogenic carbonates in late Quaternary relict soils, New Mexico, USA: an integrated record of pedogenic overprinting. J Sediment Res 72:809-822

Dutton A, Wilkinson BH, Welker JM, Bowen G, Lohmann KC (2005) Spatial distribution and seasonal variation in $^{18}O/^{16}O$ of modern precipitation and river water across the coterminous USA. Hydrological Processes 19:4121-4146

Eiler JM, Schauble E (2004) $^{18}O^{13}C^{16}O$ in Earth's atmosphere. Geochim Cosmochim Acta 68:4767-4777

Eiler JM, Garzione C, Ghosh P (2006) Response to comment on "Rapid uplift of the altiplano revealed through $^{13}C-^{18}O$ bonds in paleosol carbonates. Science 314:5800

Ekart DD, Cerling TE, Montanez IP, Tabor NJ (1999) A 400 million year carbon isotope record of pedogenic carbonate: implications for paleoatmospheric carbon dioxide. Am J Sci 299: 805-827

Fox D, Koch PL (2004) Carbon and oxygen isotopic variability in Neogene paleosol carbonates: constraints on the evolution of C_4 grasslands. Palaeogr Palaeocl Palaeoecol 207:305-329

Friedman I, Smith GI, Gleason JD, Warden A, Harris JM (1992) Stable isotope composition of waters in southeastern California 1. modern precipitation. J Geophys Res 97(D5):5795-5812

Fritz P, Suzuki O, Silva C, Salati E (1981) Isotope hydrology of groundwater in the Pampa del Tamarugal, Chile. J Hydrol 53:161-184

Garzione CN, Quade J, DeCelles PG, English NB (2000a) Predicting paleoelevation of Tibet and the Himalaya from $\delta^{18}O$ versus altitude gradients in meteoric water across the Nepal Himalaya. Earth Planet Sci Lett 183:215-229

Garzione CN, Dettman DL, Quade J, DeCelles PG, Butler RF (2000b) High times on the Tibetan Plateau: paleoelevation of the Thakkola graben, Nepal. Geology 28:339-342

Garzione CN, Dettman DL, Horton BK (2004) Carbonate oxygen isotope paleoaltimetry: evaluating the effect of diagenesis on paleoelevation estimates for the Tibetan plateau. Palaeogeog Palaeoclim Palaeoec 212:119-140

Garzione CN, Molnar P, Libarkin JC, MacFadden BJ (2006) Rapid late Miocene rise of the Bolivian Altiplano: Evidence for removal of mantle lithosphere. Earth Planet Sci Lett 241:543-556

Ghosh P, Adkins J, Affek H, Balta B, Guo W, Schauble EA, Schrag D, Eiler JM (2006a) $^{13}C-^{18}O$ bonds in carbonate minerals: A new kind of paleothermometer. Geochim Cosmochim Acta 70:1439-1456

Ghosh P, Eiler JM, Garzione C (2006b) Rapid uplift of the Altiplano revealed in abundances of $^{13}C-^{18}O$ bonds in paleosol carbonate. Science 311:511-515

Gile LH, Peterson FF, Grossman RB (1966) Morphological and genetic sequences of carbonate accumulation in desert soils. Soil Sci 101:347-360

Gonfiantini R, Stichler W, Rozanski K (1995) Standards and intercomparison materials distributed by the International Atomic Energy Agency for stable isotope measurments. *In:* IAEA-TECDOC-825, Reference and intercomparison materials for stable isotopes of light elements. IAEA, Vienna. p. 13-29

Gonfiantini R, Roche M-A, Olivry J-C, Fontes J-C, Zupi GM (2001) The altitude effect on the isotopic composition of tropical rains. Chem Geol 181:147-167

Guo W, Eiler JM (2007) Evidence for methane generation during the aqueous alteration of CM chondrites. Meteoritics and Planetary Sciences (in press)

Hardy DR, Vuille M, Bradley RS (2003) Variability of snow accumulation and isotopic composition on Nevado Sajama, Bolivia. J Geophys Res 108 (D22):4693

Harris N (2006) The elevation history of the Tibetan Plateau and its implications for the Asian monsoon. Palaeogeogr Palaeoclim Palaeoecol 24:4-21

Hays PD, Grossman EL (1991) Oxygen isotopes in meteoric calcite cements as indicators of continental paleoclimate. Geology 19:441-444

Hernández R, Jordan T, Dalenz Farjat A, Echavarría L, Idleman B, Reynolds J (2005) Age, distribution, tectonics and eustatic controls of the Paranense and Caribbean marine transgressions in southern Bolivia and Argentina. J South Am Earth Sci 19:495-512

Hillel D (1982) Introduction to Soil Physics. Academic Press, New York

Hoorn C (2006) The birth of the mighty Amazon. Scientific American 294:52-59

Horton BK, Hampton BA, Lareau BN, Baldellon E (2002) Tertiary provenance history of the northern and central Altiplano (Central Andes, Bolivia); a detrital record of plateau-margin tectonics, J Sedimentary Res 72:711-726

Hsieh JCC, Chadwick O, Kelly E, Savin SM (1998) Oxyen isotopic composition of soil water: quantifying evaporation and transpiration. Geoderma 82:269-293

Hulka C (2005) Sedimentary and tectonic evolution of the Cenozoic Chaco foreland basin, southern Bolivia (Ph. D. dissertation), Freien Universität, Berlin

Jenny H, Leonard C (1934) Functional relationships between soil properties and rainfall. Soil Sci 38:363-381

Kay RF, MacFadden BJ, Madden RH, Sandeman H, Anaya F (1998) Revised age of the Salla beds, Bolivia, and its bearing on the age of the Deseadan South American Land Mammal "Age." J Vert Paleont 18:189-199

Kennan L, Lamb S, Rundle C (1995) K-Ar dates from the Altiplano and Cordillera Oriental of Bolivia - implications for Cenozoic stratigraphy and tectonics. J South Am Earth Sci 8:163-186

Kent-Corson ML, Sherman LS, Mulch A, Chamberlain CP (2006) Cenozoic topographic and climatic response to changing tectonic boundary conditions in western North America. Earth Planet Sci Lett 252 (3-4): 453-466

Kim S-T, O'Neil JR (1997) Equilibrium and non-equilibrium oxygen isotope effects in synthetic carbonates. Geochim Cosmochim Acta 61:3461-3475

Koch PL, Zachos JC, Dettman DL (1995) Stable-isotope stratigraphy and paleoclimatology of the paleogene bighorn basin (Wyoming, USA). Palaeogeogr Palaeoclim Palaeoecol 115:61-89

Kroon D, Steens T, Troelstra SR (1991) Onset of monsoonal related upwelling in the western Arabian Sea as revealed by planktonic foraminifers. *In:* Proceedings of the Ocean Drilling Project, Scientific Results. Prell WL, Niitsuma N (eds) 117:257-263

Lakhanpal RN (1970) Tertiary floras of India and their bearing on historical geology of the region. Taxon 19(5):675-694

Larrain H, Velazquez F, Cereceda P, Espejo R, Pinot R, Osses P, Schemenauer RS (2002) Fog measurements at the site "Falde Verde" north of Chanaral compared with other fog stations of Chile. Atmos Res 64:273-284

Latorre C, Quade J, McIntosh WC (1997) The expansion of C_4 grasses and global change in the late Miocene: stable isotope evidence from the Americas. Earth Plan Sci Lett 146:83-96

Leier AL (2005) The Cretaceous evolution of the Lhasa terrane, southern Tibet, Ph.D. Dissertation, Univ. of Arizona

Liu B, Phillips FM, Campbell AR (1996) Stable carbon and oxygen isotopes of pedogenic carbonates, Ajo Mountains, southern Arizona: implications for paleoenvironmental change. Palaeogr Palaeocl Palaeoecol 124:233-246

Machette MA (1985) Calcic soils of the southwestern United States. *In:* Soils and Quaternary Geology of the Southwestern United States. Weide DL, Faber ML (eds) Geol Soc Am Spec Pap 203:1-21

MacFadden BJ, Campbell KE Jr, Cifelli RL, Siles O, Johnson NM, Naeser CW, Zeitler PK (1985) Magnetic polarity stratigraphy and mammalian fauna of the Deseadan (late Oligocene early Miocene) Salla beds of northern Bolivia. J Geol 93:223-250

Marshall LG, Swisher CC III, Lavenu A, Hoffstetter R, Curtis GH (1992) Geochronology of the mammal-bearing late Cenozoic on the northern Altiplano, Bolivia. J South Am Earth Sci 5(1):1-19

Mulch A, Teyssier C, Cosca MA, Vanderhaeghe O, Vennemann V (2004) Reconstructing paleoelevation in eroded orogens. Geology 32:525-528

Pagani M, Freeman KH, Arthur MA (1999) Late Miocene atmospheric CO_2 concentrations and expansion of C_4 grasses. Science 285:876-879

Pagani M, Zachos J, Freeman KH, Bohaty S, Tipple B (2005) Marked change in atmospheric carbon dioxide concentrations during the Oligocene. Science 309:600-603

Pearson PN, Palmer MR (2000) Atmospheric carbon dioxide concentrations over the past 60 million years. Nature 406:695-699

Poage MA, Chamberlain CP (2001) Empirical relationships between elevation and the stable isotope composition of precipitation and surface waters: consideration for studies of paleoelevation change. Am J Sci 301:1-15

Quade J, Cerling TE (1995) Expansion of C_4 grasses in the late Miocene of northern Pakistan: evidence from stable isotopes in paleosols. Palaeogr Palaeocl Palaeoecol 115:91-116

Quade J, Cerling TE, Bowman JR (1989a) Systematic variation in the carbon and oxygen isotopic composition of Holocene soil carbonate along elevation transects in the southern Great Basin, USA. Geol Soc Am Bull 101:464-475

Quade J, Cerling TE, Bowman JR (1989b) Development of the Asian Monsoon revealed by marked ecological shift in the latest Miocene in northern Pakistan. Nature 342:163-166

Quade J, Cater JML, Ojha TP, Adam J, Harrison TM (1995) Dramatic carbon and oxygen isotopic shift in paleosols from Nepal and late Miocene environmental change across the northern Indian sub-continent. Geol Soc Am Bull 107:1381-1397

Quade J, Rech J, Latorre C, Betancourt J, Gleason E, Kalin-Arroyo M (2007) Soils at the hyperarid margin: the isotopic composition of soil carbonate from the Atacama Desert. Geochim Cosmochim Acta (in press)

Ramesh R, Sarin MM (1995) Stable isotope study of the Ganga (Ganges) river system. J Hydrol 139:49-62

Rech JA, Currie BS, Michalski G, Cowan AM (2006) Neogene climate change and uplift in the Atacama Desert, Chile. Geology 34(9):761-764

Roe GH (2005) Orographic precipitation. Ann Rev Earth Planet Sci 33:645-671

Roperch P, Herail G, Fornari M (1999) Magnetostratigraphy of the Miocene Corque Basin, Bolivia; implications for the geodynamic evolution of the Altiplano during the late Tertiary. J Geophys Res 104(9):20415-20429

Rowley DB, Pierrehumbert RT, Currie BS (2001) A new approach to stable isotope-based paleoaltimetry: implications for paleoaltimetry and paleohypsometry of the High Himalaya since the Late Miocene. Earth Planet Sci Lett 188:253-268

Rowley DB, Currie BS (2006) Palaeo-altimetry of the late Eocene to Miocene Lunpola basin, central Tibet. Nature 439:677-681

Rowley DB, Garzione CN (2007) Stable isotope-based Paleoaltimetry. Annu Rev Earth Planet Sci 35:463-508

Royer D (1999) Depth to pedogenic carbonate horizon as a paleoprecipitation indicator? Geology 27:1123-1126

Rozanski K, Sonntag C (1982) Vertical distribution of deuterium in atmospheric water vapour. Tellus 34:135-141

Ruddiman WF, Raymo ME, Prell WL, Kutzbach JE (1997) The climate uplift-connection. *In:* Tectonic Uplift and Climate Change. Ruddman WL (ed) Plenum Press. p 3-15

Salomans W, Goudie A, Mook WG (1978) Isotopic composition of calcrete deposits from Europe, Africa and India. Earth Surf Processes 3:43-57

Santrock J, Studley SA, Hayes JM (1985) Isotopic analysis based on the mass-spectrum of carbon-dioxide. Anal Chem 57:1444-1448

Savin SM, Douglas RG, Stehli FG (1975) Tertiary marine paleotemperatures. Geol Soc Am Bull 86:1499-1510

Schauble EA, Ghosh P, Eiler JM (2006) Preferential formation of ^{13}C-^{18}O bonds in carbonate minerals, estimated using first-principles lattice dynamics. Geochim Cosmochim Acta 70:2510-2529

Siegenthaler U, Oeschger, H (1980) Correlation of ^{18}O in precipitation with temperature and altitude. Nature 285:314-17

Smith AG, Hurley AM, Briden JC (1981) Phanerozoic paleocontinental world maps, Cambridge, UK, Cambridge University Press, 102 p

Stern, LA, Chamberlain, CP, Reynolds, RC, Johnson GD (1997) Oxygen isotope evidence of climate changes from pedogenic clay minerals in the Himalayan molasses. Geochim Cosmochim Acta 61:731-744

Stern LA, Blisniuk PM (2002) Stable isotope composition of precipitation across the southern Patagonian Andes. J Geophys Res 107(D23): doi:10.1029/ 2002JD002509

Stewart MK (1975) Stable isotope fractionation due to evaporation and isotopic exchange of falling waterdrops—applications to atmospheric processes and evaporation of lakes. J Geophys Res 80:1133-1146

Talbot MR (1990) A review of the paleohydrological interpretation of carbon and oxygen isotopic ratios in primary lacustrine carbonates. Chem Geol 80:261-279

Talma AS, Netterberg F (1983) Stable isotope abundances in calcretes. *In:* Residual Deposits: Surface Related Weathering Processes and Materials. Wilson, RCL (ed) Oxford: Blackwell Scientific Publ., p 221-233

Tian L, Masson-Delmotte V, Stievenard M, Tao T, Jouzel J (2001) Tibetan Plateau summer monsoon northward extent revealed by measurements of water stable isotopes. J Geophys Res 106:28,081-28,088

Tian L, Yao T, White JWC, Yu W, Wang N (2005) Westerly moisture transport to the middle of Himalayas revealed from the high deuterium excess. Chinese Sci Bull 50:1026-130

Uba CE, Heubeck C, Hulka C (2006) Evolution of the late Cenozoic Chaco foreland basin, southern Bolivia. Basin Res 18:145-170

Veizer J, Ala D, Azmy K, Bruckschen P, Buhla D, Bruhn F, Carden GAF, Diener A, Ebneth S, Godderis Y, Jasper T, Korte C, Pawellek F, Podlaha OG, Strauss H (1999) $^{87}Sr/^{86}Sr$, $\delta^{13}C$ and $\delta^{18}O$ evolution of Phanerozoic seawater. Chem Geol 161:59-88

Vuille M, Bradley RS, Werner M, Healy R, Keimig F (2003) Modeling $\delta^{18}O$ in precipitation over the tropical Americas: 1. Interannual variability and climatic controls. J Geophys Res 108: doi:10.1029/ 2001JD002038

Wang ZG, Schauble EA, Eiler JM (2004) Equilibrium thermodynamics of multiply substituted isotopologues of molecular gases. Geochim Cosmochim Acta 68:4779-4797

Wolfe BB, Aravena R, Abbott MB, Seltzer GO, Gibson JJ (2001) Reconstruction of paleohydrology and paleohumidity from oxygen isotope records in the Bolivian Andes. Palaeogeogr Palaeoclimatol Palaeoecol 176(1-4):177-192

Wynn JG (2004) Influence of Plio-Pleistocene aridification on human evolution: evidence from paleosols of the Turkana Basin. Am J Phys Anthro 123:106-118

Zachos J, Pagani M, Sloan L, Thomas E, Billups K (2001) Trends, rhythms, and aberrations in global climate 65 Ma to present. Science 292:686-693

Zhang X, Nakawo M, Yao T, Han J, Xie Z (2002) Variations of stable isotopic compositions in precipitation on the Tibetan Plateau and its adjacent regions. Science in China (Series D) 25(6):481-493

Reviews in Mineralogy & Geochemistry
Vol. 66, pp. 89-118, 2007
Copyright © Mineralogical Society of America

4

Stable Isotope Paleoaltimetry in Orogenic Belts – The Silicate Record in Surface and Crustal Geological Archives

Andreas Mulch[1,2,*], C. Page Chamberlain[1]

[1]*Geological and Environmental Sciences, Stanford University, Stanford, U.S.A.*
[2]*Institut für Geologie, Universität Hannover, Hannover, Germany*

Current Address: Callinstr. 30, D-30167 Hannover, Germany
mulch@geowi.uni-hannover.de

ABSTRACT

In this chapter, we discuss the use of variations in the oxygen and hydrogen isotope compositions of hydrous silicates to paleoaltimetry studies. Currently, there are numerous isotopic (lacustrine, palustrine, and pedogenic carbonates, hydrous silicates, fluid inclusions) and paleofloral (leaf physiognomy, stomata density) proxies that provide information on climatic and topographic histories of mountain belts. Compared to lacustrine and paleosoil carbonates that frequently provide detailed and temporally extensive paleoaltimetric records, the use of hydrous silicates (such as smectite, kaolinite, chert, as well as metamorphic minerals that grow in the presence of meteoric waters) in paleoclimate and paleoaltimetry studies has important additional advantages. First, hydrous silicates provide both a hydrogen and oxygen isotope record. These two isotope systems can be used to evaluate climate change, but are also useful in understanding evaporative effects on meteoric waters prior to mineral formation. Since evaporation can cause isotopic changes that otherwise would correspond to changes on the order of kilometers of surface uplift, it is critical in any isotopic paleoaltimetry study within continental interiors to examine both the hydrogen and oxygen isotope record. Second, hydrogen isotope ratios of micas in deep-seated extensional shear zones often reflect the composition of meteoric waters that infiltrated these rocks during the late stages of orogenesis. As such, it is possible to use the combined oxygen, hydrogen and geochronological record of synkinematic micas in extensional shear zones as a paleoaltimeter. This information when combined with high-precision geochronology and paleoaltimetry studies of intermontane basins formed above these shear zones allows for the development of an integrated structural, sedimentologic, and surface elevation history of an evolving orogen.

INTRODUCTION

With the advent of stable isotope paleoaltimetry towards the turn of the millennium the stable isotope and tectonics communities have witnessed an increasing number of isotopic mineral proxies developed to address the long-term topographic histories of orogenic belts and continental plateaus. These proxies include calcite from paleosols (see for example Quade et al. 2007, this volume and references therein), fluvial and lacustrine rocks; the phosphate and carbonate component of mammal teeth (Kohn and Dettman 2007, this volume and references therein), smectite and kaolinite from paleosols, weathered sediments and volcanic ashes (e.g., Chamberlain et al. 1999; Takeuchi and Larson 2005; Mulch et al. 2006a); as well as white mica from extensional shear zones and fluid inclusions in hydrothermal veins (e.g., Mulch et al.

1529-6466/07/0066-0004$05.00
DOI: 10.2138/rmg.2007.66.4

2004, 2007; Sharp et al. 2005). The bulk of these isotope paleoaltimetry studies have focused on calcite, because of its ubiquity in semi-arid and arid settings and its relative ease in obtaining carbon ($\delta^{13}C$) and oxygen ($\delta^{18}O$) isotopic data. Fewer studies have focused on hydrous silicate minerals, although these minerals—due to their mineral compositions and genesis—often provide very robust information on the topographic and climatic evolution of orogens.

The major advantage of various hydrous silicate minerals originates from the fact that multiple (oxygen and hydrogen) isotope systems can be applied and used for stable isotope paleoaltimetry. These systems will allow evaluation of evaporative effects that may have influenced meteoric water prior to mineral formation. Second, in some cases the timing of mineral formation and/or isotopic exchange can even be directly dated using radiometric techniques. As an example, smectite and kaolinite formed at or near the Earth's surface in general retain the hydrogen and oxygen isotope compositions of the waters from which they formed (Savin and Hsieh 1998) unless subject to metamorphism. Information on the role of evaporation in the near-surface environment is essential in paleoaltimetry studies as the associated isotopic effects, in some cases, can be larger than the isotopic changes induced by the rise of the tallest mountains on Earth. These different isotopic systems respond differently to environmental parameters. As authigenic clay minerals provide two isotopic systems (hydrogen and oxygen), effects of evaporation may be separated from changes in past topography due to the different slopes of combined sample arrays in the $\delta D - \delta^{18}O$ diagram (e.g., Abruzzese et al. 2005; Davis et al. 2007). Due to their distinct retention and fractionation behavior, systematic variations in the hydrogen and oxygen isotope composition, therefore, may provide the only direct means to (semi-) quantitatively evaluate the effects of protracted evaporation on the composition of the isotopic proxy material. As a complementary approach to isotope proxies that formed at or near the Earth's surface, white mica formed in the presence of meteoric waters penetrating deep within extensional shear zones records hydrogen isotope data that can also be used to assess paleoelevation. Under favorable circumstances, the timing of recrystallization and isotopic exchange within these micas can be dated providing much-needed information for reconstruction of topographic histories and their relation to tectonic and surface processes. Although the isotopic information gained from hydrous silicate minerals represents a powerful tool for tectonic and climatic reconstructions, these data are commonly more difficult to obtain. It may be for this reason that the number of paleoaltimetry studies that exploit the isotope record from carbonates by far exceeds those that make use of silicate mineral proxies.

In this chapter we focus our discussion on paleoaltimetry studies that use hydrous silicates as isotope proxies. It is our hope that scientists from the various subdisciplines of Earth Sciences interested in paleoaltimetry and paleoclimate will take advantage of the integrative approach we propose in order to gain a more complete understanding of the long-term topographic evolution of orogens.

STABLE ISOTOPE PALEOALTIMETRY AND MOUNTAIN BUILDING PROCESSES

Quantitative paleoelevation estimates as boundary conditions for climate models

Much of what is currently known about the Earth's climate comes from the application of stable isotopes collected from ocean drill cores in marine sediments (e.g., Zachos et al. 2001). These isotopic data sets provide detailed records of how the Earth's oceans have responded to changing climate and are extremely valuable in assessing global climate histories down to millennial scales. Similar detailed isotopic records for terrestrial systems are, however, uncommon and frequently continuous terrestrial climate records that span millions to tens of millions of years are not preserved in the terrestrial geologic record. With the advent of paleoaltimetry studies targeted directly at the coupled isotopic effects of changes in climate and topography in the different geological archives amenable to stable isotopic analysis, it is now possible to collect much-needed isotopic data for many of the Cenozoic mountain belts and plateaus. These isotopic data show that, unlike isotopic data from the ocean floors, the isotopic trends are highly spatially variable (see for example the western United States given in Kent-Corson et al. 2006), as would be expected in that tall mountain ranges and continental plateaus affect terrestrial climate and both exert first-order influence on the isotopic composition of meteoric waters and ultimately proxy materials in the geologic record. In summary, the ultimate goal of stable isotope paleoaltimetry must be to provide geologic and geochemical evidence, in the field and in the laboratory, to identify the competing or sustaining effects of climate and topographic changes on the stable isotopic composition in the geologic record. Only then will we be able to make precise inferences about the effect of tall mountain ranges and continental plateaus in the past on global atmospheric circulation patterns.

The rise of the major Cenozoic mountain ranges and plateaus has exerted a strong control over the Cenozoic climate at global, regional and local scales (Kutzbach et al. 1989; Ruddiman and Kutzbach 1989; Ruddiman et al. 1989; Prell and Kutzbach 1992; Ramstein et al. 1997). Because these large mountainous regions represent key elements in understanding climate change it is imperative to reconstruct their surface uplift histories, one of the goals of stable isotope paleoaltimetry. Indeed, extrapolation of climate models back into the geologic past requires that at some level the topographic history of the Earth is known (e.g., Sewall and Sloan 2006). In general, these topographic estimates are based on geological and paleofloral arguments. At present, we know of no studies that combine stable isotope paleoaltimetry with climate models. Herein lies the problem. Climate models need information from stable isotope paleoaltimetry studies, and paleoaltimetry studies have to rely on information from climate models to be more accurate (see for example Blisniuk and Stern 2005). For example, as reviewed in Kent-Corson et al. (2006), the stable isotope record of terrestrial sediments in the western United States suggests that regional mean surface elevation has migrated to the southwest through time. In this interpretation, mean surface elevation within the hinterland of the central North American Cordillera was high in the early to middle Eocene in Montana, Idaho and British Columbia, in the late Eocene to early Oligocene in northeastern Nevada, and late Oligocene to Miocene in southern Nevada. Interpretation of the isotopic data form the basis of this work. However, interpretation of the isotopic record in terms of changes in topography requires that precipitation patterns and source regions of air masses are known to some degree. This information, in general, comes from the climate models that have been constructed for the western United States for different periods during the Cenozoic (Huber and Sloan 1999; Huber and Caballero 2003; Sewall and Sloan 2006). But these models assume a topography that may or may not be consistent with the more detailed stable isotope paleoaltimetry studies. This problem is not unique to studies of the western United States Cordillera, but affects all stable isotope paleoaltimetry studies.

To solve this problem will require collaboration of the climate modeling and stable isotope paleoaltimetry communities. Iterative climate models that incorporate stable isotopes applied to regional isotopic data sets, like the one generated for the western United States (compiled in Kent-Corson et al. 2006), should provide sets of solutions that allow better assessment of past topography and more accurate climate models.

Quantitative paleoelevation estimates as boundary conditions for tectonic models

Over the last decade considerable effort of the tectonics community has been directed towards the development of thermomechanical models that describe the collisional history and the internal dynamics of orogenic belts and continental plateaus (e.g., Beaumont et al. 2001, 2004; Koons et al. 2002). These models are commonly tested against thermobarometric, thermochronologic, and geochronologic data. However, by definition, these data sets only provide constraints on rates of rock uplift or exhumation; the surface response to tectonic processes remains elusive. Unfortunately, the topographic evolution of mountain ranges during orogeny is one -if not the- most important boundary condition for geodynamic and tectonic models (Clark 2007, this volume).

At present, paleoaltimetric data for the major continental plateaus provide previously unavailable constraints on a determination of the relative importance of the various mechanisms including: (a) continental subduction, crustal shortening and thickening (Dewey and Burke 1973), (b) thickening of crust and mantle lithosphere and syn- to postorogenic removal of lithospheric mantle by delamination (e.g., Molnar et al. 1993; Beaumont et al. 2004), as well as (c) lateral escape (Tapponnier et al. 1982; Taponnier et al. 2001) and (d) extrusion at the flanks of the orogen (Royden et al. 1997). Further, these data allow us to place constraints on how these mechanisms accommodate the distribution of strain within the orogenic belt. It is no surprise then, that the advent of paleoaltimetric data sets for the major orogenic belts and continental plateaus on Earth has stimulated renewed debate not only about the timing and magnitude of elevation change but also about the mechanism(s) responsible for elevation change and the inherent geodynamic implications of continental plateaus (e.g., Chase et al. 1998, Garzione et al. 2000, 2006; Molnar et al. 2006; Rowley and Currie 2006; Mulch and Chamberlain 2006; Mulch et al. 2007) or orogenic belts (Jones et al. 2004; Mulch et al. 2006a).

As one example, the mechanisms and processes governing the enigmatic evolution of the deep crust and mantle lithosphere during continental plate convergence and collision but also during the late stages of orogeny when broad mountain ranges and orogenic plateaus may undergo extension and gravitational collapse are still highly debated. Geological and geophysical evidence has been interpreted that under certain conditions the lower part of the mantle lithosphere under evolving orogens or mountain ranges has been tectonically removed (e.g., Bird 1978; Houseman et al. 1981; England and Houseman 1989; Molnar et al. 1993; Houseman and Molnar 1997; Ducea and Saleeby 1998, Jull and Kelemen 2001; Saleeby et al. 2003; Zandt 2003). Such a removal of dense lithospheric material should result in a complete reorganization of the internal dynamics of the orogen and ultimately be manifest in significant (and rapid) surface uplift. The implications of such a process should be testable using paleoaltimetric data.

A case has been made that rapid late Miocene rise of the Bolivian Altiplano can be attributed to delamination of mantle lithosphere where cold and negatively buoyant mantle material was replaced by hot asthenosphere, thus creating the necessary thermal and rheological conditions to sustain rapid surface uplift on the order of 2-3 km (Garzione et al. 2006; Ghosh et al. 2006). This conclusion, however, has not remained unchallenged (e.g., Sempere et al. 2006). Garzione et al. (2006) base their argument on a ~3.3‰ decrease in the oxygen isotope composition of paleosol carbonate in synorogenic deposits of the central Andean belt in Bolivia that is attributed to decreasing temperatures of carbonate formation (Ghosh et al. 2006) and $\delta^{18}O$ values of precipitation during surface uplift and hence this decrease in $\delta^{18}O$ values of precipitation marks the timing when the plateau region attained modern elevations. The main support for this model of Miocene surface uplift due to delamination of the lithospheric mantle component comes from the proposed rapid (~1-3 km /Ma) rate at which a relatively large amount of surface uplift is supposed to have occurred (Garzione et al. 2006). In contrast to other modes/mechanisms that could be held responsible for surface uplift of the Central Andean orogenic plateau such as for example crustal shortening and thickening where shortening at plate tectonic rates could allow uplift rates of ~0.3 km /Ma or changing the thermal structure of the central Andean belt through internal heating, delamination of mantle lithosphere provides a plausible mechanism for rapid surface rise if rapid surface uplift indeed occurred over very short time scales. Given the strong variability of climate and the potential feedback on the process of mountain building during the critical time interval of uplift in the Andes (Molnar 2004), more regionally extensive stable isotopic data sets are needed to account for spatially and temporally varying rates of surface uplift and to ultimately solve the question of when and how major surface uplift occurred in the Puna-Altiplano region.

In the case of the Tibetan plateau, Rowley and Currie (2006) presented oxygen isotope data from lake carbonates of the southern and central Tibetan plateau that in concert with previous data sets (Garzione et al. 2000; Rowley et al. 2000; Currie et al. 2005; Cyr et al. 2005) suggest progressive northward growth of the high-elevation (>4 km) plateau since at least 35 Ma. This model is in stark contrast to proposed renewed Miocene (8-10 Ma) surface uplift of the plateau region (e.g., Harrison et al. 1992; Molnar et al. 1993, 2006). In brief, both examples document that the major disagreement(s) center on the relative role(s) of thickening of the crust and/or mantle lithosphere during continental collision and convergence and whether or not subsequent convective destabilization and delamination of mantle lithosphere significantly contributed to the modern high elevations in the two major orogenic plateaus on this planet.

What can the geodynamic and tectonic modeling community expect from future stable isotope paleoaltimetry studies?

Stable isotope paleoaltimetry in general should be considered as a method that is still in its developing stages. In contrast to an evolving body of stable isotope paleoaltimetry studies that focus on the surface record of orogens, there are currently only a few integrated tectonic and paleoaltimetric datasets (e.g., Mulch et al. 2004, 2007) that incorporate stable isotope measurements from (formerly) tectonically active internal parts of evolving orogens. However, over the past few years, progress is continually made in understanding how the interactions of surface uplift, orogen-scale tectonics, as well as erosion and sediment transport create dynamic feedbacks between the biosphere, the atmosphere and the crust and upper mantle. As outlined in more detail below, integrated multi-isotope, multi-proxy techniques from a variety of tectonic and sedimentary environments can greatly enhance the ability to make assertions about the elevation history of orogenic belts and continental plateau regions with the ultimate goal of reconstructing first-order topographic variations in evolving orogens. These approaches, which take account of both surface and deeper Earth processes, are necessary to complement tectonic and modeling studies of orogens and will greatly enhance our predictive capabilities in such a task.

STABLE ISOTOPE PALEOALTIMETRY USING SILICATE PROXIES

Cherts and clay minerals

In most cases terrestrial stable isotope records are derived from the analysis of $\delta^{18}O$ in carbonates ($\delta^{18}O_{cc}$) from either fluvial or lacustrine deposits or paleosol successions. These carbonates capture long-term paleoclimatic trends via the temperature-dependant oxygen isotope calcite-water fractionation if carbonates formed in oxygen isotopic equilibrium with either lake, fluvial or soil waters (e.g., Drummond et al. 1993; Garzione et al. 2000, 2004, 2006; Rowley et al. 2000; Morrill and Koch 2002; Dettman et al. 2003; Horton et al. 2004; Cyr et al. 2005; Currie et al. 2005; Graham et al. 2005; Blisnuik et al. 2005; Horton and Chamberlain 2006; Ghosh et al. 2006; Rowley and Currie 2006; Kent-Corson et al. 2006; DeCelles et al 2007). Despite the relatively large number of successful applications, $\delta^{18}O_{cc}$ values in intermontane (and eventually even internally-drained) basins may be compromised by: 1) the combined effects of evaporation of the soil or lake water prior to carbonate formation that shift $\delta^{18}O_{cc}$ towards more positive values (e.g., Garzione et al. 2004; Cyr et al. 2005), an isotopic effect that can be relatively large (up to several per mil in $\delta^{18}O_{cc}$); 2) the fact that carbonate is particularly prone to diagenetic recrystallization postdating original carbonate formation (e.g., Morrill and Koch 2002) (in contrast to the effects of evaporation, recrystallization at higher burial temperatures commonly results in a decrease in the $\delta^{18}O_{cc}$ values relative to the primary values); 3) carbonate growth rates (Morse 1983) can be considerably faster than that of many clay minerals (Egli et al. 2003; Price et al. 2005), and, thus, they may record higher frequency isotopic shifts not associated with tectonism; and 4) calcite formed on an uplifting plateau is more sensitive to temperature variations that would be attendant with uplift of mountains, resulting in a dampened isotopic signal as compared to kaolinite (Poage and Chamberlain, 2001). For these reasons, it is often useful to examine the $\delta^{18}O_{cc}$ record in concert with oxygen and hydrogen isotope compositions of hydrous silicates, although these analyses are considerably more difficult to obtain (see below).

A historical note on stable isotope paleoaltimetry

Some of the very early work that used stable isotopes to make inferences about past topographic changes indeed used oxygen and hydrogen isotopes of (hydrous) silicate minerals as proxies for past elevations (e.g., O'Neil and Silberman 1974; Lawrence and Rashkes Meaux 1993; France-Lanord et al. 1988). Within this context it is worthwhile to note that the general concept of stable isotope paleoaltimetry—the systematic decrease in $\delta^{18}O$ or δD of precipitation with altitude and how it could be preserved in the geological record—has long been recognized and that at least the stable isotope community was well aware of potential applications since at least the late 1960's. One anecdotal remark within this context was that Caltech Professor S. Epstein set up an experiment where he collected rain water on the roof of the Caltech laboratory during a rain storm and then drove his car up the adjacent hills to sample precipitation from the same storm front at various elevations. The results of this experiment are hidden in a poorly circulated abstract but at least the general principle had been demonstrated and presented to the stable isotope community (HP Taylor and JR O'Neil pers. comm. 2006). Ironically, it was not until the late 1990's that integrative approaches bridging the fields of tectonics and stable isotope geochemistry started to systematically exploit the terrestrial stable isotope record for paleoaltimetric studies.

As one example, the regional studies of Lawrence and Rashkes Meaux (1993) on the oxygen and hydrogen isotope composition of kaolinite from Cretaceous to Tertiary weathered horizons showed isotopic gradients across the Rocky Mountains and the authors suggested this was due to ancient high elevation and topography. Chamberlain et al. (1999) used the oxygen isotope composition of kaolinite from weathered horizons on the east side of the Southern Alps of New Zealand to constrain the timing of uplift of this mountain belt. Since these studies, the combined oxygen and hydrogen isotope records of hydrous silicates to examine topographic histories of mountain belts have grown considerably, with applications to the Bohemian Massif in Europe (Gilg 2000), Sierra Nevada of California (Chamberlain and Poage 2000; Poage and Chamberlain 2002; Mulch et al. 2006a), Basin and Range (Horton et al. 2004), Rocky Mountains (Sjostrom et al. 2006), and Cascade Mountains (Takeuchi and Larson 2005), the hinterland of the North American Cordillera (Mulch et al. 2004, 2007) and the European Alps (Vennemann et al. 2000; Sharp et al. 2005; Kocsis et al. 2007). At present, hydrous silicates from depositional or weathering environments that have been used to reconstruct paleoelevations include: 1) smectite from paleosols and weathered ashes (Stern et al. 1997; Chamberlain and Poage 2000; Poage and Chamberlain 2002; Horton et al. 2004; Takeuchi and Larson 2004; and Sjostrom et al. 2006), 2) kaolinite from paleosols and weathered horizons (Lawrence and Rashkes Meaux 1993; Chamberlain et al. 1999; Gilg 2000; Mulch et al. 2006a); and 3) freshwater chert (Abruzzese et al. 2005).

All of these paleoaltimetry studies require that the isotopic record of these hydrous silicates reflects past climatic conditions and that the measured isotopic compositions can be retained within the proxy mineral over geologic time. There is excellent evidence that oxygen isotope compositions of kaolinite, smectite, and chert do record the isotopic composition of surface waters from which they formed and that, baring extreme diagenesis of up to 200 °C for $\delta^{18}O$ and <100 °C for δD, retain their original oxygen and hydrogen isotope values (see review by Savin and Hsieh (1998) and laboratory experiments by O'Neil and Kharaka (1976) for clay minerals; and Knauth and Epstein (1976) for chert). This work is substantiated in paleoaltimetry studies that show that oxygen isotope trends observed in calcite are also present in coexisting smectite (Stern et al. 1997; Horton et al. 2004), and chert (Horton et al. 2004; Abruzzese et al. 2005) from the same stratigraphic sections. As would be expected, hydrogen isotopes of kaolinite and smectite do exchange more readily than oxygen isotopes (O'Neil and Kharaka 1976), but unless burial temperatures are relatively high (above ~100 °C) the hydrogen isotope compositions of these clay minerals are retained unless mineral phases undergo dissolution and reprecipitation reactions (see review by Savin and Hsieh 1998).

Although hydrous silicates are excellent isotopic proxies for paleoaltimetry studies due to their different retention behavior of $^{18}O/^{16}O$ ratios, their advantage over other single isotopic proxies, such as calcite, is the possibility of obtaining both hydrogen and oxygen isotopes from the same sample. Combined hydrogen and oxygen isotope analysis of clay minerals and cherts allows one to evaluate the effects of evaporation. Evaporation can cause relatively large positive shifts in oxygen isotopes (over 10‰ in the samples discussed below), such that studies that solely rely on calcite as the isotopic proxy could be significantly compromised and the resultant data could cause serious errors in tectonic and paleoclimatic interpretations. Because the uplift of mountain ranges and large plateaus can cause severe aridity in the developing rain shadow, evaporation will be common in most intermontane basins. Thus, assessing whether any large isotopic effects are due to changes in surface elevation or simply reflect changing precipitation patterns that resulted in increased evaporation is imperative.

Two studies (Horton et al. 2004; Abruzzese et al. 2005) of the Great Basin of eastern Nevada and western Utah may serve as an example of the general utility of clay minerals and chert in paleoaltimetry. In several intermontane basins in this region these authors measured oxygen and hydrogen isotope compositions of authigenic minerals from lacustrine and fluvial sedimentary deposits ranging in age from the Eocene to the Pleistocene. These data show a decrease in $\delta^{18}O$ values from various proxy materials from the middle Eocene to early Oligocene followed by an increase in $\delta^{18}O$ values since the middle Miocene (Fig. 1), trends that have been interpreted to be caused by the direct or indirect effects of regional surface uplift (Eocene) and a subsequent decrease in mean surface elevation (Miocene) (Horton et al. 2004). For the purpose of this paper, there are several aspects of this data set that are particularly illustrative. First, oxygen isotopic compositions of calcite, smectite, and chert follow the same isotopic trends (Fig. 1), supporting the general utility of these systems for paleoaltimetry studies in terrestrial basin and paleosoil systems. Second, the $\delta^{18}O$ values of calcite, smectite, and chert are roughly identical for a given time period. If these minerals were in oxygen isotope equilibrium then the $\delta^{18}O$ values should systematically increase from smectite to calcite to chert as a result of different water-mineral fractionation factors between these minerals. This disequilibrium, however, is common in terrestrial depositional systems and has been demonstrated in previous studies that have used multiple isotope mineral proxies (Stern et al. 1997; Poage and Chamberlain 2002). It has been tentatively described that the disequilibrium between calcite and smectite is the result of different seasons of formation (Stern et al. 1997). Furthermore, some fraction of the chert-calcite disequilibrium can be explained by chert forming during diagenesis at higher temperatures (40 °C for chert and 20 °C for calcite) (Abruzzese et al. 2005). Third, close inspection of the data reveals that there is a period during the Eocene (Fig. 1) where there is a larger variation in the oxygen isotopic data, which for reasons given below is attributed to evaporative effects.

To assess the role of evaporation on the isotopic trends shown in Fig. 1, Horton et al. (2004) measured oxygen and hydrogen isotopic compositions of smectite coexisting with chert and calcite in individual sedimentary horizons. When plotted on a δD versus $\delta^{18}O$ diagram samples relatively unaffected by evaporation will lie along a slope of ~8 (the slope of the global meteoric line as defined by Craig (1961) and Dansgaard (1964). In contrast, samples strongly affected by evaporation will lie along shallower positive slopes (Dansgaard 1964). Horton et al.

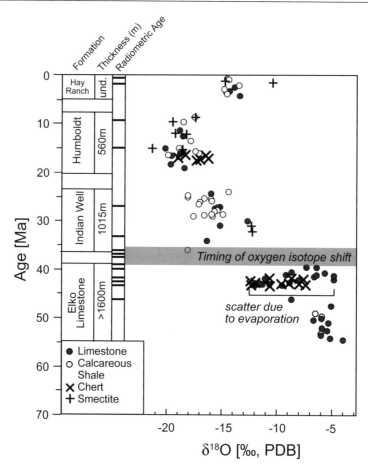

Figure 1. $\delta^{18}O$ values for the Eocene to Pleistocene stratigraphic record in the Elko Basin (Carlin-Pinon Range, Nevada). A significant (ca. -8 to $-10\%_0$) decrease in $\delta^{18}O$ values of calcite at ca. 35-40 Ma is consistent with complete reorganization of precipitation patterns in central Nevada, likely due to increasing mean elevation and associated depletion in ^{18}O of precipitation (Horton et al. 2004). $\delta^{18}O$ values for lacustrine limestone (filled circles), calcareous shale (open circles), lacustrine chert (cross-hatch marks), and smectite (crosses) indicate oxygen isotope disequilibrium most likely as a result of different temperatures (and timescales) of formation. The general consistency of large scale oxygen isotope variations and oxygen isotopic trends in materials that form under different temperature conditions and over a wide range of timescales, however, indicates that terrestrial sedimentary sequences reliably record the long-term evolution of correlated changes in climate and elevation during the evolution of orogens. See Horton et al. (2004) for further details.

(2004) showed that δD and $\delta^{18}O$ values of smectite for the northern and central Great Basin lie along a line parallel to the meteoric water line with a slope of approximately 8 (Fig. 2). Thus, for most of the presented smectite data evaporation has played, at best, a minor role and most likely did not significantly affect the isotopic trends shown on Fig. 1. However, for the Eocene samples that display the largest degree of scatter (at ~40 Ma) evaporation most likely exerted the dominant control on the oxygen isotope compositions of the various authigenic minerals.

    Combined oxygen and hydrogen isotope analyses of Eocene cherts from this particular time interval define linear arrays with slopes significantly less than 8 (1.6 for cherts in central

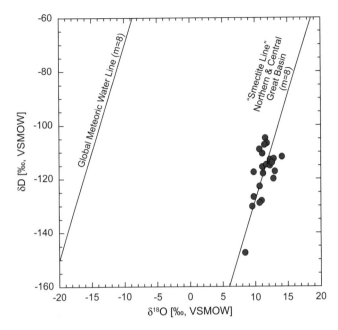

Figure 2. δD vs. δ^{18}O of volcanic ash derived smectite in the northern Great Basin of Nevada. A regression line through the smectite data parallels the global meteoric water line (δD = 8×δ^{18}O +10; Craig 1961) indicating that the oxygen and hydrogen isotopic compositions of meteoric water from which those clay minerals formed had not been altered severely by evaporative processes prior to crystallization of the clays. After Horton et al. (2004).

Utah and 2.7 for cherts in northern Nevada) (Abruzzese et al. 2005) (Fig. 3). These shallow slopes indicate that during the Eocene considerably large lake systems in this part of the North American Cordillera experienced a period of intense evaporation (Abruzzese et al. 2005), a conclusion supported by trace element studies of coexisting calcite, that has high Mg/Ca and Sr/Ca ratios (Horton et al. 2004) and geologic evidence from the nearby Green River basin that this particular time interval during the Eocene was characterized by anomalously high temperatures, large amounts of precipitation, and extreme evaporation (Sewall and Sloan 2006). Without the additional information from hydrogen isotopic compositions it would be difficult to interpret the relatively large (>10‰) oxygen isotope variations observed in these rocks.

Such studies illustrate the usefulness of using combined hydrogen and oxygen isotopic compositions of hydrous silicates in paleoaltimetry and paleoclimatic studies. That said, however, there are disadvantages of this approach as compared to the analysis of carbonate minerals. Combined oxygen and hydrogen isotope analyses are considerably more time consuming and laborious. The difficulty in obtaining reliable and accurate data lies not with mass spectrometry, but rather in obtaining high purity mineral separates that clearly reflect one single generation of mineral growth. In many cases, paleosols and weathered horizons contain multiple hydrous silicates. Separation and purification of clay minerals, therefore, often requires multiple steps in off-line chemical purification and quantitative X-ray analysis (see Stern et al. 1997; Chamberlain et al. 1999). However, because the effects of evaporation on the δ^{18}O of the analyzed proxy material can be larger than effects of uplift on the isotopic composition of authigenic minerals it should be reiterated that these more detailed studies are highly encouraged as they may ultimately provide important insight into the rates and timing of surface weathering as well as conditions of mineral growth.

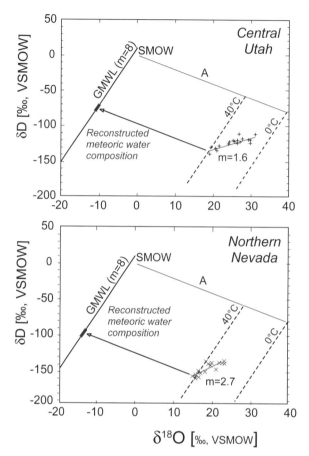

Figure 3. Oxygen and hydrogen isotope data for Eocene freshwater cherts from central Nevada and Utah (after Abruzzese et al. 2005). Samples with the most negative $\delta^{18}O$ values in the Utah and Nevada sections define an array parallel to the global meteoric water line (GMWL) corresponding to temperatures of formation of ~40 °C. Deviations from lines parallel to the GMWL (for samples with higher $\delta^{18}O$) along lower slopes reflect lake basins characterized by seasonal (or long-term) desiccation and evaporation and intermittent lake recharge. If variations in the oxygen and hydrogen isotopic compositions of the freshwater cherts were caused by variations in formation (and burial) temperature, the data should plot along a line with negative slope parallel to A. Reconstructed meteoric water compositions least affected by evaporation are shown as bold segments along the GMWL.

Shear zone silicates as paleoelevation proxies

Recently, the approach of using the stable isotope records from continental basins as paleoaltimetry proxies has been extended to normal fault and detachment systems (Mulch et al. 2004, 2007) as these are known to be pathways for meteoric fluids in actively extending upper and middle crust (e.g., Fricke et al. 1992; Morrison 1994; Morrison and Anderson 1998; Mulch et al. 2005, 2006b; Person et al. 2007) (Fig 4). This approach of reconstructing the isotopic composition of ancient meteoric waters is certainly less straightforward than the approach of exploiting the often (quasi-) continuous stratigraphic and isotopic record in long-lived orogenic and intermontane basins. However, the transport and mineralization processes as well as the source areas of meteoric waters within the orogen can be very different from the processes affecting the isotopic composition of mineral deposits in synorogenic basins

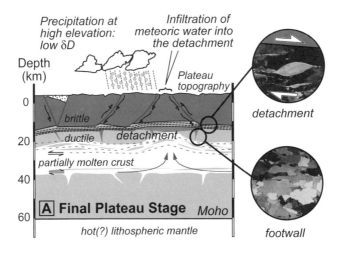

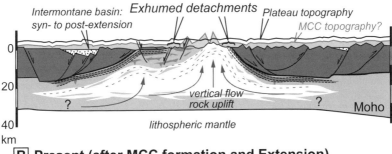

Figure 4. Extensional detachment systems as archives for meteoric fluid compositions. Following thickening of orogenic crust, normal fault and detachment systems develop that dissect the plateau morphology. Once crustal-scale detachment systems control the internal dynamics of the orogen, precipitation sourced at high elevations starts to convect in the detachments and associated fluids may percolate down to the brittle-ductile transition. Here, neocrystallized synkinematic hydrous silicates equilibrate with evolved meteoric water. The hydrogen isotope composition of these minerals, therefore, provides a proxy for meteoric water compositions that are sourced in the internal domains of the evolving orogen. Following exhumation and erosion of the metamorphic core complex (MCC) overall mean elevation and relief have adapted to changing tectonic boundary conditions (after Mulch et al. 2004).

and isotopic data from such reservoirs are, therefore, highly complementary to any study of paleoclimatic records.

Synkinematic minerals that either crystallize from meteoric fluids in the brittle segment of extensional detachment systems (e.g., Morrison and Anderson 1998) or recrystallize during protracted deformation in ductile mid-crustal shear zones of these detachments provide critical isotopic and geochronologic markers that link surface processes with those governing the internal dynamics of orogens. Therefore, the major advantage of this technique is that stable isotope paleoaltimetry data can be temporally and kinematically linked to the evolution of extensional detachment systems that during the late(st) stages of orogeny control surface and rock uplift within the orogen. So far, this approach has been restricted to mylonitic detachment systems of the metamorphic core complexes in the North American Cordillera (Fig. 4; Mulch et al. 2004, 2007). However, the same principle can be generalized and applied to any neo- or recrystallized fault material if the temperature and timing of mineral formation (and trapping of meteoric water) can be adequately constrained and the isotopic composition of the proxy material can be measured with reasonable precision.

As in any stable isotopic study of meteoric-hydrothermal systems, water-rock isotopic exchange can play a significant role when trying to reconstruct the primary isotopic composition of evolved meteoric water at depth (for a review see Criss and Taylor 1986). Although the mechanisms of fluid transport across the brittle-ductile transition in extensional detachment systems are not completely understood, stable isotopic studies have established that meteoric water can penetrate to significant depths in normal fault and detachment systems and that these waters occur at the interface between the brittle hanging wall and ductile segments of the footwall of the detachment (e.g., Wickham and Taylor 1987; Fricke et al. 1992; Morrison 1994; Morrison and Anderson 1998; Bebout et al. 2001; Mulch et al. 2004, 2007; Person et al. 2007). The main challenge within such rapidly evolving and dynamic systems is to establish the timing and duration of fluid flow and water-mineral isotope exchange and the temperature at which this exchange occurred.

Current paleoaltimetry studies exploiting the hydrogen and oxygen stable isotopic record in high(er) temperature extensional detachment systems (Mulch et al. 2005b, 2007) have concentrated on compositionally and isotopically relatively simple detachments where extension related strain is localized within well-defined, highly deformed quartzitic shear zones commonly several 10's to 100's of m thick. Within these shear zones, water is partitioned into either the hydrous mineral phase (white mica) or can be found as primary or secondary fluid inclusions in recrystallized quartz. In contrast to other approaches aimed at determining paleoaltimetry based on the stable isotope record in fluid inclusions (e.g., Sharp et al. 2005) the approach presented here has the additional benefit of providing kinematic, stable isotopic, and thermometric data that can be directly correlated with geochronological methods, thus allowing the temporal reconstruction of meteoric fluid flow in the crust and ultimately paleoelevation. Although more complex rock compositions do not *a priori* prohibit the successful application of the technique, tracer isotope (e.g., hydrogen and oxygen) partitioning between multiple mineral phases may add additional complexities to the successful application of the method.

Hydrogen and oxygen isotopes in hydrous minerals that crystallized in the presence of fluids provide two independent tracers of the primary composition and ultimately the origin of meteoric waters. In silicate rocks, oxygen is the dominant constituent of the main rock-forming minerals and hence protracted rock-water interaction will commonly affect the oxygen isotope record of the water and erase any information about the initial oxygen isotopic composition (e.g., Criss and Taylor 1986; Person et al. 2007). In contrast, hydrogen represents a trace element in most silicate rocks with concentrations commonly below 2000 ppm (e.g., Sheppard 1986). Therefore, in extensional shear zones that receive significant amounts of external, meteoric fluids during deformation, water represents the major hydrogen reservoir and the primary hydrogen isotopic composition of these meteoric waters will determine the D/H ratios of neo- or recrystallizing minerals within the shear zone.

White mica that recrystallized in the presence of evolved meteoric waters in extensional shear zones provides a suitable target for paleoaltimetric studies as: 1) it records the oxygen and hydrogen isotopic composition of water present during deformation, 2) it is known to be resistant to post-deformational alteration and low-temperature isotopic exchange, especially within rapidly cooling rocks, 3) the (microstructural) deformation history in the mylonite can be correlated with the fluid flow history, and 4) the timing and duration of recrystallization and associated isotope exchange can be dated precisely using e.g., texturally-controlled $^{40}Ar/^{39}Ar$ geochronology (e.g., Müller et al. 2000; Sherlock et al. 2003; Goldstein et al. 2005; Mulch and

Cosca 2004; Mulch et al. 2005, 2006b). The utility of synkinematic white mica in mylonitic quartzite for paleoaltimetry is, therefore, based on the coupled stable isotope ($^{18}O/^{16}O$, D/H) and radiogenic isotope ($^{40}Ar/^{39}Ar$) record.

Temperature-dependent isotope fractionation can be used to determine the temperatures of oxygen isotope exchange between quartz and muscovite. One of the prerequisites for obtaining reliable oxygen isotope exchange temperatures is that for example quartz-muscovite oxygen isotope fractionation within the shear zone mylonite is measured between mineral pairs that either formed (neocrystallized) or recrystallized synkinematically from the same fluid. Even though direct oxygen isotope fractionation data from recrystallized mineral assemblages are sparse, recent data indicate that reaction products are likely to be in oxygen isotopic equilibrium even in relatively "dry" metamorphic environments (e.g., Kohn 1999; Müller et al. 2004). In a mylonitic rock sample where mineral phases are formed either by synkinematic mineral growth (by a solution-precipitation mechanism) or synkinematic recrystallization it is expected that we see (local) oxygen isotopic equilibrium on the length scales of the spacing of the mylonitic foliation that is assumed to control the transport and availability of fluids in the shear zone. Based on these temperatures, the hydrogen isotopic composition of the proxy material (e.g., muscovite), δD_{proxy}, can then be converted to that of water, δD_{water}, assuming proxy-water hydrogen isotope equilibrium during recrystallization. For the purposes of this study, δD_{water} designates the hydrogen isotopic composition of aqueous fluids present in the shear zone during deformation. These fluids are derived from near-surface groundwaters that in turn reflect a precipitation-weighted average of the hydrogen isotopic composition of precipitation, δD_{ppt}.

The long time scales (10^4 to 10^6 m.y.) involved in deformation and synkinematic isotope exchange in white mica provide a robust, long-term average of meteoric water-rock interaction, characteristic for the time scales of major readjustments in surface elevation. Thus the hydrogen isotope record in recrystallized muscovite provides a direct link between the tectonic and fluid flow history in the shear zone and temporal variations in meteoric water composition due to changing surface elevation.

We want to reiterate that determination of the timing and temperature of hydrogen isotope exchange in the ductile segment of the detachment is of prime importance for any paleoaltimetry study that aims at relating surface processes to those occurring contemporaneously at depth. It is, therefore, important to document that the geochronologic (e.g., $^{40}Ar/^{39}Ar$) and thermometry (e.g., quartz-muscovite $^{18}O/^{16}O$ fractionation) data record neo- or recrystallization of white mica during deformation in the shear zone rather than e.g., diffusional isotopic exchange during retrograde metamorphism.

The coupled basin-detachment approach

The $\delta^{18}O$ and δD of water in intermontane basins, whose areal extent is likely to be controlled by regional variations in topography and the distribution of zones of rock uplift, provide a record of water compositions within the entire catchment area of a basin and thus integrate over the different hypsometries of the various drainages feeding into the basin. As the basin and drainage geometries at any time in the geologic past are usually poorly constrained, detailed analysis of the sedimentation history, the interconnectivity with adjacent lake basins, and the exact timing of major facies changes are important tools in converting oxygen and hydrogen stable isotope data from lake deposits or authigenic minerals formed within them to some measure of paleoelevation. For example, Kent-Corson et al. (2006) and Carroll et al. (2007) present an example of combined sedimentological, geochronological, and isotopic studies that document a contemporaneous negative (−8 to −10‰) oxygen isotopic shift in the Medicine Lodge/Sage Creek (MT) and Green River Basins (WY) at ~48-49 Ma concurrent with the development of major detachments bounding the metamorphic core complexes in the northern and central part of the North American Cordillera (e.g., Teyssier et al. 2005; Mulch et al. 2006b; Foster et al. 2007). The correlation in magnitude and timing of these isotopic shifts in basins that extend for more than 600 km along the orogen suggests that the oxygen isotopic record of lacustrine and palustrine carbonates within these intermontane basins is controlled by a common mechanism and most likely reflects drainage reorganization due to increased mean elevation and enhanced topographic relief in the upstream sections of the basins (Kent-Corson et al. 2006; Carroll et al. 2007). However, unless there is direct stable isotope evidence from within the orogenic belt the locus and tectonic significance of this increase in elevation and/or topographic relief remain speculative. It is at this point where combined isotopic, sedimentological and structural data sets from within the uplifting crystalline core of the orogen as well as the developing intermontane basins provide critical information on the spatial and temporal evolution of surface elevation in orogenic belts (Fig. 5).

APPLICATIONS OF STABLE ISOTOPE PALEOALTIMETRY

Tracking of orogenic rain shadows

The most straightforward approach to understanding the topographic history involves tracking the development of a rain shadow through time. Many of the world's mountain belts lie along the edges of continents and have relatively simple climatic regimes, in which precipitation is provided by a single dominant moisture source and (quasi-) constant atmospheric circulation patterns (on decadal to millennial time scales). The transport of water vapor across these mountain ranges results in a rain shadow with orographic precipitation on the windward side of the range and drastically reduced precipitation rates on the leeward side. In these "simple" mountain ranges (such as e.g., the Sierra Nevada of California) the asymmetric patterns of precipitation cause relatively large differences in the hydrogen and oxygen isotopic compositions of surface waters on the windward and leeward sides of a given mountain range (e.g., Ingraham and Taylor 1991). Using isotopic proxies for surface waters in stratigraphic sections on the leeward side of the mountain range, it is possible to reconstruct the topographic evolution of the orographic barrier through time. This approach using stable isotope ratios of meteoric water in the rainshadow of orographic barriers has shown to be most reliable in 1) extremely warm climates, 2) in topographic settings with comparatively simple (unidirectional and stable) atmospheric circulation patterns, and 3) in the direct leeward rain shadow within close proximity (10's to 100's of kilometers) of the orographic barrier (see e.g., Chamberlain et al. 1999; Blisniuk and Stern 2005; Quade et al. 2007, this volume). It has therefore been increasingly used in paleotopography studies in the Austral Andes of southern Patagonia (Blisniuk et al. 2005), the Southern Alps of New Zealand (Chamberlain et al. 1999), the Oregon and Washington Cascades (Kohn et al. 2002; Takeuchi and Larson 2005) and the Sierra Nevada of California (Chamberlain and Poage 2000; Poage and Chamberlain 2002; Koch and Crowley 2005; Mulch et al. 2006a) of western North America.

This approach of detecting the presence or absence of an orographic barrier in the stable isotope record of basin deposits on the leeward side of an orogen was first-tested in the Southern Alps of New Zealand (Chamberlain et al. 1999). The Southern Alps lie along the west-side of the South Island of New Zealand and moisture to this region is largely provided by southern hemisphere Westerlies. Moisture transport across these mountains results in a nearly tenfold decrease in precipitation across the mountain belt, producing a strong rain shadow on the east side of the Southern Alps. The $\delta^{18}O_{water}$ values of surface waters decrease by ~4.5‰ and δD_{water} by about 40‰ from east to west across the mountain range (Fig. 6). By tracking the development of this isotopic rain shadow through time it is possible to constrain the long-term evolution of surface elevation of the Southern Alps (Chamberlain et al. 1999).

Chamberlain et al. (1999) measured the oxygen isotope composition of authigenic kaolinite, smectite and illite in weathered horizons from intermontane basins on the east side of the Southern Alps. Neither smectite nor illite produced useful results, as reflected in the

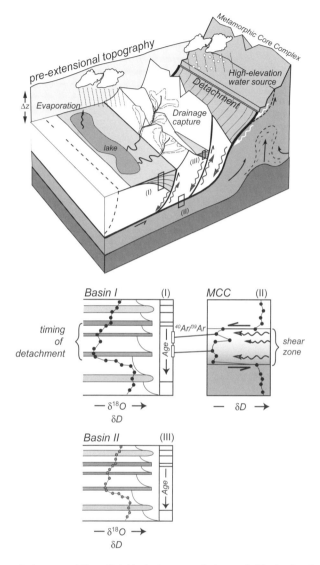

Figure 5. Conceptual representation of stable isotope records in coupled-basin detachment systems. Syntectonic basins develop along the orogenic front and within the foreland and hinterland regions of the orogen. Depending on their location with respect to the main orographic barrier the isotopic record of changing climate and topographic conditions may be dampened (Basin III) or controlled by evolution of the drainage system in the upstream sections of the mountain rivers feeding the lake basin (Basin I). River capture and drainage reorganization directly respond to the larger-scale changes in topography and frequently are difficult to distinguish from the latter. Extensional detachment systems control the overall geometry of basins in the evolving upper plate as well as the exhumation and rock uplift in the lower plate. Waters percolating in the fault and detachment systems are likely to be sourced in the ancestral high elevation regions of the orogen where the combined effects of changing tectonic boundary conditions (contraction to extension) and vertical and lateral flow in the middle and lower crust (see Fig. 10 below) are propagated to the surface. Recrystallized syntectonic minerals that form at the brittle-ductile transition provide temporal constraints on the timing and duration of deformation and related fluid flow. The isotopic signal recorded in these minerals, therefore, represents an integrated long-term record of meteoric fluids that together with the basin record permits the direct correlation between elevation history and orogen-scale tectonic processes.

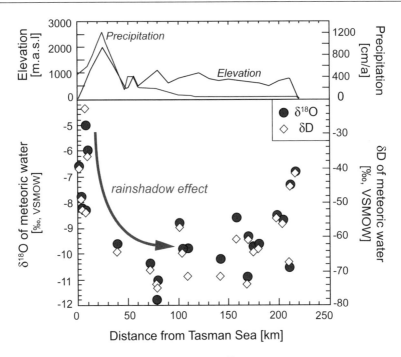

Figure 6. Topography (solid), precipitation (gray), and $\delta^{18}O$ and δD values along a transect across the southern South Island of New Zealand. The modern Southern Alps create a strong rain shadow on the eastern side of the mountain range that is characterized by a reduction in rainfall from up to 3000 mm/a on the western side to values <300 mm/a on the eastern side of the range. Similarly, hydrogen and oxygen isotopic compositions of meteoric water are strongly depleted in D and ^{18}O, and decrease by ca. −40‰ and −4.5‰, respectively. Circles: $\delta^{18}O$ values, diamonds: δD values. Data from Chamberlain et al. (1999) and Stewart et al. (1983). After Chamberlain et al. (1999).

considerable scatter of the oxygen isotope compositions (Fig. 7). This scatter resulted from: 1) the inability to obtain pure smectite separates; and 2) illite forming as an incomplete weathering product of detrital, metamorphic white mica, and thus the oxygen isotope values are inherited, in part, from the white mica. In contrast, the oxygen isotope values of authigenic kaolinite decreased from $\delta^{18}O_{kaolinite}$ = +10‰ to $\delta^{18}O_{kaolinite}$ < +5‰ between the late Miocene and early Pliocene (Fig. 7). After accounting for isotopic effects due to known changes in climate and the oxygen isotope composition of ocean water, they attributed approximately 3 to 4‰ of this shift as due to the rise of the Southern Alps, by an elevation of up to 2 km. This rise of the Southern Alps during the Late Miocene to Pliocene is consistent with geodynamic (Walcott 1998) and sedimentologic (Sutherland 1996) studies that show the plate motion vectors for the South Island of New Zealand to have changed at ~5 Ma and in turn increased convergence and transpressional motion along the Alpine Fault (the boundary between the Australian and Pacific Plates that forms the western edge of the Southern Alps).

Other studies of isotopic rain shadows in the Washington Cascades (Takeuchi and Larson 2005) and the Sierra Nevada of California (Chamberlain and Poage 2000; Poage and Chamberlain 2002) used smectite in paleosols and altered ashes as an isotopic proxy. In these studies, the oxygen isotope composition of smectite from east of the Washington Cascades decreased by 3 to 4‰ from the middle Miocene to recent, whereas smectite from east of the Sierra Nevada increased by 2 to 3‰ from the middle Miocene to late Pliocene. The two studies argued that these isotopic results indicate an elevation increase in the Washington Cascades (Takeuchi and Larson

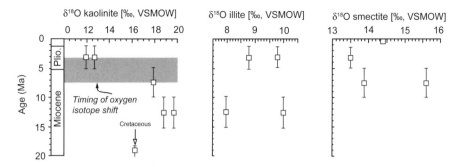

Figure 7. $\delta^{18}O$ values of kaolinite, illite, and smectite from weathered basin deposits on the leeward side of the modern Southern Alps (New Zealand). Vertical bars represent the possible formation and growth periods for the individual authigenic mineral phases. Oxygen isotope analysis of authigenic kaolinite shows a ca. 5-6‰ decrease in the early Pliocene; from 18.2‰ in Miocene and older samples to 12.3‰ in the Pliocene kaolinite. This decrease in $\delta^{18}O$ values of kaolinite forming in equilibrium with meteoric water at or near the surface records the topographic development of the rising Southern Alps during this time interval. The suggestion of a Miocene-to-Pliocene strengthening of the rain shadow is further supported by the increased abundance of smectite in the younger stratigraphic record that indicates a change to a generally drier climate south of the crest of the Southern Alps. After Chamberlain et al. (1999).

2005) and decreased or constant elevation in the Sierra Nevada (Poage and Chamberlain 2002; Horton and Chamberlain 2006) during this time period. The fact that these two distinct trends are from regions only about 800 kilometers apart strongly indicates that the oxygen isotope compositions of smectite are recording local climate conditions caused by the adjacent mountain ranges, rather than larger scale global climatic changes. This is further supported by oxygen isotope studies of calcite (Poage and Chamberlain 2002) and fossil mammal teeth (Kohn et al. 2002; Koch and Crowley 2005; Kohn and Fremd 2007) that display similar isotopic trends.

Reconstructing paleo-river slopes

Apart from the quality of the available proxy materials, isotopic, sedimentologic or tectonic data, the success of any stable isotopic paleoaltimetry study strongly depends upon the knowledge of past climatic and tectonic boundary conditions. One promising approach to stable isotope paleoaltimetry is to study modern high elevation continental interiors with the question of when these orogens or orogenic plateaus attained their current stature. However, since these areas are frequently remote from the original moisture source and characterized by arid or semi-arid climates such paleoaltimetric studies need to take into account the effects of evaporation of surface waters and paleoaltimetric results are subject to additional uncertainties arising from complex or even unknown atmospheric circulation and precipitation patterns.

Ancient river systems provide a different framework for paleoaltimetry. Here, weathering products or authigenic minerals that form within the fluvial deposits of the river system represent valuable information on the combined effects of river gradients and river discharge. One example of well-preserved paleo-river deposits is the Eocene Yuba River, one of the major drainage systems of the northern Sierra Nevada (California), which due to its important economic role during the 1849 gold rush, most likely represents one of the most intensely studied and best characterized Early Cenozoic drainage systems in the western United States (e.g., Lindgren 1911; Yeend 1974). The ancient Yuba River deposited vast amounts of granitic detritus that, during a period of intense Eocene weathering, was rapidly transformed into kaolinite clay. These clays formed *in situ* in overbank deposits of the river system and—where preserved— provide a record of near-surface alteration and weathering by meteoric waters. Additionally, these weathering products can be retrieved continually from Eocene sea level upstream within three different drainages of the river system, thus permitting reconstruction of not only the course of the ancient river but also the variation of oxygen and hydrogen isotope ratios within an ancient mountain river system along the windward side of a mountain range (Fig. 8).

In contrast to the vast majority of paleoaltimetric studies that in one way or the other recover the isotopic composition of precipitation after atmospheric moisture has crossed an orographic barrier, the approach of tracing the effects of progressive rain out along an ancient mountain range provides a much more direct way of estimating the height of a mountain range by reconstructing the slope and headwater elevation of the ancient river system. Hydrogen isotopic compositions of Eocene kaolinite clays that were either formed in situ at the site of deposition of the river gravels or that were subsequently transported downstream by the Yuba river itself demonstrate that ~40 million years ago the topography of the western flank of the northern Sierra Nevada was very similar to that observed today (Mulch et al. 2006a). As expected, hydrogen isotopic compositions of kaolinite (δD_{kaol}) decrease systematically with distance from the Eocene marine shoreline (Fig. 9). Due to the unique preservation of the strongly weathered sedimentary sequences a bias towards more strongly D-depleted kaolinite at any given elevation by either slope transport processes or interaction with down-slope flow of groundwater can be excluded from the interpretation of the hydrogen isotope pattern. The observed systematic decrease in δD_{kaol} found in all major drainages of the Eocene Yuba River together with different estimates of isotopic lapse rates, therefore, indicates that already during the Eocene the northern Sierra Nevada provided a major orographic barrier for atmospheric flow and most likely created a rain shadow to the west (Mulch et al. 2006a). When combined with river incision and erosion data, the results of hydrogen isotope paleoaltimetry require headwater elevations of the ancient Yuba River in excess of 1.6 km and peak elevations that were at least 2.2 km above Eocene sea level. Even though the set of boundary conditions for the

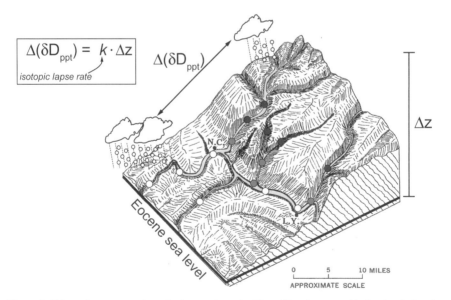

Figure 8. Schematic representation of rain out along the Sierra Nevada orographic barrier during the Eocene and associated change in isotopic composition of surface waters. Sampling of weathering products along the course of the ancestral Yuba River (circles) provides first order constraints on the change in elevation from the Eocene shore line upstream to the headwaters of the river. This change in elevation (Δz) is related to the change in the hydrogen isotopic composition of precipitation ($\Delta\delta D_{ppt}$) by an isotopic lapse rate (k). Topography from Yeend (1974).

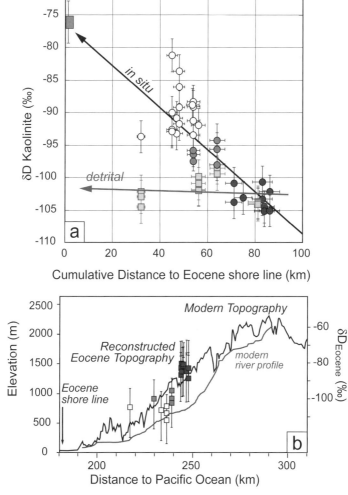

Figure 9. (a) δD_{kaol} along the Eocene Yuba and American rivers (circles) versus distance to Eocene shoreline of the marine embayment in the modern Great Valley (California). The Yuba and American rivers represent the major westward-directed drainage systems of the ancestral northern Sierra Nevada. Linear regression through δD_{kaol} from kaolinite formed in situ along the Yuba River indicates a sea level δD_{kaol} of $-76 \pm 5\%o$ ($r^2 = 0.76$) within error identical to kaolinite in isotopic equilibrium with modern precipitation at the Pacific coast (gray square). Detrital kaolinite that was transported downstream by the Eocene Yuba River (squares) overlaps with most D-depleted in situ kaolinite. The presence of large differences in δD_{kaol} between detrital and in situ kaolinite at the same sampling sites indicates that post-depositional processes did not significantly affect the hydrogen isotopic composition of the kaolinite clays. The 25 $\pm5\%o$ decrease in δD_{kaol} from sea level upstream reflects rainout during uplift of moisture along the western flank of the Eocene Sierra Nevada (modified after Mulch et al. 2006a). (b) Orographic effect of Eocene northern Sierra Nevada on δD of precipitation along the ancestral Yuba river can be translated into information on the Eocene slope of the Yuba River. Decrease in δD of kaolinite and ultimately δD of precipitation indicates that modern and Eocene western slopes of the northern Sierra Nevada were similar.

case of the Eocene Yuba River (knowledge of sea level with respect to the sampling site, river geometry, upstream-downstream relationships, windward-facing slope of the orographic barrier) are not common, ancient river systems in general provide important targets for paleoaltimetric studies, and we strongly advocate for integrated studies that aim at reconstructing the erosional, depositional, and topographic histories of such systems.

Reconstructing paleoelevation in eroded mountain ranges

Eroded mountain ranges and orogenic plateaus expose the combined structural complexities of surface deformation and crustal and lithospheric flow on the architecture of orogens. Deeply exhumed crustal (and even lithospheric) cross sections, may serve as a guide to understanding the internal dynamics governing surface uplift as well as the superimposed demise of a plateau region through mechanisms such as lithospheric extension, orogenic collapse or lateral escape. For the purposes of stable isotope paleoaltimetry it is important to note that in deeply dissected and eroded mountain ranges the topographic starting conditions for paleoaltimetric studies are usually either model-derived or largely unknown. In contrast to modern plateau regions or rain shadow environments, where studies aim at reconstructing the timing when the current topographic configuration was attained, paleoaltimetric studies in partially collapsed orogenic belts and plateaus need to be tested against thermomechanical or tectonic models that make predictions about the surface response to , for example, various collapse mechanisms (Fig. 10).

Late-orogenic extensional detachment systems represent first-order tectonic elements that control the internal architecture of orogens and strongly influence redistribution of mass and heat during orogenic collapse (e.g., Vanderhaeghe and Teyssier 2001; Rey et al. 2001; Teyssier et al. 2005; Mulch et al. 2006b). Crustal sections in the exhumed footwalls of such detachments provide an opportunity to compare thermochronological (e.g., Stockli 2006 and references therein), stable isotopic (e.g., Fricke et al. 1992; Morrison and Anderson 1998; Mulch et al. 2006b) and (micro-) structural data sets in rocks that have experienced conceivably simple exhumation and cooling histories. One example of a tectonically exhumed extensional detachment system is the Columbia River Detachment in southern British Columbia and northern Washington that represents the eastern termination of the Shuswap-Okanagan metamorphic core complex(es) (e.g., Orr and Cheney 1987; Vanderhaeghe et al. 1999; Teyssier et al. 2005). Geochronologic and thermochronologic data document rapid cooling and exhumation at around 48-50 Ma in the upper and middle crust postdating decompressional melting and subsequent crystallization of partially molten middle crust at around 52-55 Ma (Vanderhaeghe et al. 1999, 2003; Teyssier et al. 2005; Mulch et al. 2006b). Teyssier et al. (2005) interpret the Columbia

---

Figure 10. *(ON FACING PAGE)* (a) Schematic cross-section of a Cordilleran-type orogen with a fold-thrust belt foreland, intermontane terrane thrust sheet and crustal duplex that probably characterized the thickening processes during the contractional stages of orogeny of the North American Cordillera. Right column: Expected surface response to deeper crustal processes. Modified after Teyssier et al. (2005). (b) Development of a continental plateau following thickening. Thickened crust in the hinterland region may be supported by a zone of partially molten crust. Intracrustal flow is shown to be determined by a channel geometry on the hinterland side and the development of a foreland-dipping detachment of the foreland side. High mean surface elevation and low relief may characterize the internal domains of the orogen. (c) The foreland dipping detachment develops into a rolling-hinge structure (Teyssier et al. 2005) during incipient collapse. Note that at this point plate boundary conditions do not permit free collapse and lithospheric-scale extension. Due to thinning of upper and middle crust mean elevation is lowered compared to the plateau stage. The evolution of relief at this stage depends on the magnitude, relative timing, and location of vertical displacements and erosion within the internal parts of the orogen. (d) With further extension, the rolling-hinge detachment now controls vertical and horizontal material trajectories within the orogen; this lateral displacement is a direct measure of extension. Decompression of deep crustal rocks enhances partial melting and triggers a diapiric instability and associated rapid (sub-) vertical rock uplift. Depending on the relative rates of thinning in the crust and lithospheric mantle mean elevation may first increase then decrease or continually decrease as a net result of lithospheric stretching. Ultimately, mean elevation will decrease after significant extension has occurred in the upper and middle crust. It is very likely that at this stage dramatic reorganization of relief and drainage patterns occurs. These changes should be reflected in the stable isotopic record of synextensional basins as well as detachment systems that provide pathways for meteoric fluids.

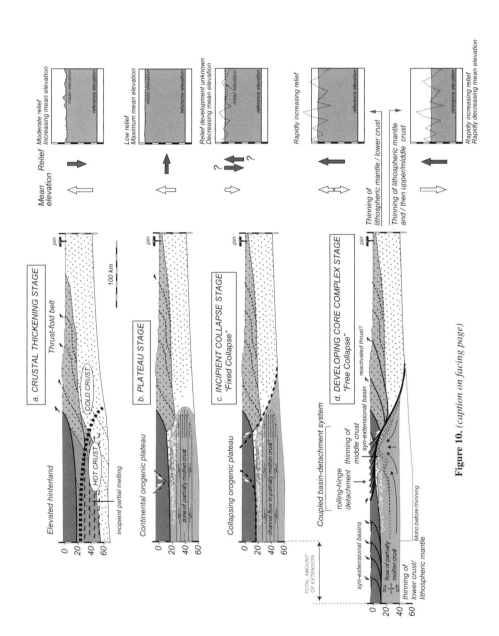

Figure 10. *(caption on facing page)*

River Detachment as a relatively shallowly dipping 'rolling-hinge' detachment on the foreland side of the metamorphic core complex whose geometry is governed by strain localization at the critical interface between two crustal segments with contrasting thermal histories. In contrast, it was conceivable that on the hinterland side of the evolving core complex (Fig. 10) differential displacement between upper crust and deeper portions of the lithosphere is accommodated by channelized flow along a long-lived interface that separated partially molten crust and perhaps even lithosphere from a rigid upper crustal lid (e.g., Nelson et al. 1996; Rey et al. 2001; Vanderhaeghe and Teyssier 2001; Beaumont et al. 2001). Combined hydrogen, oxygen, and $^{40}Ar/^{39}Ar$ isotopic data document that during extensional deformation and exhumation meteoric fluids infiltrated the ductile segment of the detachment and that the hydrogen isotope composition of these meteoric waters can be used to reconstruct the paleoelevation within the North American Cordillera during core complex formation at these latitudes (Mulch et al. 2004; Mulch et al. 2007; Person et al. 2007). These data indicate that hydrogen isotope exchange in the mylonitic segment of the detachment occurred at temperatures of 420 ± 40 °C (Shuswap metamorphic core complex, British Columbia at ~52°N) and 380 ±40 °C (Kettle Dome, Washington) with evolved meteoric fluids that attained δD_{water} values lower than −135‰ (Fig. 11). Together with estimates of Early-to-Mid-Eocene climate and precipitation conditions these low δD_{water} values indicate that these meteoric waters originated at elevations of around 3500-4500 m, some 1000-1500 m higher than today's highest peaks in the respective areas (Fig. 12). Variations in the hydrogen isotopic composition of meteoric waters found at ~48°N (−110 to −115‰; Kettle Dome) and ~52°N (−135‰; Shuswap; Thor-Odin Dome) of the Columbia River detachment suggest that at the time of detachment faulting the southern termination of the Shuswap-Okanagan metamorphic core complex attained elevations that were measurably lower than its northern counterparts in British Columbia. This inference is based on a very limited set of data, but such studies may enhance our predictive capabilities when interpreting the results of numerical and thermomechanical models describing the relationships between the internal dynamics and the surface evolution during orogeny.

Even though paleoaltimetric data from internal structural elements of orogenic belts and plateaus represent much needed complementary information to those derived from surface deposits or weathering products we caution about the uncritical use of stable isotopic data from deeper Earth environments in paleoaltimetric studies. It is highly desirable to obtain reliable thermometric, structural, and isotopic tracer data before attempting any paleoaltimetric reconstruction in such environments, as uncertainties exist about the fluid pathways and mechanisms responsible for fluid transport into the ductile crust. Maybe more importantly, it is imperative to document that the timing of meteoric water-rock interaction can be dated precisely, especially within thermally and kinematically rapidly evolving tectonic environments such as extensional detachment systems.

CONCLUSIONS AND FUTURE DIRECTIONS

In this chapter we emphasize the importance of using hydrous silicate minerals in paleoaltimetery studies. The bulk of the isotopic data for paleoaltimetry studies to date come from the analysis of authigenic or pedogenic calcite, primarily because calcite is relatively easy to analyze using modern continuous flow mass spectrometry. As such, relatively large oxygen isotope data sets can -and should- be generated using calcite as a proxy material for terrestrial paleoclimate and paleotopography. These large data sets are imperative in interpreting isotopic profiles as isotopic shifts in terrestrial basins can vary spatially and temporally within orogenic belts.

We further propose that studies of coupled systems such as major orogen-scale shear zones and coeval basins permit previously unavailable interpretations of the history of surface elevation. In addition, approaches that integrate surface processes (sedimentation and erosion

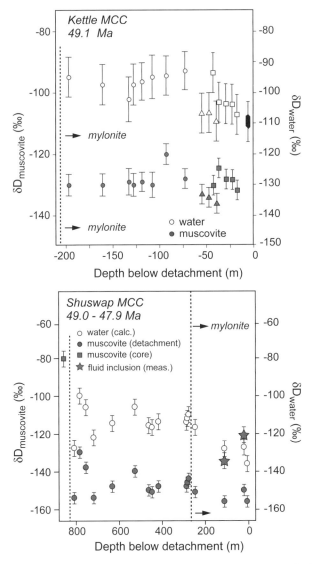

Figure 11. δD values of recrystallized muscovite (δD$_{muscovite}$) from mylonitic quartzite in the footwall of the Columbia River Detachment at two latitudes of the Shuswap-Okanagan Metamorphic core complex. (a) Composite section at the latitude of the Kettle dome (Washington): δD$_{muscovite}$ values (solid symbols) are largely constant throughout the detachment and attain minimum values closest to the hanging wall. Calculated water compositions in equilibrium with muscovite (open symbols) and amphibole (black rectangle) are δD$_{water}$ = −110 ± 4‰ and −112 ± 4‰, respectively. Muscovite-water hydrogen isotope fractionation was calculated using temperatures from quartz-muscovite oxygen isotope thermometry. (b) Hydrogen isotope transect across ca. 800 m of mylonitic quartzite of the Columbia River detachment at the latitude of Thor-Odin dome (Shuswap metamorphic core complex, British Columbia): Water in equilibrium with muscovite at 420 ±40 °C has δD values of −135 ± 4‰, typical for high-latitude/high altitude meteoric waters. The calculated water compositions (open symbols) based on measured δD$_{muscovite}$ (solid symbols) and temperature information from quartz-muscovite oxygen isotope thermometry are in good agreement with measured δD values of fluid inclusion water (stars). Muscovite hydrogen isotopic composition of quartzite in the core of the complex (δD$_{muscovite}$ = −80‰) indicates that infiltration of meteoric water shifted δD$_{muscovite}$ by up to −75‰. After Mulch et al. (2007).

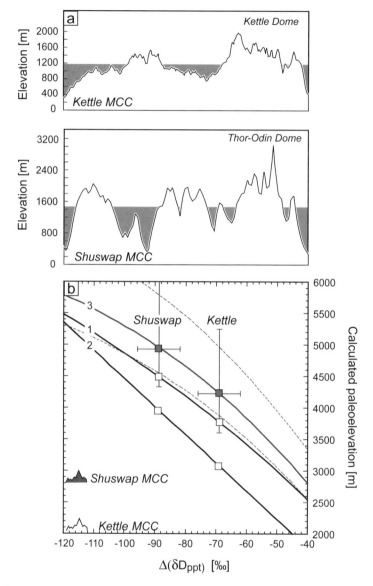

Figure 12. (a) Modern east-west topographic profile and estimate of isostatically supported mean elevation at different latitudes (Kettle dome and Thor-Odin dome) of the Shuswap-Okanagan metamorphic core complex in Washington and British Columbia. (b) Predicted Eocene (ca. 49-47 Ma) paleoelevations based on detachment-derived meteoric water compositions using the lapse rates of (1) Garzione et al. (2000), (2) Poage and Chamberlain (2001), (3) Rowley et al. (2001). For the latter, Mulch et al. (2007) assumed zonal atmospheric flow with moisture derived from 13° to 32° N and used climate conditions from Huber and Caballero (2003) as input parameters to the model). In all cases, calculated paleoelevations are regarded as minimum estimates as down-gradient hydrogen isotopic exchange in the shear zone may have shifted δD to more positive values. The paleoaltimetric reconstruction for the time of initial extension and core complex formation suggests that following a period of crustal thickening the hypsometry of the surface area that provided the meteoric water for the extensional detachment systems was at least 1000 m higher than present-day's highest peaks (Mulch et al. 2007). Insets show modern maximum peak elevations. After Mulch et al. (2007).

rates, drainage evolution, changing local and regional climate patterns) and processes controlling the internal architecture of the orogen (crustal and lithospheric scale faulting, rock uplift and exhumation, vertical and lateral flow) aid greatly to understand the role of various mechanisms proposed for the formation and demise of orogenic belts and continental plateaus.

Given the importance of paleoaltimetric data for our understanding of orogen-building processes, we advocate that, when possible, stable isotope paleoaltimetry studies should rely on multiple proxies from both silicates and carbonates for the following two reasons. First, evaporative isotopic effects can be as large or larger than the effect of uplift of major mountain ranges and these effects can best be assessed through the combined hydrogen and oxygen isotope analysis of hydrous silicates. Second, the growth rates of the different types of calcite found in terrestrial basins can vary by several orders of magnitude. Whereas carbonate nodules in paleosols may form over time scales of 10^3 to 10^4 years, growth of calcite laminae in lake sediments or calcite cements/crusts is much faster than the growth rates of clay minerals and as a result rapidly forming calcite may record isotopic shifts that are unrelated to climate changes associated with tectonic processes (see also Quade et al. 2007, this volume for use of carbonates in stable isotope paleoaltimetry). Where oxygen isotope compositions of both calcite and hydrous minerals have been determined, calcite always shows more scatter than clay minerals, arguably, in part, due to differences in growth rates and seasonality of carbonate formation.

Based on the examples presented in this chapter we are cautiously optimistic that stable isotope paleoaltimetry will be able to address fundamental problems in the interaction between climate and tectonics. However, before stable isotope paleoaltimetry can be used to its full potential, a number of key problems should be examined:

1) There is a need for the application of multiple isotope proxies in paleoaltimetry studies. In the few studies that have used multiple isotopic proxies the same isotopic trends are generally obtained, but for the reasons stated above it is critical to examine as many isotopic systems as possible.

2) Better information is needed regarding growth rates of authigenic minerals and what indeed these minerals record. Understanding growth rates and seasonality of mineral growth in soils and weathering profiles is critical to evaluate the types of surface waters present during formation of the analyzed mineral proxies. Since there can be large seasonal variations in the isotopic composition of surface waters, this information is key to paleoaltimetry studies.

3) It is essential to understand the isotopic variability within authigenic minerals that form the basis of terrestrial paleoclimate records and between such samples that are isochronous within an outcrop of a stratigraphic formation. Except for variability studies in biogenic materials (teeth, bone, shells) (e.g., Kohn et al. 2002), we know of no paleoaltimetry studies that have examined this natural variability in authigenic or pedogenic proxy materials. Such variability studies should be conducted before any quantitative estimates of paleoelevation can be made.

4) We anticipate that much can be learned from the combination of isotope paleoaltimetry studies with climate models. As stated above, extrapolation of climate models to past conditions requires information on paleotopography and interpretation of stable isotopic data require information from paleoclimate models.

5) Any interpretation of stable isotope paleoaltimetry data must be placed in the proper structural, sedimentological, and geologic context. As such, we encourage multidisciplinary studies that bring all of these disparate data sets to bear on the past evolution of Earth's topography.

ACKNOWLEDGMENTS

We would like to dedicate this chapter to our friend and colleague James R. O'Neil, one of the most influential stable isotope geochemists of the past decades who through his fundamental work on stable isotopes contributed largely to the current success of stable isotope paleoaltimetry.

We acknowledge the thorough and helpful reviews of C. Garzione, T. Vennemann, and D. Foster and the editorial comments of M. Kohn. We thank Christian Teyssier for introducing the first author to the challenges and fun of studying tectonic processes in metamorphic core complexes and how these may be reflected at the Earth's surface. Much of the research presented here was supported by the National Science Foundation through the EAR and SGER programs.

REFERENCES

Abruzzese MJ, Waldbauer JR, Chamberlain CP (2005) Oxygen and hydrogen isotope ratios in freshwater chert as indicators of ancient climate and hydrologic regime. Geochim Cosmochim Acta 69:1377-1390

Beaumont C, Jamieson RA, Nguyen MH, Lee B (2001) Himalayan tectonics explained by extrusion of a low-viscosity crustal channel coupled to focused surface denudation. Nature 414:738–742

Beaumont C, Jamieson RA, Nguyen MH, Medvedev S (2004) Crustal channel flows: 1. Numerical models with applications to the tectonics of the Himalayan—Tibetan orogen. J Geophys Res 109:doi:10:1029/2003JB002809

Bebout GE, Anastasio DJ, Holl JE (2001) Synorogenic crustal fluid infiltration in the Idaho-Montana thrust belt. Geophys Res Lett 28:4295-4298

Blisniuk PM, Stern LA (2005) Stable isotope paleoaltimetry: A critical review. Am J Sci 305:1033-1074

Blisniuk PM, Stern LA, Chamberlain CP, Idleman B, Zeitler PK (2005) Climatic and ecologic changes during Miocene surface uplift in the southern Patagonian Andes. Earth Planet Sci Lett 230:125-142

Bird P (1978) Initiation of intracontinental subduction in the Himalaya. J Geophys Res 83:4975-4987

Carroll AR, Doebbert AC, Booth AL, Chamberlain CP, Rhodes-Carson MK, Smith ME, Johnson CM, Beard BL (2007) Capture of high altitude precipitation by a low altitude Eocene lage, western United States. Geology (in press)

Chamberlain CP, Poage M, Craw D, Reynolds R. (1999) Topographic development of the Southern Alps recorded by the isotopic composition of authigenic clay minerals, South Island, New Zealand, Chem Geol 155:279-294

Chamberlain CP, Poage MA (2000) Reconstructing the paleotopography of mountain belts from the isotopic composition of authigenic minerals. Geology 28:115-118

Chase CG, Gregory-Wodzicki KM, Parrish JT, DeCelles PG (1998) Topographic history of the Western Cordillera of North America and controls on climate. *In:* Tectonic Boundary Conditions for Climate Reconstruction. Crowley TJ, Burke KC (eds) Oxford Univ. Press, New York, p 73–97

Clark MK (2007) The significance of paleotopography. Rev Mineral Geochem 66:1-21

Craig H (1961) Isotopic variations in meteoric waters. Science 133:1702-1703

Criss RE, Taylor HP (1986) Meteoric-hydrothermal systems. Rev Min Geochem 16:373-424

Currie BS, Rowley DB, Tabor NJ (2005) Middle Miocene paleoaltimetry of southern Tibet: Implications for the role of mantle thickening and delamination in the Himalayan orogen. Geology 33:181-184

Cyr AJ, Currie BS, Rowley DB (2005) Geochemical evaluation of Fenghuoshan Group lacustrine carbonates, north-central Tibet: implications for the paleoaltimetry of the Eocene Tibetan Plateau. J Geol 113:517–533

Dansgaard W (1964) Stable isotopes in precipitation. Tellus 16:436-468

Davis SJ, Mulch A, Carroll AR, Chamberlain CP (2007) Paleogene landscape evolution of the central North American Cordillera: Developing topography and hydrology in the Laramide foreland. GSA Bull (in review)

DeCelles PG, Quade J, Kapp P, Fan MJ, Dettman DL, Ding L (2007) High and dry in central Tibet during the Late Oligocene. Earth Planet Sci Lett 253:389-410

Dettman DL, Fang X, Garzione CN, Li J (2003) Uplift-driven climate change at 12 Ma: a long $\delta^{18}O$ record from the NE margin of the Tibetan plateau. Earth Planet Sci Lett 214:267-277

Dewey JF, Burke KCA (1973) Tibetan, Variscan, and Precambrian basement reactivation: products of continental collision. J Geol 81:683-692

Drummond CN, Wilkinson BH, Lohmann KC, Smith GR (1993) Effect of regional topography and hydrology on the lacustrine isotopic record of Miocene paleoclimate in the Rocky Mountains. Paleogeo, Palaeoclim, Palaeoecol 101:67-79

Ducea MN, Saleeby JB (1998) A case for delamination of the deep batholithic root beneath the Sierra Nevada, California. Int Geology Rev 133:78-93

England P, Houseman GA (1989) Extension during continental convergence, with application to the Tibetan Plateau: J Geophys Res 94:17,561–17,579

Egli M, Mirabella A, Fitze P (2003) Formation rates of smectites derived from two Holocene chronosequences in the Swiss Alps. Geoderma 117:81-98

Foster DA, Doughty PT, Kalakay TJ, Fanning CM, Coyner S, Grice WC, Vogl J (2007), Kinematics and timing of exhumation of Eocene metamorphic core complexes along the Lewis and Clark fault zone, Northern Rocky Mountains, USA. *In:* Till A, Roeske S Foster DA, Sample J (eds) Exhumation Along Major Continental Strike-Slip Systems: Geological Society of America Special Paper 434, Geological Society of America, Boulder, (**in press**)

France-Lanord C, Sheppard SMF, LeFort P (1988) Hydorgen and oxygen isotope variations in the High Himalaya peraluminous Manaslu leucogranite: Evidence for heterogeneous sedimentary source. Geochim Cosmochim Acta 52:513-526

Fricke HC, Wickham SM, O'Neil JR (1992) Oxygen and hydrogen isotope evidence for meteoric water infiltration during mylonitization and uplift in the Ruby Mountains-East Humboldt Range core complex, Nevada. Contrib Mineral Petrol 111:203-221

Garzione CN, Quade J, DeCelles PG, English NB (2000), Predicting paleoelevation of Tibet and the Himalaya from $\delta^{18}O$ vs. altitude gradients in meteoric water across the Nepal Himalaya. Earth Planet Sci Lett 183:215-229

Garzione CN, Dettman DL, Horton BK (2004) Carbonate oxygen isotope paleoaltimetry: evaluating the effect of diagenesis on paleoelevation estimates for the Tibetan plateau. Paleogeo, Palaeoclim, Palaeoecol 212:119-140

Garzione CN, Molnar P, Libarkin JC, MacFadden BJ (2006) Rapid late Miocene rise of the Bolivian Altiplano: Evidence for removal of mantle lithosphere. Earth Planet Sci Lett 241:543 556

Ghosh P, Garzione CN, Eiler JM (2006) Rapid uplift of Altiplano revealed through ^{13}C-^{18}O bonds in paleosol carbonates. Science 311:511-515

Gilg HA (2000) D-H evidence for the timing of kaolinization in northeast Bavaria, Germany, Chem Geol 170:5-18

Goldstein A, Selleck B, Valley JW (2005) Pressure, temperature, and composition history of syntectonic fluids in a low-grade metamorphic terrane. Geology 33:421-424

Graham SA, Chamberlain CP, Yue YJ, Ritts BD, Hansons AD, Horton TW, Waldbauer JR, Poage MA, Feng X (2005) Stable isotope records of Cenozoic climate and topography, Tibetan plateau and Tarim basin. Am J Sci 305:101-118

Harrison TM, Copeland P, Kidd WSF, Yin A (1992) Raising Tibet. Science 255:1663-1670

Horton TW, Sjostrom DJ, Abruzzese MJ, Poage MA, Waldbauer JR, Hren M, Wooden J, Chamberlain CP (2004) Spatial and temporal variation of Cenozoic surface elevation in the Great Basin and Sierra Nevada. Am J Sci 304:862-888

Horton TE, Chamberlain CP (2006) Stable isotopic evidence for Neogene surface downdrop in the central Basin and Range Province. Geol Soc Am Bull 118:475–490

Houseman GA, McKenzie DP, Molnar P (1981) Convective instability of a thickened boundary layer and its relevance for the thermal evolution of continental convergent belts: J Geophys Res 86:6115–6132

Houseman GA, Molnar P (1997) Gravitational (Rayleigh-Taylor) instability of a layer with nonlinear viscosity and convective thinning of continental lithosphere. Geophys J Int 128:125–150

Huber M, Sloan LC (1999) Warm climate transitions: A general circulation modeling study of the Late Paleocene Thermal Maximum. J Geophys Res 104:16,633-16,655

Huber M, Caballero R (2003) Eocene El Nino: Evidence for robust tropical dynamics in the "hothouse". Science 299:877-881

Ingraham NL, Taylor BE (1991) Light stable isotope systematics of large-scale hydrologic regimes in California and Nevada, Water Resources Res 27: 77-90

Jones CH, Farmer CL, Unruh J (2004) Tectonics of Pliocene removal of lithosphere of the Sierra Nevada, California. Geol Soc Am Bull 116:1408-1422

Jull M, Kelemen PB (2001) On the conditions for lower crustal instability. J Geophys Res 106:6423-6446

Kent-Corson ML, Sherman LS, Mulch A, Chamberlain CP (2006), Cenozoic topographic and climatic response to changing tectonic boundary conditions in Western North America. Earth Planet Sci Lett 252:453-466

Knauth LP, Epstein S (1976) Hydrogen and oxygen isotope ratios in nodular and bedded cherts. Geochim Cosmochim Acta 40:1095-1108

Koch PL, Crowley BE (2005) The isotopic rain shadow and elevation history of the Sierra Nevada Mountains, California. Eos Transactions Am Geophys Union 86(52), Fall Meeting Suppl

Kocsis L, Vennemann TW, Fontignie D (2007) Migration of sharks into freshwater systems during the Miocene and implications for Alpine paleoelevation. Geology 35:451-454

Kohn MJ (1999) Why most "dry" rocks should cool "wet". Am Mineral 84:570-580

Kohn MJ, Dettman DL (2007) Paleoaltimetry from stable isotope compositions of fossils. Rev Mineral Geochem 66:119-154

Kohn MJ, Miselis JL, Fremd TJ (2002) Oxygen isotope evidence for progressive uplift of the Cascade Range, Oregon. Earth Planet Sci Lett 204:151-165

Kohn MJ, Fremd TJ (2007) Tectonic controls on isotope compositions and species diversification, John Day Basin, central Oregon. PaleoBios (in press)

Koons PO, Zeitler PK, Chamberlain CP, Craw D, Meltzer AS (2002) Mechanical links between erosion and metamorphism in Nanga parbat, Pakistan Himalaya. Am J Sci 302:749-773

Kutzbach JE, Guetter PJ, Ruddiman WF, Prell WL (1989) Sensitivity of climate to Late Cenozoic uplift in southern Asia and the American West: Numerical Experiments. J Geophys Res 94:18,393 – 18,407

Lawrence JR, Rashkes Meaux J (1993) The stable isotope composition of ancient kaolinites of North America. *In*: Climate Change in Continental Isotopic Records. Swart, PK, Lohmann KC, McKenzie J, Savin S (eds) Geophysical Monograph 78:249-261

Lindgren W (1911) The Tertiary gravels of the Sierra Nevada of California. USGS Prof Paper 73: 226 p

Molnar P, England P (1990) Late Cenozoic uplift of mountain ranges and global climate change: chicken or egg? Nature 346:29-34

Molnar P, England P, Martinod J (1993) Mantle dynamics, uplift of the Tibetan Plateau, and the Indian monsoon. Rev Geophys 31:357-396

Molnar P (2004) Late Cenozoic increase in accumulation rates of terrestrial sediment: How might climate change have affected erosion rates? Annu Rev Earth Planet Sci 32:67-89

Molnar P, Houseman GA, England PC (2006) Paleoaltimetry of Tibet. Nature 444:E4 doi:10.1038/nature05368

Morrill C, Koch PL (2002) Elevation or alteration? Evaluation of isotopic constraints on paleoaltitudes surrounding the Eocene Green River Basin. Geology 30:151-154

Morrison J (1994) Meteoric water-rock interaction in the lower plate of the Whipple Mountain metamorphic core complex, California. J Metam Geol 12:827-840

Morrison J, Anderson JL (1998) Footwall refrigeration along a detachment fault: Implications for the thermal evolution of core complexes. Science 279:63-66

Morse JW (1983) The kinetics of calcium carbonate dissolution and precipitation. Rev Mineral 11:227-264

Müller W, Aerden D, Halliday AN (2000) Isotopic dating of strain fringe increments: Duration and rates of deformation in shear zones. Science 288:2195-2198

Müller T, Baumgartner LP, Foster CT, Vennemann TW (2004) Metastable prograde mineral reactions in contact aureoles. Geology 32:821-824

Mulch A, Cosca MA (2004) Recrystallization or cooling ages? *In situ* UV-laser $^{40}Ar/^{39}Ar$ geochronology of muscovite in mylonitic rocks. J Geol Soc London 161:573-582

Mulch A, Teyssier C, Cosca MA, Vanderhaeghe O, Vennemann T (2004) Reconstructing paleoelevation in eroded orogens. Geology 32:525-528

Mulch A, Cosca MA, Fiebig J, Andresen A (2005a) Time scales of mylonitic deformation and meteoric fluid infiltration during extensional detachment faulting: an integrated in situ $^{40}Ar/^{39}Ar$ geochronology and stable isotope study of the Porsgrunn-Kristiansand Shear Zone (Southern Norway), Earth Planet Sci Lett 233:375-390

Mulch A, Teyssier C, Chamberlain CP (2005b) Stable isotope-based paleoelevation reconstructions in coupled basin-detachment systems: Cenozoic topography of the North American Cordillera. Geol Soc Am Abstracts with Programs 37:273

Mulch A, Chamberlain CP (2006) The rise and growth of Tibet, Nature 439:670-671

Mulch A, Graham SA, Chamberlain CP (2006a) Hydrogen Isotopes in Eocene River Gravels and Paleoelevation of the Sierra Nevada. Science 313:87-89

Mulch A, Teyssier C, Cosca MA, Vennemann TW (2006b) Thermomechanical analysis of strain localization in a ductile detachment zone. J Geophys Res 111:B12405 doi:10.129/2005JB004032

Mulch A, Teyssier C, Chamberlain CP, Cosca MA (2007) Stable isotope paleoaltimetry of Eocene Core Complexes in the North American Cordillera. Tectonics doi:10.1029/2006TC001995.

Nelson KD et al. (1996) Partially molten middle crust beneath southern Tibet: Synthesis of Project INDEPTH results. Science 274:1684–1688

O'Neil JR, Silberman ML (1974) Stable Isotope Relations in Epithermal Au-Ag Deposits. Economic Geology 69:902-909

O'Neil JR, Kharaka YK (1976) Hydrogen and oxygen isotope exchange reactions between clay minerals and water. Geochim Cosmochim Acta 40:241-246

Orr KE, Cheney ES (1987) Kettle and Okanogan domes, northeastern Washington and British Columbia. Washington Division of Geology and Earth Resources Bulletin 77:55-71

Person M, Mulch A, Teyssier C, Gao Y (2007) Isotope transport and exchange within metamorphic core complexes. Am J Sci 307:555-589

Poage MA, Chamberlain CP (2001) Empirical relationships between elevation and the stable isotope composition of precipitation and surface waters: Considerations for studies of paleoelevation change. Am J Sci 301:1-15

Poage MA, Chamberlain CP (2002) Stable isotopic evidence for a Pre-Middle Miocene rain shadow in the western Basin and Range: Implications for the paleotopogaphy of the Sierra Nevada. Tectonics 21:16-1 – 16-10

Prell WL, Kutzbach JE (1992) Sensitivity of the Indian monsoon to forcing parameters and implications for its evolution. Nature 360: 647-652

Price JR, Velbel MA, Patino LC (2005) Rates and time scales of clay-mineral formation by weathering in saprolitic regoliths of the southeren Appalachians from geochemical mass balance. Geol Soc Am Bull 117: 783-794

Quade J, Garzione C, Eiler J (2007) Paleoelevation reconstruction using pedogenic carbonates. Rev Mineral Geochem 66:53-87

Rey P, Vanderhaeghe O, Teyssier C (2001) Gravitational Collapse of the continental crust: definition, regimes and modes. Tectonophysics 342:435-449

Ramstein G, Fluteau F, Besse J, Joussame S (1997) Effect of orogeny, plate motion and land-sea distribution on Eurasian climate change over the past 30 million years. Nature 386:788-795

Roehler HW (1993) Eocene climates, depositional environments, and geography in the greater Green River Basin, Wyoming, Utah and Colorado. USGS Prof Paper 1506-F

Royden LH, Burchfiel BC, King RW, Wang E, Chen Z et al. (1997) Surface deformation and lower crustal flow in eastern Tibet. Science 276:788-790

Rowley DB, Pierrehumbert RT, Currie BS (2001) A new approach to stable isotope-based paleoaltimetry: implications for paleoaltimetry and paleohypsometry of the High Himalaya since the Late Miocene. Earth Planet Sci Lett 188:253–268

Rowley DB, Currie BS (2006) Palaeo-altimetry of the late Eocene to Miocene Lunpola basin, central Tibet. Nature 439:677–681

Ruddiman WF, Prell WL, Raymo ME (1989) Late Cenozoic uplift in southern Asia and the American West: Rationale for general circulation modeling experiments. J Geophys Res 94:18,379-18,391

Ruddiman WF, Kutzbach JE (1989) Forcing of late Cenozoic Northern Hemisphere climate by plateau uplift in southern Asia and the American west. J Geophys Res 94:18,409-18,427

Saleeby J, Ducea M, Clemens-Knott (2003) Production and loss of high-density batholithic root, southern Sierra Nevada, California. Tectonics 22, doi: 10.1029/2002TC001374

Savin SM, Hsieh JCC (1998) The hydrogen and oxygen isotope geochemistry of pedogenic clay minerals: principles and theoretical background. Geoderma 82:227-253

Sempere T, Hartley A, Roperch P (2006) Comment on "Rapid uplift of the Altiplano revealed through ^{13}C-^{18}O bonds in paleosol carbonates". Science 314:760b

Sewall JO, Sloan LC (2006) Come a little bit closer: A high-resolution climate study of the early Paleogene Laramide foreland. Geology 34:81-84

Sharp ZD, Masson H. Lucchini R.(2005) Stable isotope geochemistry and formation mechanisms of quartz veins; extreme paleoaltitudes of the Central Swiss Alps in the Neogene. Am J Sci 305:187-219

Sheppard SMF (1986) Caharcterization and isotopic variations in natural waters. Rev Mineral 16:165-184

Sherlock SC, Kelley SP, Zalasiewicz JA, Schofield DI, Evans JA, Merriman RJ, Kemp SJ (2003) Precise dating of low-temperature deformation: Strain-fringe analysis by $^{40}Ar/^{39}Ar$ laser microprobe. Geology 31:219-222

Stern LA, Chamberlain CP, Reynolds RC, Johnson GD (1997) Oxygen isotope evidence of climate change from pedogenic clay minerals in the Himalayan molasses. Geochem Cosmochim Acta 61:731-744

Sjostrom DJ, Hren MT, Horton TW, Waldbauer JR, Chamberlain CP (2006) Stable isotope evidence for an early Tertiary elevation gradient in the Great Plains-Rocky Mountain region. *In:* Tectonic, Climate and Landscape Evolution. Willet S, Hovius N, Fisher D, Brandon M (eds) Geol Soc Am Spec Pap 398: 309-319

Stockli DF (2006) Application of low temperature thermochronometry to extensional tectonic settings. Rev Mineral Geochem 58:411-448

Stewart MK, Cox MA, James MR, Lyon GL (1983) Deuterium in New Zealand rivers and streams, Wellington, New Zealand. Inst of Nuclear Sciences Report INS-R-320, 32 pp (1983)

Sutherland R (1996) Transpressional development of the Australian-Pacific boundary through the southern South Island, New Zealand: constraints from Miocene-Pliocene sediments, Waiho-1 borehole South Westland. New Zealand J Geol Geophys 39:251-264

Takeuchi A, Larson PB (2005) Oxygen isotope evidence for the late Cenozoic development of an orographic rain shadow in eastern Washington. Geology 223:127-146

Tapponnier P, Peltzer G, LeDain AY, Armijo R, Cobbold P (1982) Propagating extrusion tectonics in Asia: new insights from simple experiments with plasticine. Geology 10:611-616

Tapponnier P, Zhiqin X, Roger F, Meyer B, Arnaud N, Wittlinger G, Jinsui Y (2001) Oblique stepwise rise and growth of the Tibet plateau. Science 294:1671-1677

Teyssier C, Ferre EC, Whitney DL, Norlander B, Vanderhaeghe O, Parkinson D (2005) Flow of partially molten crust and origin of detachments during collapse of the Cordilleran Orogen. Geol Soc London Spec Pub 245:39-64

Vanderhaeghe O, Teyssier C (2001) Crustal scale rheological transitions during late-orogenic collapse. Tectonophysics 335:211–228

Vanderhaeghe O, Teyssier C, Wysoczanski R (1999) Structural and geochronological constraints on the role of partial melting during the formation of the Shuswap metamorphic core complex at the latitude of the Thor–Odin Dome, British Columbia. Can J Earth Sci 36:917–943

Vanderhaeghe O, McDougall I, Dunlap WJ, Teyssier C (2003) Cooling and exhumation of the Shuswap metamorphic core complex constrained by $^{40}Ar/^{39}Ar$ thermochronology. Geol Soc Am Bull 115: 200–216

Vennemann TW, Vdovic R, Thoral S (2000) Clay minerals of the north Alpine molasses sediments: archives of Alpine upliftment and climate change? Goldschmidt conference 2000, Journal of Conference Abstracts V5, 2:1049

Walcott RI (1998) Models of oblique compression: Late Cenozoic tectonics of the South Island of New Zealand. Rev Geophys 36:1-26

Wickham SM, Taylor HP (1987) Stable isotope constraints on the origin and depth of penetration of hydrothermal fluids associated with Hercynian regional metamorphism and crustal anatexis in the Pyrenees. Contrib Mineral Petrol 95:255-268

Yeend WH (1974) Gold-bearing gravel of the ancestral Yuba River, Sierra Nevada, California. USGS Prof Pap 722: 44 pp

Zachos J, Pagani M, Sloan LC, Thomas E, Billups K (2001) Trends, rhythms, and aberrations in global climate since 65 Ma to present. Science 292:686-692

Zandt G (2003) The southern Sierra Nevada drip and the mantle wind direction beneath the southwestern United States. Int Geology Rev 45:213-223

Reviews in Mineralogy & Geochemistry
Vol. 66, pp. 119-154, 2007
Copyright © Mineralogical Society of America

5

Paleoaltimetry from Stable Isotope Compositions of Fossils

Matthew J. Kohn*

Department of Geological Sciences
University of South Carolina
Columbia, South Carolina, 29208, U.S.A.

*present address and e-mail:
Department of Geosciences, Boise State University, Boise, Idaho, 83725, U.S.A.
mattkohn@boisestate.edu

David L. Dettman

Department of Geosciences
University of Arizona
Tucson, Arizona, 85721, U.S.A.
dettman@mail.arizona.edu

ABSTRACT

Stable isotope systematics of phosphatic (vertebrate) and carbonate (invertebrate) fossils are reviewed, emphasizing external vs. biological controls on isotope compositions and their variation. External controls include elevation and isotopic lapse rates, temperature, atmospheric circulation patterns, aridity, and changes to precipitation seasonality, e.g., development of monsoons. Biological controls include temperature regulation and temperature-dependent biological activity, water balance, behavior, and rates of hard-part secretion and maturation. Some key factors include what sources of water an animal samples, how those sources respond isotopically to elevation, and the isotopic sensitivity of a biologic tissue to environmental changes. General research design criteria are enumerated, including numbers of fossils and analyses required, as well as uncertainties in the interpretability of isotopic shifts. Isotope zoning from fossil teeth and shells from the Indian foreland demonstrate comparable seasonal monsoon signals at ~11 Ma vs. today, implying a high and broad plateau at that time. Mean isotope compositions of fossil teeth from the lee of the Cascade Range, central Oregon, show a pronounced decrease in $\delta^{18}O$ initiating ~7 Ma, signaling topographic rise associated with impingement of Basin and Range extension on the arc. Mean compositions and zoning in fossil shells from the northern Rocky Mountains indicate high elevations and topographic relief in the late Cretaceous and early Paleogene. Paleoaltimetry will principally benefit from better estimates of temperature, because temperature lapse rates are less variable than isotopic lapse rates, and because the combination of temperature and isotopic studies may help distinguish between local isotopic changes and distal ones caused by catchment or rain shadow effects.

INTRODUCTION

As discussed in the seminal paper by England and Molnar (1990) paleoelevation reflects the combined chemical and physical state of the lithosphere, including thicknesses, thermal structure and bulk chemistry. Tectonic processes such as lithospheric delamination and growth of mountain ranges through either collisional orogenesis or arc evolution may gradually or abruptly

1529-6466/07/0066-0005$05.00 DOI: 10.2138/rmg.2007.66.5

change the thickness, temperature, and/or chemistry of the crust and lithospheric mantle. Just as a history of basin depth ("negative paleoelevation") constrains extensional processes through its implications for lithosphere thickness and thermal structure (McKenzie 1978; Royden and Keen 1980), so too do paleoelevations constrain convergent processes. Consequently, the last ~15 years have witnessed increasing effort to develop methods for inferring paleoelevation, ranging from chemical proxies that include stable isotopes (Rowley 2007; Quade et al. 2007; Mulch and Chamberlain 2007; this chapter) and cosmogenic isotopes (Riihimaki and Libarkin 2007), to physical proxies that include leaf morphology (Meyer 2007), leaf structure (Kouwenberg et al. 2007), and basalt vesicularity (Sahagian and Proussevitch 2007).

In this chapter, we discuss paleoaltimetric application of stable isotopes in fossils. Fossils share some features of other stable isotope proxies, for example in their link to atmospheric circulation patterns and meteoric water compositions. However they differ in other key respects, for example in their link to biology, and in their preservation of seasonal isotope variations, reflecting seasonal climate. These variations complicate some interpretations, but also provide alternative approaches to constrain elevations.

Note that fossils may contain several chemical components that allow measurement of multiple stable isotopes, e.g., H in apatite, C or N in collagen, C in carbonate, but these either are not preserved on timescales of interest to tectonics, or are usually investigated for biological or ecological reasons. Thus, we focus our discussion of stable isotopes in biogenic tissues on oxygen isotopes, with some additional consideration of carbon isotopes, when they help to interpret oxygen isotope trends.

LINKS BETWEEN ELEVATION AND OXYGEN ISOTOPE COMPOSITIONS

As in pedogenic and authigenic materials considered in other chapters (Mulch and Chamberlain 2007; Quade et al. 2007), oxygen isotope compositions principally reflect local water compositions. In the case of invertebrates and diagenetically altered bone, temperature of precipitation or recrystallization is an additional factor influencing compositions, much as for paleosol carbonates and other authigenic minerals. Conventionally, geochemists have linked original biogenic compositions of fossils to local water compositions (often assumed to reflect precipitation) that are then linked to elevation. General principles of isotopes in precipitation are discussed elsewhere (Dansgaard 1964; Rozanski 1993; Rowley et al. 2001; Rowley 2007; Rowley and Garzione 2007), but we reiterate two of these here—isotopic lapse rates and monsoons—to focus on those signals that have been used to infer paleoelevations from oxygen isotopes in fossils, and complexities that affect applications.

Isotopic lapse rates

$\delta^{18}O$ values generally decrease with increasing elevation (e.g., Garzione et al. 2000; Poage and Chamberlain 2001), due to rainout from ascending and cooling airmasses (Dansgaard 1964; Rowley et al. 2001, Fig. 1); this fundamental process yields lower $\delta^{18}O$ values in precipitation both at high elevations and in any orographic rainshadows relative to precipitation compositions of the originating air mass (see Rowley and Garzione 2007 and Rowley 2007 for recent reviews). The west coast and western interior of North America provide good examples of both types of isotopic depletions (Fig. 1). The $\delta^{18}O$ of meteoric water decreases dramatically across the Sierra Nevada, Cascades, and Coast Ranges because they oppose prevailing (east-directed) weather patterns, and their high elevations force ascent and cooling of airmasses. The temperature lapse rate is a classical measure of the decrease in temperature with vertical ascent. The isotopic lapse rate derives from this concept and reflects the decrease in $\delta^{18}O$ with elevation, usually expressed in ‰/km (paleoaltimetry literature) or ‰/100 m (paleoclimate modeling literature). An increasing literature is devoted to characterizing isotopic lapse rates, either empirically from precipitation records (e.g., the IAEA database) or theoretically (Rowley et al. 2001). Often

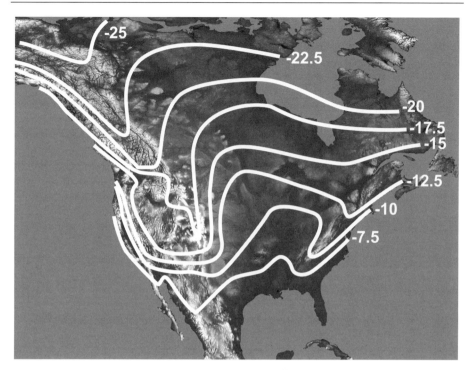

Figure 1. Geographic map of North America contoured for $\delta^{18}O$ of precipitation relative to V-SMOW showing strong depletions across north-south – trending ranges, and broadly similar compositions throughout the Basin and Range province. From Sheppard et al. (1969), adjusted for data from Coplen and Kendall (2000) and Friedman et al. (1992, 2002b). Basemap from NOAA.

it is assumed that a single isotope lapse rate describes all water in a range, and that the low precipitation $\delta^{18}O$ at the range crest establishes all subsequent compositions downwind. In fact, hydrologic and atmospheric processes are much more variable, and 3 sources of variation produce substantial uncertainty that must be considered in most paleoelevation estimates.

(1) Measured isotopic lapse rates in different regions are much more variable than might be implied by Rayleigh distillation models (Rowley et al. 2001) or the consistency of datasets from a single region (e.g., Garzione et al. 2000 2006). For example, tabulated data of the $\delta^{18}O$ shift with elevation show a mean lapse rate of −2.9 ‰/km (Poage and Chamberlain 2001) with greater values at higher elevation (Garzione et al. 2000 2006). These observations are consistent with Rayleigh fractionation attending adiabatic cooling (Dansgaard 1964; Rowley et al. 2001). However, even ignoring extreme values of measured lapse rates (e.g., −6.2 ‰/km in Greenland and −11.2 ‰/km in Antarctica: Johnson et al. 1989, Masson et al. 2000), the ± 1σ confidence interval ranges from −1.8 ‰/km to −4.0 ‰/km. This result reminds us that, in the absence of direct estimates of isotopic lapse rates in the specific area and time of study, paleoelevation estimates can readily vary by a factor of 2.

(2) Isotope compositions in the lees of ranges or across plateaux do not always remain constant and low but can increase downwind irrespective of elevation (Smith et al. 1979; Ingraham and Taylor 1991; Ingraham 1998; Tian et al. 2001, 2003; Dettman 2002; Stern and Blisniuk 2002; Blisniuk et al. 2005). This fact contrasts with many discussions that presume that the isotopic composition of rain on the leeward side of a range or in the interior of a plateau remains unchanged at a very negative $\delta^{18}O$ value set by the minimum achieved at the range

crest or plateau edge (Chamberlain and Poage 2000; Poage and Chamberlain 2001; Rowley et al. 2001; Stern and Blisniuk 2002; Currie et al. 2005; Cyr et al. 2005; Blisniuk and Stern 2006; Rowley and Currie 2006). This latter assumption may hold for some ranges (e.g., see the Cascades example below), but cannot be applied to all ranges.

For the high Sierra Nevada and Canadian Rockies of western North America, the increase in precipitation $\delta^{18}O$ on leeward slopes reflects mixing of air masses moving parallel to the ranges, evaporative enrichment of ^{18}O in arid rain shadows, and occasional transport of moisture against prevailing winds (Yonge et al. 1989; Friedman et al. 2002a). Further east, in the interior, isotope compositions remain generally low and relatively uniform, sometimes irrespective of local elevations, because previous rainout already depleted ^{18}O, and because water recycling over large areas in the western interior redistributes moisture without much further depletion (Ingraham et al. 1998). These generally low values do reflect high upwind elevations and in principle may be usefully applied in altimetry studies. But no theoretical framework exists that relates them quantitatively both to the elevation and leeward lapse rate of these high western ranges.

(3) Lapse rates are commonly measured for precipitation, but the materials analyzed for isotope composition do not ordinarily sample precipitation directly. For fossils, isotope compositions are instead related to local surface water, e.g., rivers or lakes. Yet the catchment for a river may extend to much higher elevations than the local sampling site, i.e., the composition of local water is not necessarily linked to local elevation. Rather, the elevation estimate obtained reflects the hypsometric mean of the catchment (Rowley et al. 2001).

In sum, factors other than elevation affect the change in oxygen isotope ratio across an orogen. Although quantifying and removing such factors may be possible in young settings, ancient settings present greater difficulties. Regardless, error analysis of calculated elevations must take into account assumptions regarding the isotopic lapse-rate, as well as what water source an isotopic proxy is sampling.

Monsoons

Plateaux drive monsoons, which are characterized by strongly seasonal wind and precipitation patterns (Fig. 2): heating over a plateau in summertime causes an atmospheric pressure low that draws in moisture from nearby water sources, whereas cooling in wintertime causes a pressure high that drives moisture away. The resulting variations in precipitation and temperature cause large seasonal variations in oxygen isotope compositions. New Delhi, in northern India, provides a good example of such large variations in seasonal isotope and precipitation patterns, which are driven by monsoonal flow associated with the Tibetan Plateau (Figs. 2, 3). These large variations contrast with low-lying regions in continental interiors that are either less strongly influenced by monsoonal precipitation (Wuhan, China), or essentially independent of monsoons (Mississippi-Arkansas, United States; Fig. 3). The presence or absence of a monsoon does not quantitatively measure plateau elevations, but rather a combination of elevation plus geographic extent, i.e., the plateau must be sufficiently "big" to drive a monsoon. Although we recognize that there are many complications in interpreting seasonal isotope signals, seasonality remains an important component of regional climate, and fossils provide one of very few quantitative sources of seasonal data in the geologic record.

MATERIALS AND METHODS

Choosing a problem

As with other stable isotope proxies, to infer paleoelevations or the presence/absence of a plateau, the isotope geochemist analyzing fossils has several options: 1) "Look high." Use a fossil sequence collected at high modern elevations (e.g., in a high mountain range

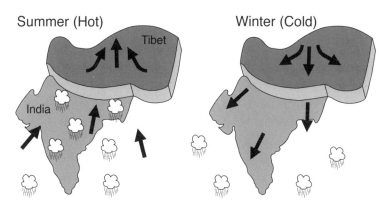

Figure 2. Sketch of Indian subcontinent and neighboring Tibet, showing seasonal changes in rainfall in the Indian subcontinent and atmospheric pressure over the plateau.

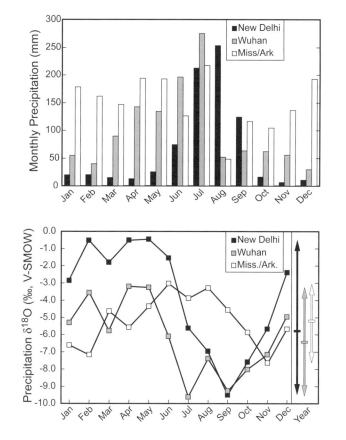

Figure 3. Comparison of isotope and precipitation seasonality for a region strongly influenced by a monsoon (New Delhi) and 2 other regions at comparable latitudes (Arkansas-Mississippi, United States; Wuhan, China) that are less influenced by monsoons. These patterns underscore the strong precipitation and isotope variations associated with monsoons.

or plateau) and determine when that sample site was at low elevation. 2) "Look low." Use a fossil sequence collected in a low elevation rainshadow and determine when that rainshadow began to form. 3) "Look seasonally." An increase in isotope seasonality near a plateau may reflect the time when the plateau became sufficiently high and broad to initiate a monsoon. In the first two cases, mean isotope compositions are sufficient, but in the third case seasonality must be inferred from zoning analysis along or across biogenic tissues. These choices affect both what samples are needed and how they are analyzed. In some cases, the relative rarity of specific fossils targeted for research (e.g., teeth from a particular taxon) does not permit routine collection of materials in a few short field seasons. Rather, concerted collecting effort over years or decades may be required, necessitating use of museum collections. Museum curators normally request specific advance information about numbers of samples and degree of destruction before agreeing to collaborate or approve release of materials for analysis.

Vertebrate fossils

By vertebrate fossils, we usually mean fossilized bones and teeth, although fish scales and otoliths may also be analyzed (e.g., Fricke et al. 1998; Wurster and Patterson 2001). Bone consists of a dense, outer layer, referred to as cortical or compact bone, that surrounds an inner porous core, referred to as trabecular, cancellous or spongy bone. Teeth are a composite of one or more layers of enamel, which is highly crystalline, interfolded and/or surrounding dentine (or dentin), which is less crystalline and more organic rich. Stable isotopes of teeth and diagenetic alteration of fossil bones and teeth were reviewed by Koch (1998, 2006), Kohn and Cerling (2002) and Trueman and Tuross (2002), so here we provide a brief summary of some key points they raise, updated for research in the last ~5 years.

Mineralogy and structure. Modern phosphatic tissues consist of dahllite, or CO_3^{2-} – substituted hydroxy-apatite, with an average composition for enamel of ~$[Ca_{4.3}Na_{0.15}Mg_{0.05}]$ $[(PO_4)_{2.5}(HPO_4)_{0.2}(CO_3)_{0.25}][(OH)_{0.5}Cl_{0.05}]$ (e.g., Driessens and Verbeeck 1990, p. 107 and 128) where CO_3^{2-} and HPO_4^{2-} substitution for PO_4^{3-} is charge-balanced by Na substitution for Ca and vacancies in the Ca and OH sites. Bone and dentine differ from enamel in having much finer crystal sizes (at most a few nm in thickness; see review of Elliott 2002), much greater organic contents, and less hydroxyl in the dahllite, possibly none (Pasteris et al. 2004). CO_3^{2-} concentrations are ~3.5, ~4.5, and ~5 wt% in enamel, dentine and bone respectively (Driessens and Verbeeck 1990, Wright and Schwarcz 1996; Koch et al. 1997). Either the PO_4^{3-} or the CO_3^{2-} component may be separated chemically and analyzed independently for $\delta^{18}O$. The most commonly used methods of preparation and analysis of the CO_3^{2-} and PO_4^{3-} components are detailed elsewhere (Koch et al. 1997; Dettman et al. 2001; Kohn et al. 2002; Vennemann et al. 2002). In modern materials, $\Delta(CO_3^{2-}-PO_4^{3-}) = 8-9‰$ (Iacumin et al. 1996; Bryant et al. 1996; Zazzo et al. 2004), and in comparing them in this paper we use a Δ value of 8.5‰. Because of susceptibility to diagenetic alteration, components in the OH site have not been used for inferring biogenic compositions in any fossil materials.

After burial, microorganisms degrade the abundant organic constituents in bone and dentine, leading to massive recrystallization of crystallites and chemical transformation to francolite (CO_3^{2-} – substituted fluor-apatite) on timescales of years to a few tens of thousands of years (Tuross et al. 1989; Millard and Hedges 1999; Trueman and Tuross 2002; Trueman et al. 2004; Kohn and Law 2006; Kohn 2006). Although the PO_4^{3-} component may preserve original biogenic compositions in some bone samples (Barrick and Showers 1994, 1995, 1999; Barrick et al. 1996, 1998; Barrick 1998), the CO_3^{2-} component appears to adopt a pedogenic composition (Kohn and Law 2006) dependent on soil water composition and temperature; that is, bone CO_3^{2-} behaves analogously to paleosol CO_3^{2-}, with all its attendant advantages and disadvantages (Quade et al. 2007). In contrast, tooth enamel's highly crystalline structure and sparse organic compounds confer strong resistance to recrystallization and oxygen isotope alteration (Ayliffe et al. 1992, 1994; Wang and Cerling 1994; Kohn and Cerling 2002), although its trace element composition can change radically due to exchange with Ca and OH (Kohn et al. 1999). Both the PO_4^{3-} and CO_3^{2-} components have been analyzed reliably for original biogenic compositions in tooth enamel as old as the Eocene (Kohn et al. 2007; Zanazzi et al. 2007). Mammals provide most of the fossil teeth analyzed for oxygen isotopes in paleoclimate studies (although see Fricke and Rogers 2000; Stanton Thomas and Carlson 2004; Straight et al. 2004; Botha et al. 2005; Amiot et al. 2006), and because mammals thermoregulate at ~37 °C, temperature effects in enamel isotopes can be ignored.

Modern oxygen isotope trends. Oxygen isotopes in modern mammal bones and teeth correlate strongly with local water compositions (Fig. 4) because drinking plus water in food contributes most (~80%) of the oxygen intake of mammals (Kohn 1996; Kohn and Cerling 2002). Note that although some might assume this water is precipitation, in fact it is whatever water the animal ingests, including local drinking water sources (rivers, lakes, ponds, etc.) plus water in plants derived from soil water. These sources are generally correlated with precipitation, but not equivalent. Data for the subfamily Bovinae (*Bos, Bison, Syncerus,* and *Ovibos*; cattle, bison, water buffalo, and musk ox) illustrate the mammal-local water dependence well (Fig. 4), with a strong correlation ($r^2 \sim 0.95$) and slope slightly less than one (because of air O_2 intake, as the $\delta^{18}O$ of air O_2 is invariant worldwide; Kroopnick and Craig 1972). A dependence of $\delta^{18}O$ on aridity, normally expressed in terms of relative humidity or water deficit, is apparent from positive deviations from the global correlation line both for drought-tolerant animals and for water-dependent animals that live in arid ecosystems (Ayliffe and Chivas 1990; Luz

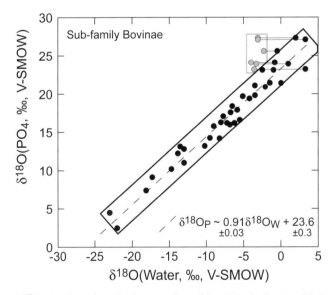

Figure 4. Plot of $\delta^{18}O$ of modern bioapatite from members of the subfamily Bovinae (black dots) vs. local water composition, showing strong correlation (data from Luz and Kolodny 1984; D'Angela and Longinelli 1990; Koch et al. 1991; Bocherens et al. 1996; Hoppe 2006; Levin et al. 2006). Compositions are for either PO_4^{3-}, or $CO_3^{2-} - 8.5‰$ (to account for *in vivo* fractionation between these two components: Bryant et al. 1996; Iacumin et al. 1996; Zazzo et al. 2004). Water compositions are mainly from local rainfall or direct measurement of local water sources. Some local water compositions (Koch et al. 1991; Bocherens et al. 1996; Levin et al. 2006) were estimated from hippo compositions, assuming hippo body water is enriched relative to local water by a constant ~1.5‰ (data from Koch et al. 1991 and Koch pers. comm. 2007). Regional rainwater compositions for data of Levin et al. (2006) from relatively dry ecosystems (gray dots) connect to estimated local water compositions (thin horizontal lines) and illustrate evaporative enrichment of local water ^{18}O. Water compositions for two datapoints of Hoppe (2006; Wind Cave and Ordway Prairie) were adjusted to −13‰, based on regional trends (Coplen and Kendall 2000).

et al. 1990; Delgado Huertas et al. 1995; Kohn and Cerling 2002). For example, if it assumed that animals ingest regional (unevaporated) meteoric water (Levin et al. 2006), then data from *Syncerus* are elevated by as much as 5‰ from the trends defined by all other data for their subfamily (gray dots, Fig. 4). Zebra and elephant values in arid environments are similarly skewed to positive values (not shown). If instead local water compositions are estimated from the $\delta^{18}O$ of hippos (because they live in water and eat foods adjacent to water; see discussion of Bocherens et al. 1996), then the "deviant" data are reconciled (Fig. 4, thin horizontal lines), further illustrating that animals to not commonly ingest meteoric water.

The physiology of plants is predicted to cause an increase in $\delta^{18}O$ of herbivores with increasing aridity (Ayliffe and Chivas 1990; Luz et al. 1990; Kohn 1996), because $\delta^{18}O$ values of both leaf water and cellulose increase with decreasing humidity (Craig and Gordon 1965; Sternberg 1989; Roden and Ehleringer 1999). Macropods (kangaroos and wallabies), deer, and rabbits show pronounced enrichments in arid environments (Ayliffe and Chivas 1990; Luz et al. 1990; Delgado Huertas et al. 1995), although the relative contributions of plant physiology vs. evaporative enrichment of local waters are not resolved (Kohn 1996).

Zoning. Like many biogenic tissues, enamel forms progressively, nucleating at the dentine-enamel interface at the occlusal (or wear) surface, and growing downward and outward with time. For typically analyzed herbivores, the duration of growth for a tooth may span a few months (Kohn et al. 1998) to over a year (Balasse 2002; Passey and Cerling 2002; Hoppe 2006; see also Kohn 2004 for a review), or even decades for proboscideans (Koch et al. 1989). Tooth enamel almost always exhibits isotopic zoning (Fricke and O'Neil 1996), because isotope compositions of local water vary during the year, in response to seasonal changes in precipitation regime and/or temperature (Kohn et al. 1998). This zoning means that fossil tooth enamel cannot usually be sampled in bulk unless the timing and rate of tooth formation are already known. Otherwise, the isotope composition may be biased to a particular season. Also, specific teeth form at different times in different individuals. Zoning profiles characterize seasonal isotope variations and allow more accurate inference of yearly mean compositions, but of course require far more analyses per tooth than a bulk analysis.

Four factors dampen enamel isotope variations relative to seasonal variations in meteoric or local water compositions: a non 1:1 correspondence between the compositions of body water (tooth enamel) vs. meteoric water (Fig. 4, Kohn 1996; Kohn and Cerling 2002), a non-negligible residence time of oxygen in the animal (Kohn et al. 2002; Kohn and Cerling 2002), a non-negligible maturation interval for tooth enamel at any point along the length of the tooth (Balasse 2002; Passey and Cerling 2002), and environmental averaging of compositions in water sources used by herbivores. The first three factors reflect animal physiology, whereas the last reflects environment. Models of isotope variation that consider animal physiology and seasonal changes in water composition and humidity closely match measured isotope variations in enamel (Kohn et al. 1998), demonstrating that climate seasonality drives isotope zoning in enamel. Even in cases where damping may be significant, restricting analysis to one taxon, subfamily, or family, in which physiologies were likely similar, eliminates the first three problems from consideration, and focuses more directly on changes to environmental variability (Zanazzi et al. 2007).

Physiology. Many studies now demonstrate the differences in isotope compositions of different animals living in the same areas (e.g., Koch et al. 1991; Bocherens et al. 1996; Kohn et al. 1996; Levin et al. 2007), related physiologically to diet, water dependence, and water turnover processes (Kohn 1996). This fact stresses that the comparability of fossil compositions from different strata depends strongly on (paleo-)biology, and stable isotope geochemists contemplating analysis of fossils should either have a background in paleobiology or work closely with paleobiologists. For example, digestive physiologies differ strongly between modern artiodactyls (an order that includes pigs, camels, deer, cattle, and antelopes) vs. perissodactyls (an order that includes horses, rhinos and tapirs); artiodactyls are foregut fermenters and most are ruminants, whereas perissodactyls are hindgut fermenters. These physiological differences affect water requirements (McNab 2002), which in turn determine sensitivity to water compositions (Kohn 1996; Kohn et al. 1996; Kohn and Fremd 2007). For example, digestive physiology of perissodactyls requires high water turnover, whereas that of artiodactyls can confer but does not demand drought tolerance (McNab 2002). Thus, comparing different species, genera, or even families within an order is far more reliable for perissodactyls than for artiodactyls. For example, although all members of the families Equidae and Rhinocerotidae probably have had quite similar and high water requirements, members of the Bovidae exhibit remarkable variations in water requirements (Kohn et al. 1996). In general, it is better to analyze a single taxon if possible through a sequence to ensure constant physiology and water requirements. If this is not possible, use of perissodactyls is encouraged by their digestive physiology.

Migration. Unlike all other climate proxies that do not generally "move around" as they form, most commonly analyzed mammals are sufficiently large to be capable of seasonal migrations during tooth growth. Although the popular press has highlighted herbivore migrations over hundreds of km, long lateral migrations are not especially common for terrestrial mammals, and tend to occur only in larger-bodied forms (e.g., Estes 1991). More common patterns involve local summer migrations to higher elevations where conditions are generally cooler and wetter, and winter migrations to lower elevations where conditions are warmer. We expect such migrations to dampen seasonal isotope variations, rather than shift mean values. This consideration does affect interpretation of isotope seasonality, if changes to topographic relief occur over the timespan of investigated strata, but not mean elevations.

Numbers of samples. Multiple samples of teeth are required for any particular time to account for intra-populational variability and isotope seasonality, and the number of teeth and analyses required to assure interpretable results is of great interest to many scientists (perhaps most importantly to museum curators and funding agency managers). However, the optimal number in part depends on the size of the teeth, which generally reflects how much of a year is recorded (Kohn 2004), and the two approaches used to measure isotope compositions, i.e., bulk sampling vs. zoning analysis. Very generally, bulk analysis is less destructive but requires more teeth, whereas zoning analysis is more destructive but requires fewer teeth. Clementz and Koch (2001) provide the only statistically rigorous estimates on numbers of teeth required for bulk sampling. Although their study focused on marine mammals, paleoaltimetry studies generally consider terrestrial herbivores, so we discuss their results for M3 teeth (third, or rearmost, molars) for black-tailed deer from central California.

Clementz and Koch collected bulk samples by removing enamel with a drill. Their oxygen isotope data from 47 teeth show a variation of $\pm 1.3‰$ ($\pm 1\sigma$). This result generally implies that resolution of a 1‰ difference between two sites or times at 95% confidence would require 10-15 teeth per species investigated from each stratigraphic level (although resolving the composition to $\pm 1‰$ at any one level requires only ~6 teeth, twice as many are required per level to resolve a 1‰ change). In our experience, few if any museums could honor such a large request because fossil teeth are taxonomically so important. However, bulk sampling does not resolve seasonal isotope variations or differences in the times of tooth formation among individuals. Deer teeth are relative small and mineralize quickly, so they may not be well-suited to a bulk sampling approach. As mentioned previously, it is important to identify numbers of samples and a preferred analytical approach prior to requesting samples from museums, so we believe relatively detailed discussion of this issue is warranted.

Deer M3's are ~10 mm long, and the rates of tooth formation and maturation in deer and other similarly sized ungulates are 30-60 mm/yr (Kohn 2004); therefore each M3 probably encodes only 2-4 months of time. Moreover, M3 eruption times for ~100 kg deer range broadly from 9 to 26 months after birth (Kierdorf and Kierdorf 2000). Thus the M3's analyzed by Clementz and Koch may well represent an assemblage of 2-4 month time slices spread out over all seasons of the year. The $\delta^{18}O$ of precipitation in northern coastal California ranges seasonally over 7‰ (on-line isotope data from GNIP and USNIP), and deer teeth from northern California indeed show zoning in excess of 5‰ (Kohn, unpubl. data). If yearly seasonal variability in tooth compositions has a ~5‰ amplitude and a sinusoidal form, then a random sample of teeth would exhibit a ~ ± 1.5‰ variation (± 1σ), even as averaged over several months. Thus, Clementz and Koch's data cannot discount the hypothesis that all deer actually had the same yearly isotope trends, but that differences in the *season* of M3 tooth growth among individuals resulted in the observed M3 compositional variability of ± 1.3‰. Because of this ambiguity, Clementz and Koch's recommendations regarding numbers of teeth, while completely accurate for bulk samples of rapidly mineralized teeth, cannot be applied to other sampling methods or to teeth that mineralize slowly. Larger teeth that mineralize over a year or more (generally from larger animals) may be better suited for bulk analysis. Data for bison teeth (Hoppe 2006), although not quite sufficient for rigorous error analysis, do show smaller standard deviations of ± 0.9‰ (1σ) for the larger sample sizes (*n* = 8 to 10). Thus, only ~6 bison teeth per site would be required to resolve an isotopic shift of ~1‰ between sites, although of course larger numbers would be required for better isotopic resolution.

An alternative approach is to measure isotope zoning in many teeth, and determine how many teeth are required to resolve median compositions and variations. Note that we prefer medians rather than means in statistical comparisons because they are less susceptible to outliers resulting from isotopically anomalous seasons. A rigorous number of samples depends not only on the spatial and temporal resolution of the analyses, but also on the magnitude of the isotope zoning and the amount of time encoded by a tooth, which may not be known *a priori*. Although by no means completely rigorous, we generally request at least 3 teeth from each stratigraphic level, and optimally analyze 5-6. Combined with zoning analysis, these numbers ordinarily resolve medians to ± 0.5‰ at 95% confidence (e.g., see data from Kohn et al. 2002; Zanazzi et al. 2007). Including more teeth generally does not influence median compositions or estimates of isotope seasonality. Interestingly, the total numbers of analyses collected for the zoning vs. bulk approaches do not differ that much; each requires several tens of analyses for a resolution of ~ ± 0.5‰. However, the direct investigation of zoning not only allows assessment of seasonal isotope variations, but also minimizes the number of fossils required.

Invertebrate fossils

Although many different groups of invertebrate organisms have been investigated to infer the isotopic composition of terrestrial surface waters, only two, mollusks and ostracodes, have commonly been used in continental settings. We focus primarily on mollusks in this paper because they are the most common fossils that have been used for paleoelevation studies in ancient terrestrial systems. Ostracodes will also be discussed because they are potential targets for elevation studies, although they carry the added complication of being associated with lacustrine environments. There is no general review of the stable isotope systematics in freshwater mollusks. The use of stable isotope ratios of ostracodes in paleolimnology has been reviewed by Schwalb (2003), but we reconsider some key points below.

Mineralogy and structure. Freshwater mollusk shell is made up of calcium carbonate, $CaCO_3$, with a minor component of organic matter. All freshwater bivalve taxa produce aragonite shells (Carter 1980). Freshwater gastropods also produce aragonite shell, except for a few genera that have developed a calcitic outer layer, and the calcitic opercula of some *Pila sp.* and *Bythnia sp.* (see Bandel 1990 for further discussion). These carbonate shells are built of various crystal forms (nacreous, prismatic, crossed-lamellar, etc.) and layering arrangements;

these may be useful in assessing diagenesis or the collection of sclerochronological data, but will not be discussed here (see Carter 1980 for an introduction). The overwhelming predominance of aragonite offers a simple test for diagenesis, in that alteration of the shell would most likely result in stabilization to calcite. Therefore X-ray diffraction or careful use of Feigl's solution, a stain for aragonite (Friedman 1977), are important tools in assessing the preservation of fresh water shell. Cathodoluminescence microscopy is another common observational technique for assessing alteration, partial calcitization, or intergrowth of later phase calcite into shell. Although more commonly used to image alteration in marine carbonates, the contrast between shell aragonite (green-gold luminescence) and later-stage calcite (orange-red luminescence) can be very clear in freshwater shells (Nickel 1978).

Ostracodes are micro-crustaceans that produce a series of calcitic shells during their growth (De Deckker 1981). Typical size is around 1mm, although shells can range in size up to ~4 mm. Unlike the accretionary growth of mollusk shells, the ostracode shell is produced in a very short time, a matter of days (Turpen and Angell 1971). The shell, therefore, provides only a snapshot of the local environmental conditions. The shells are low-Mg calcite, usually less than 4 mole% Mg, and a good deal of effort has been expended in understanding the relationship between shell Mg/Ca and temperature (see e.g., Dettman et al. 2002; Holmes and Chivas 2002). These shells are therefore very stable at earth-surface conditions and there are no simple chemical tests for diagenesis. Assessment for alteration therefore must rely on petrographic, SEM, or CL microscopy.

Modern oxygen isotope relationships in shells. Calibration studies of mollusk shell across a wide range of taxonomy and habitat have observed little difference between the temperature-fractionation relationship for oxygen isotopes measured by Grossman and Ku (1986) and calibration data (Wefer and Berger 1991). This relationship apparently holds true for freshwater mollusks, although few calibration studies have been done (e.g., Dettman et al. 1999; White et al. 1999; Ricken et al. 2003; Shanahan et al. 2005). This implies that there are only two environmental factors that affect the $\delta^{18}O$ of shell: the isotopic composition and the temperature of the water in which the shell is grown. In order to calculate the isotopic composition of the surface waters in which these mollusks lived, therefore, one must somehow constrain the temperature at which shell was grown. As with vertebrate fossils, a basic understanding of mollusk physiology and ecology is required, particularly as non-marine mollusks inhabit a continuum from fully terrestrial to fully aquatic. The shells of land snails can be much more positive than predicted by the temperature and rainfall $\delta^{18}O$ of a location due to strong evaporative enrichment of the body water of the snail (Goodfriend 1992), although in some regions this offset is not large (LeColle 1985). But even among aquatic snails caution must be employed, as some are only semi-aquatic. For example, pulmonates, which use a lung to breath instead of a gill structure, are often found on plants above the pond surface and their $\delta^{18}O$ values may be more positive compared to fully aquatic snails (Shanahan et al. 2005). Therefore it is important to understand the habitat of the mollusk species used in a study and to use fully aquatic taxa if possible for the reconstruction of water $\delta^{18}O$.

The Grossman and Ku (1986) fractionation relationship for oxygen isotope ratios in aragonite has been rewritten because the original paper represents the fractionation between water and aragonite as a simple difference ($\delta_c - \delta_w$) where δ_c is the $\delta^{18}O_{VPDB}$ of the carbonate and δ_w is defined as the $\delta^{18}O$ of the water (V-SMOW) minus 0.2‰. While the temperature dependent fractionation can be calculated from this difference within the error of measurement for water compositions between -10‰ and $+10$‰, a systematic error is introduced when the water is less than -10‰ (V-SMOW). The re-written relationship (Dettman et al. 1999) is:

$$1000\ln(\alpha) = 2.559(10^6 T^{-2}) + 0.715 \tag{1}$$

where T is degrees Kelvin and α is the fractionation between water and aragonite:

$$\alpha = \frac{\left(1000 + \delta^{18}O_{aragonite-VSMOW}\right)}{\left(1000 + \delta^{18}O_{water-VSMOW}\right)} \qquad (2)$$

To use this relationship to calculate the oxygen isotope ratio of river or lake water from a mollusk shell requires constraints on the temperature of the water body. Sometimes the temperature can be estimated using other lines of evidence, such as plant macrofossils or pollen. In the absence of external constraints, average shell $\delta^{18}O$ values can be used thereby averaging shell grown across a range of temperatures. Although the uncertainty in this approach is significant, note that an uncertainty of ± 4.5 °C leads to an uncertainty of only ± 1‰ in the back-calculation of water $\delta^{18}O$. Therefore if $\delta^{18}O$ variations through time (or in space) are large, an uncertainty of ± 9 °C in the Mean Annual Temperature (MAT) leads to a (perhaps tolerable) uncertainty of ± 2‰ in water $\delta^{18}O$.

Biology can also be used to constrain the temperature at which shell grows in certain groups of mollusks. Modern unionids or naiads, large freshwater bivalves of the superfamily Unionacea, have a restricted temperature range of shell growth. Shell growth is highly restricted or absent below ~10 °C (Howard 1921; Negus 1966; Dettman et al. 1999). Also, because these animals grow most rapidly in the late spring and early summer, there is a temperature bias in shell formation. Both the limited range of growth temperature (10 to 30 °C) and the bias (most rapid growth from 20 to 25 °C) result in similar growth temperatures for a bulk shell $\delta^{18}O$ measurement. This is demonstrated in Figure 5, where the average $\delta^{18}O$ of river water is compared to the average $\delta^{18}O$ value in unionid shell for 28 rivers in the USA. The locations are from a wide variety of climates: coastal, continental interior, Mediterranean, high elevation and seasonal-temperate. MATs range from 3 °C in northern North Dakota to 23 °C in southern Florida (Easterling et al. 1999). This simple linear relationship shows that, for unionids, bulk shell $\delta^{18}O$ values can be used to calculate average river water $\delta^{18}O$ directly without estimating

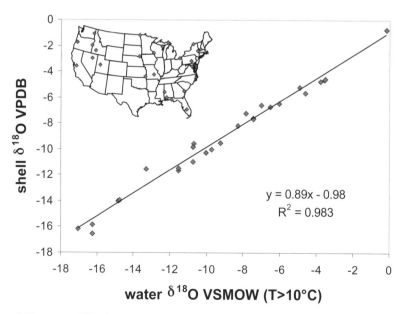

Figure 5. The average $\delta^{18}O$ of river water (when water T is greater than 10 °C) compared to bulk shell $\delta^{18}O$ for samples from 28 rivers across the USA. MAT ranges from 3 °C to 23 °C at sampling locations (inset).

temperature. Two features combine to make this such a strong correlation: the temperature limits and biases of unionid shell growth and the fact that rivers across North America go through very similar temperature cycles regardless of their mean climates. The monthly maximum T for the Red River of the North (North Dakota) river is 29 °C (Harkness et al. 2000), while it is 30 °C for Fisheating Creek, central Florida (U.S. Geol. Survey 1982). Shell growth stops or slows due to cold temperatures in both river systems. The effects of temperature can be seen in the slope of the relationship, 0.89, which shows that more shell material is grown at warmer temperatures in areas with river water $\delta^{18}O$ values close to 0‰ and more shell is grown at cooler temperatures for $\delta^{18}O$ values in the more negative portion of this plot. There seems to be no difference in the oxygen isotopic fractionation in unionid shell across a wide range of taxonomy, allowing different species to be combined in studies (Dettman et al. 1999).

Many of the same uncertainties described for mollusks exist for ostracodes when trying to use ostracode shell material as a measure of the $\delta^{18}O$ of water. Although ostracodes are present in river environments, they are not well preserved in fluvial sediments. They are often preserved in lake sediments and they are primarily used as indicators of lacustrine paleoenvironments. In both culturing and natural environment calibration studies of oxygen isotope fractionation in ostracodes, all species examined so far are more positive in $\delta^{18}O$ than the value predicted by the inorganic calcite relationship of Friedman and O'Neil (1977). Both Holmes and Chivas (2002) and Schwalb (2003) in reviews of oxygen isotope studies using ostracodes have argued that this offset is between +2 and +3‰. Dettman et al. (2005) expanded the data set used in these reviews and, due to questions of possible disequilibrium conditions in the inorganic calcite study of Kim and O'Neil (1997, discussed in Chivas et al. 2002), re-targeted the comparison to $\delta^{18}O$ values based on Friedman and O'Neil (1977). This comparison suggests that the offset ranges from +2.5 to +0.3‰ with a mean of +1.3‰ (Dettman et al. 2005). This data set also implies that there are significant taxonomic differences in the oxygen isotope systematics of ostracodes, and calibration studies or mono-specific approaches are needed for this kind of work.

Because ostracode shells are produced in a matter of hours or days, a single valve represents only a snapshot of the local environment. Two approaches have been used to trace long-term changes in the $\delta^{18}O$ of water. First, by analyzing many valves as a single sample, the full spectrum of ambient temperatures in which ostracodes grew may be averaged, thus allowing the average valve $\delta^{18}O$ to be used as a proxy for the $\delta^{18}O$ in the water (e.g., Schwalb et al. 1994; Curtis et al. 1998). The uncertainties present in this approach are difficult to quantify. Changes in MAT could affect the average valve $\delta^{18}O$ data with little change in the $\delta^{18}O$ of ambient water. However, because ostracode species may have a preferential season for growth and molting of valves, there may be a temperature bias in growth and valve production which could remove the effects of a change in MAT. Another approach has been to run a moderate number of valves individually from a given stratigraphic horizon and use the most positive $\delta^{18}O$ values as valves produced at the coldest temperatures in the seasonal cycle. Dettman et al. (1995) used this technique to track changes in early Holocene Lake Huron by associating a cold temperature (3 ± 2 °C) to the most positive ostracode $\delta^{18}O$ values and calculating water $\delta^{18}O$ with uncertainties on the order of ± 0.5 ‰. Using a very cold temperature was feasible for this study because *Candona subtriangulata,* the most abundant species, can grow and reproduce in the permanently cold waters of deep Lake Huron.

Seasonal records in shells. Freshwater mollusk shells, whether gastropods or bivalves, provide an opportunity to examine seasonal records in ancient environments. Sclerochronological techniques can be used to independently establish annual markers in the shells and thereby document $\delta^{18}O$ change between spring, summer, fall, and (perhaps) winter. This seasonal information can be useful in paleo-elevation studies by revealing the presence of "monsoon-like" seasonal cycles in shells and in the $\delta^{18}O$ of surface water (Dettman et al. 2001). It can also be useful in documenting snow-melt or glacial runoff from high elevation regions in a river's catchment (Dettman and Lohmann 2000). Examples of both uses are discussed in the examples below.

High resolution oxygen isotope records from mollusk shells are readily extracted by using current micro-milling techniques. When necessary, sample resolutions of 10 to 50 μm are achievable, resulting in weekly to subweekly resolution of oxygen isotope ratios (Dettman and Lohmann 1995). The accretionary nature of these shells provides a time series of environmental change on a sub-annual basis (although not all of the year may be present). Seasonal records are rare in the geologic record, but these kinds of records can be very useful in, for example, tracing the history of the Asian monsoon, a seasonal phenomenon established by a high elevation plateau. In the example discussed below seasonal records demonstrate a very arid dry season leading to strong evaporation in surface waters on a seasonal basis. In other cases there may be less evidence for monsoon-like wet/dry seasonality, but seasonal patterns in surface water $\delta^{18}O$ may indicate the presence of monsoon circulation in other ways. In the monsoon region the isotopic composition of summer rainfall is more negative than that in winter (Figs. 3 and 6, Araguas-Araguas et al. 1998) and this pattern leads to unusually large seasonal $\delta^{18}O$ cycles in carbonates. Warm summer temperatures (smaller carbonate-water fractionation) coupled with low $\delta^{18}O$ surface waters enhance negative $\delta^{18}O$ values in carbonate, whereas cool winter temperatures (larger carbonate-water fractionation) coupled with high $\delta^{18}O$ surface waters enhance positive $\delta^{18}O$ values in shell. In contrast, in temperate climates, isotopic records are dampened because surface water are usually more positive in the summer when fractionations are smaller, and more negative in the winter when fractionations are larger (Fig. 7). Although little work has been done on seasonal cycles in shells, seasonal amplitude in carbonates may prove to be an important indicator of monsoon climates.

Surface waters in paleo-elevation studies. The isotopic compositions of all the fossils discussed here are strongly tied to the oxygen isotope ratio of surface waters, either rivers or lakes, although some organisms might take up a significant percentage of water from their

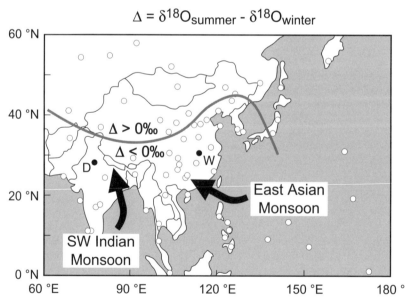

Figure 6. Oxygen isotope seasonality in Asia. Strong monsoonal precipitation exists in areas where summer precipitation is more negative than winter precipitation (modified from Araguas-Araguas 1998 based on data from IAEA/WMO 2004).

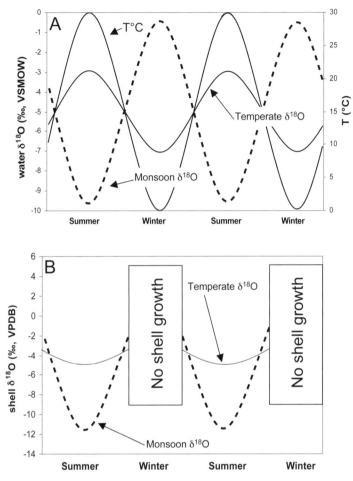

Figure 7. Simple model of seasonal temperature and $\delta^{18}O$ change in river water and how the phase relationship of these could affect the $\delta^{18}O$ amplitude of shell. (A) Ambient conditions for model. Temperature ranges from 0 to 30 °C for both monsoon and temperate cases. Temperate climate water $\delta^{18}O$ ranges from -3 to $-7‰$ VSMOW; monsoon climate water $\delta^{18}O$ ranges from -0.5 to $-9.5‰$, as in Figure 3. (B) Calculated shell $\delta^{18}O$ amplitudes for temperate and monsoon conditions in A. No shell growth occurs below 10 °C.

food, e.g., particularly small-bodied, arid adapted forms. It is important to keep in mind that these types of water bodies are not always local water. Rivers can carry water from distant sources and $\delta^{18}O$ values are controlled by the precipitation in the entire catchment. When the catchment is large, the $\delta^{18}O$ of river water can be partially disconnected from the $\delta^{18}O$ of local rainfall. Examples of this are numerous, mostly in large river systems such as the Colorado River in the southwestern USA. The $\delta^{18}O$ of Colorado River water near Yuma, Arizona, at 62 m above sealevel, MAT 23 °C, is approximately $-12‰$ V-SMOW today, while local rain water is -2 to $-6‰$ (Dickinson et al. 2006). Before the dams on the Colorado River were emplaced, the $\delta^{18}O$ of Colorado River water reached values less than $-17‰$ during the summer season (Dettman et al. 2004). The reason for this rainwater – river water discrepancy is that the Colorado River catchment is very large, extending north into Wyoming. Snowmelt from the Rocky Mountains created a large annual flood event in the early summer and the isotopic composition of this meltwater reflects the elevation of the Rocky Mountains plus distillation due to moisture transport into the interior of the continent. The issue of local vs. distal water in paleoelevation studies is not necessarily a drawback, however. The comparison of materials associated with local water, such as soils or mylonites, with distal water, which may have sources in highlands, can provide information on paleotopography.

Because mammals and aquatic invertebrates derive the majority of their drinking water from surface waters one could perhaps argue that relationships between local environment and the isotopic composition of *river* waters should be used to interpret ancient systems (Fricke and Wing 2004). This approach assumes the catchments of all rivers somehow scale equivalently, which becomes problematic when some rivers obtain their water locally whereas others transport water long distances between two very different climates. The Yuma example above demonstrates this well—the mid summer $\delta^{18}O$ of Colorado River water prior to damming was $\sim-17\%o$ SMOW (indicative of sub-freezing conditions), whereas the local temperature in July and August exceeds 40 °C and the altitude is <100 feet above sea level (Dettman et al. 2004). The very negative values of river water were established by the winter snows in the Rocky Mountains (elevations 2000-4000 m), which melt in the summer and became a large flood water event. We suggest that ancient conditions can be better interpreted based on the modern relationship between the isotopic composition of precipitation and climatic parameters than on the relationship between river water $\delta^{18}O$ and local climate. However we emphasize that one must consider the possibility that a river's catchment may encompass quite different climates or elevations (Rowley et al. 2001).

An additional complication in samples derived from lake sediments is that lake water may be evaporatively enriched in ^{18}O. Lacustrine fossils have not yet been employed in paleo-elevation studies, but they have potential. Careful assessment of the isotopic signatures must be used to evaluate the open/closed status or potential for evaporation of the lake. This could involve comparison with coeval soil carbonates or other isotopic records tied directly to local precipitation. Note that one simple test often used to distinguish the open/closed status of a lake is clearly not valid: the test for co-variance between $\delta^{18}O$ and $\delta^{13}C$ in carbonates. Many have used this isotopic co-variance to distinguish between open and closed lakes (Talbot 1990). Many closed lakes do show co-variance in oxygen and carbon isotope ratios, and many open lakes do not. But open lakes not subject to significant evaporative enrichment can also show strong co-variance (Drummond et al. 1995). This behavior can readily result from seasonal cycles in temperature that affect the $\delta^{18}O$ of lake carbonates simultaneous with seasonal changes to lake water $\delta^{13}C$, leading to significant co-variance independent of the open/closed status of the lake.

EXAMPLES AND APPLICATIONS

Tibetan Plateau

The Tibetan Plateau is sufficiently large that it affects not only its own climate (high elevations are colder), but also the climate of neighboring regions through development of a monsoon and farther reaching atmospheric perturbations. As described previously, heating over the plateau during the summer creates an atmospheric low that draws in moisture from the Indian and Pacific oceans and causes intense (low $\delta^{18}O$) rainfall and wet conditions; cooling during the winter creates an atmospheric high that causes relatively arid (high $\delta^{18}O$) conditions. Arguably, the geomorphically-induced atmospheric patterns affect climates not only in southeast Asia but also in Australia, Africa, and even North America (via atmospheric teleconnections; e.g., McCabe et al. 2004). Several studies have used mean isotope values from modern high elevation sites on the plateau to infer past elevations assuming isotopic lapse rates either from modern measurements or theory (e.g., Garzione et al. 2000; Currie et al. 2005; Cyr et al. 2005; Rowley and Currie 2006). However, fossils provide an alternative perspective regarding uplift of the plateau through preservation of seasonal isotope patterns that might be related to a monsoonal climate. Specifically, Dettman et al. (2001) used seasonal cycles in the $\delta^{18}O$ of unionid bivalves, "horse" teeth, rhino teeth, and one bovid tooth from the southern margin of the Himalaya to infer the seasonal magnitude of the monsoon in South Asia. The importance of this study lay in its inference regarding timing of the development of the monsoon. Previous interpretations indicated initiation at ~8 Ma (Harrison et al. 1992; Molnar et al. 1993), whereas the fossils that were analyzed ranged from Recent to as old as 10.7 Ma, i.e., almost 3 Myr older than the presumed rise of Tibet.

Well preserved fossil shells were collected from fluvial and flood-plain sediments of low elevation southward-flowing rivers tributary to the paleo-Ganges and Indus. Eleven different stratigraphic levels in the Siwaliks of Nepal were represented and two to four annual $\delta^{18}O$ cycles were measured from each shell bed using 1 to 3 shells. An example of one of the shells is shown in Figure 8. The maximum and minimum $\delta^{18}O$ values repeat very well in the two years of growth.

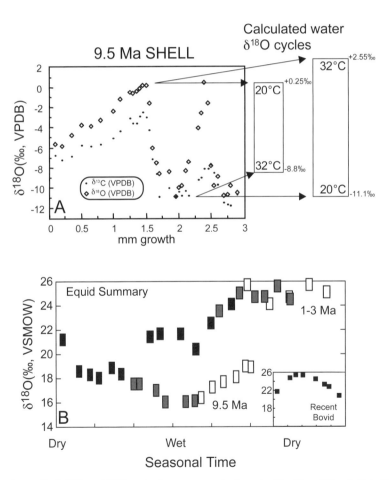

Figure 8. (A) Shell $\delta^{18}O$ and $\delta^{13}C$ cycle from one sample from the Siwaliks of Nepal. Note that the calculated amplitude of $\delta^{18}O$ change in the river water is 10‰ if cool temperature is associated with the most positive shell $\delta^{18}O$ value. The amplitude is larger (13.65‰) if the temperature association is reversed. Modified from Dettman et al. 2001. (B) Tooth enamel $\delta^{18}O$ cycle from the Siwaliks of Pakistan, showing similar amplitude variations for older vs. younger teeth.

The total range in shell $\delta^{18}O$ values for each stratigraphic interval is shown in Figure 9a. Fossil teeth were collected from fluvial deposits in the Siwaliks of Pakistan, and were first analyzed for carbon isotopes to infer paleodiets (Quade et al. 1992). One archeological and four geological horizons were analyzed. Partial annual cycles are obvious in the equid and bovid teeth (Fig. 10a), and total ranges in compositions for samples pre- vs. post-8 Ma are similar (Fig. 10b).

The $\delta^{18}O$ of river waters can be calculated from the isotope compositions of the shells, but a temperature must be assumed, and was based on fossil floras from the Siwalik sediments (Awashi and Prasad 1989; Sarkar 1989). The older part of the record (before 9 Ma) has a flora similar to Mandalay, Burma where temperatures range from 20 to 32 °C. The middle portion of the record has a flora similar to Patna, India with a temperature range of 17 to 32 °C. The post 4.5 Ma portion has floras similar to New Delhi with temperatures of 14-34 °C. The question of how to apply these temperature ranges to the shell $\delta^{18}O$ ranges can be answered by using the most conservative relationship, the coolest temperature associated with most positive $\delta^{18}O$

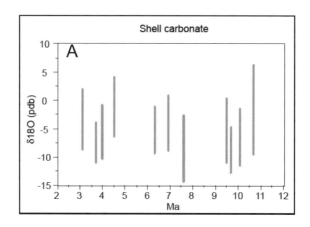

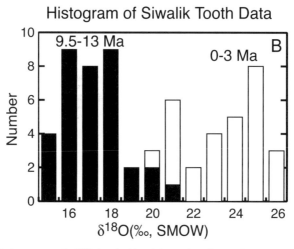

Figure 9. (A) Maximum range in $\delta^{18}O$ (vertical bars) for each shell bed. Two to four annual cycles were measured for each shell bed, using one to three shells. Note the large range of isotope compositions for shells from the Siwaliks of Nepal and high maximum $\delta^{18}O$ values. [Used by permission of the Geological Society of America, from Dettman et al. (2001), Geology, Vol. 29, Fig. 2, p. 31-34.] (B) Histogram of fossil tooth compositions, illustrating similar ranges of composition and shift to higher $\delta^{18}O$ values for younger teeth.

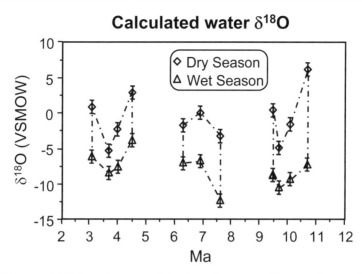

Figure 10. Estimated $\delta^{18}O$ for surface waters during dry and wet seasons. Using maximum and minimum $\delta^{18}O$ values for shell carbonate and temperatures based on paleobotanical data, water $\delta^{18}O$ is calculated using fractionation relationship for molluscan aragonite (see text). Estimated temperatures are different for three enclosed areas: oldest employs 20–32 °C; middle uses range of 17–32 °C; and youngest interval uses 14–34 °C. Error bars represent a ± 5 °C uncertainty in temperature used. [Used by permission of the Geological Society of America, from Dettman et al. (2001), Geology, Vol. 29, Fig. 2, p. 31-34.]

value and the warmest temperature with the most negative shell value. This choice yields the smallest seasonal change in the $\delta^{18}O$ of the water (Fig. 8). Any shift in the phase of this temperature cycle would increase the calculated $\delta^{18}O$ of the local waters.

The results of these calculations show that the $\delta^{18}O$ values of surface waters experienced large seasonal cycles throughout most of the record (Fig. 9b) and we have argued that these are monsoon-like. The strongest evidence for this is based on the dry-season oxygen isotope ratios, which are 0‰ VSMOW or greater in 5 of the 11 intervals studied, and are greater than −2.5‰ in 3 more cases. For a body of water in a continental interior to have a $\delta^{18}O$ value approaching 0‰ SMOW implies significant evaporation, especially when the body of water is recharged and shifts to much more negative $\delta^{18}O$ values on a seasonal basis (Ingraham et al. 1998). This is very similar to the modern South Indian Monsoon, where the dry season is very arid and lasts for 6 months or more. Two or three of the shells do not show such strongly positive dry season $\delta^{18}O$ value, but we expect this even if a strong monsoon existed at this time. Larger rivers in the monsoon region do not go through a large seasonal cycle in $\delta^{18}O$ because they transport too much water too quickly for evaporation to have a strong effect (Dettman et al. 2001).

The similar range in $\delta^{18}O$ values for the fossil teeth suggest that no difference in the seasonality of local water occurred between 9.5 and <4 Ma. Thus, if a seasonal monsoon climate is responsible for the magnitude of the isotope variation present for samples <4 Ma, then a similarly seasonal climate is inferred for samples 9.5 Ma and older. Subsequent studies have found similar magnitudes of isotopic variation in teeth from non-monsoonal climates (e.g., see Kohn et al. 2002). Nonetheless, although the teeth data might not uniquely identify the monsoon, they are consistent with the interpretations of bivalve isotopes.

The tooth data also clearly delineate a pronounced shift to more positive $\delta^{18}O$ values between 9.5 and 3 Ma (Fig. 10a,b). Although noisy, the wet-season $\delta^{18}O$ values calculated from mollusks also suggest more negative $\delta^{18}O$ values in the older river systems. Paleosol carbonate data show that this shift occurred between 7 and 8 Ma and slightly preceded the well-known shift to more positive $\delta^{13}C$ values in paleosols and fossil teeth (Quade et al. 1989, 1992, 1995; Quade and Cerling 1995). Low $\delta^{18}O$ values of precipitation in the Indian subcontinent today attend an intense summertime rainy season, which is driven by an atmospheric low over the Tibetan plateau (Fig. 3). One interpretation of the stronger depletion prior to ~8 Ma is that warmer temperatures during the late Miocene caused a more intense summertime low over the plateau, greater rainout over the Indian subcontinent, and lower summertime $\delta^{18}O$ values. Subsequent global cooling caused a decrease in the intensity of the summertime atmospheric low and rainout intensity, causing summertime low $\delta^{18}O$ values to increase.

In sum, seasonal isotope data from shells and teeth indicate a strong monsoon in the Indian subcontinent through the period 10.7 to 2.5 Ma, with a likely decrease in intensity after 7-8 Ma. Because the monsoon is driven by a high plateau, a plateau region must have been present as early as 10.7 Ma, 2-3 Myr earlier than was presumed at the time. This result in turn implies that thermal-mechanical models of plateau formation must account for a plateau already by ~11 Ma. Placing more precise constraints on the elevation or extent of the plateau is left to climate modelers who can perhaps describe the type of plateau needed to establish a monsoon with a very arid season. This dry season resulted in strong evaporation of surface waters on a seasonal basis, alternating with wet season recharge and flushing of evaporated waters.

Cascades

Many studies focus on paleoelevations attending continent-continent collisions, yet arc processes are fundamentally important to tectonics, both because arcs precede such collisions and because paleoelevation provides information about an arc lithosphere's thermal and chemical state. Clearly different arcs behave differently with respect to elevations, as even long-lived arcs such as the Aleutians (40-50 Myr of activity) can have relatively little topographic expression outside of individual volcanoes, whereas, as we describe below, events within the last 7 Myr appear to have elevated the Oregon Cascades.

The Oregon Cascades are well-suited to paleoaltimetry analysis because they produce an impressive rainshadow extending over central and eastern Oregon, which in turn includes some of the world's pre-eminent Oligocene to late Miocene fossil vertebrate sequences, specifically in basins in central Oregon (John Day basin; Fremd et al. 1997) and scattered throughout southeastern Oregon (Shotwell 1968). Note that these fossils were collected at modern elevations of no more than 800 m, the central Oregon Cascades rise to a minimum ridge height of 1600 m, and the oxygen isotope composition of meteoric water east of the Cascades is nearly constant and depleted by~7‰ relative to coastal Oregon and Willamette Valley (−8‰ vs. −15‰; Fig. 11). These data imply an apparent isotopic lapse rate of −4.3 ‰/km (=−7‰/1.6 km). This lapse rate is apparent because it is determined from isotopic differences across the range, rather than elevation transects. Nonetheless, this value may also be the true lapse rate because some data indicate that isotope compositions from winter precipitation near the crest of the range are similar to those of the interior (M Kohn and J Welker, unpubl data). Additional key fossils of Pliocene and early Pleistocene age occur in southwestern Idaho, where modern meteoric water compositions are ~−16‰ (Kohn and Fremd 2007). Storms uniformly track eastward, obviating complications of multiple tracks or north-south water transport that can occur in the southwestern US. Also evaporative enrichment is less pronounced than further south, both because winter precipitation predominates, and because interior Oregon receives more rainfall. Consequently precipitation compositions in the lee of range do not increase with distance, but rather gradually drop as expected simply from continental effects.

The Cascades are additionally interesting geologically because their protracted igneous history is well understood yet heterogeneous both temporally and spatially. Briefly, an active arc from ~40 to ~7 Ma gave rise to the western Cascades, but this volcanism was more recently supplanted by the modern eastern Cascades, which hosts the highest peaks. Important

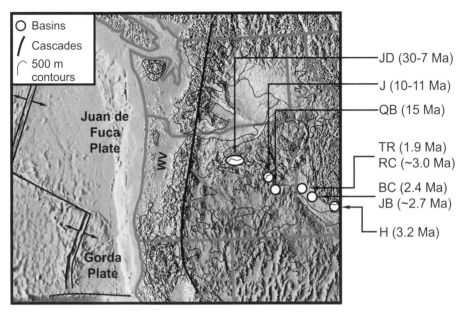

Figure 11. Topographic map contoured at 500 m intervals showing location of fossiliferous basins in Oregon and western Idaho. JD = John Day, J = Juntura, QB = Quartz Basin, TR = Tyson Ranch, RC = Rabbit Creek,BC = Birch Creek, JB = Jackass Butte, H = Hagerman. Note that the John Day basin includes the John Day Formation (30.0 – 19.4 Ma), Mascall Formation (15.8 – 15.2 Ma) and Rattlesnake Formation (7.8 – 7.2 Ma). Age constraints are summarized in Kohn and Law (2006) and Kohn and Fremd (2007).

changes to igneous activity are summarized by Conrey et al. (2004) and include eruption of the Columbia River Plateau basalts—comprising a mid-Miocene (mainly 17-14 Ma; Cummings et al. 2000) large igneous province in NE Oregon, SW Washington, and western Idaho—and a shift at 7-8 Ma from calc-alkaline andesite-dominated lavas to more varied compositions with a MORB-signature. Conrey et al. (2004) argue for a major change in the composition and presumably temperature of the sub-arc mantle at 7-8 Ma, as Basin and Range extension propagated into the arc. Although intra-arc extensional faults are dated at 5.4 Ma, rifting in the eastern Cascades probably started earlier, c. 7.4 Ma, when deposition of rift-related volcanic rocks and sediments initiated (Smith et al. 1987). Presumably, gradual accumulation of a volcanic edifice (between ~40 and ~7 Ma), eruption of the Columbia River basalts (17-14 Ma), development of the eastern high Cascades (post 7.4 Ma), and extensional tectonics (post 7.4 Ma) could all have affected lithospheric properties through changes to mantle chemistry and temperature, diversion of melts, and changes to the thickness of the crust. Direct investigation of paleoelevations provides insights into how these varied arc processes affect the lithosphere.

Three primary datasets have been collected from central and eastern Oregon, and southwestern Idaho (Fig. 11), and include oxygen isotope compositions of tooth enamel (PO_4^{3-} and CO_3^{2-} components; Kohn et al. 2002; Kohn and Fremd 2007) and fossil bone (CO_3^{2-} component; Kohn and Law 2006). Kohn and Law (2006) focused on calibrating bone $\delta^{18}O$ systematics in Oregon assuming a known isotopic record from tooth enamel. Although their results broadly corroborate the other two studies, the bone interpretations are necessarily somewhat derivative of the tooth enamel data. Tooth enamel analysis was principally restricted to equids and rhinos, as they should have had a strong dependence on water compositions, a weak dependence on humidity, and comparable isotope compositions because they share similar digestive physiologies (Kohn and Fremd 2007).

Individual teeth show large isotopic variability, reflecting smooth isotopic zoning associated with $\delta^{18}O$ highs during summer and $\delta^{18}O$ lows during winter (Fig. 12). As discussed by Kohn et al. (2002), these variations are expected and accommodated by sampling protocols (see discussion of Passey and Cerling 2002; Kohn 2004). Unlike other proxies that provide more averaged isotopic measures, such as paleosol carbonate or authigenic clays, teeth simply record a different climate proxy that includes seasonality as a main signal. Even including the standard error of mean values (which is statistically quite conservative), mean compositions have ± 1 s.e. uncertainties of only 0.2-0.3‰, mainly reflecting the large numbers of data used to reconstruct isotope seasonality. Modern compositions are estimated either from modern cattle teeth, whose compositions are indistinguishable from horses in the modern global dataset, or from known meteoric water compositions and the global dataset for horses (see Kohn and Fremd 2007).

Looking at the entire record (Fig. 13), compositions show no significant differences from ~30 Ma to 7.2 Ma, then a large, ~4‰ drop to modern compositions. Compositions from Idaho suggest that the isotope drop occurred gradually over the last ~7 Myr, although within the time resolution and uncertainties of the data there could have been steps. As discussed by Kohn et al. (2002), this record reflects climate change due to both global trends and local tectonics. Deconvolving global vs. local factors requires two models. The first describes how meteoric water compositions respond to changes in temperature; the second describes how perissodactyl compositions respond to local water composition and humidity. These models are described in detail elsewhere (Kohn 1996; Kohn et al. 1998, 2002, 2004; Kohn and Fremd 2007). In general global climate change is predicted to have little effect on perissodactyl compositions, because meteoric water compositions are relatively insensitive to temperature and are largely offset by synchronous decreases in relative humidity. Consequently, the post 7 Ma, pronounced drop in $\delta^{18}O$ values mainly reflects local tectonics, i.e., uplift of the central Cascades.

The drop in tooth enamel $\delta^{18}O$ in excess of the global climate prediction (i.e., the $\delta^{18}O$ residual) can be converted into an apparent increase in elevation (Fig. 14). Quantitatively, the

Rattlesnake samples (~7.2 Ma)

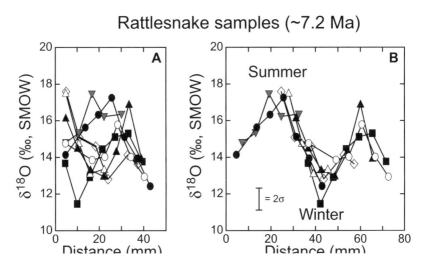

Figure 12. Oxygen isotope zoning from fossil equid teeth from the Rattlesnake Formation. (A) Raw zoning profiles, showing significant variation along the length of each tooth and differences in zoning patterns among teeth. Distances are measured from the occlusal (wear) surface to base of crown. (B) Zoning profiles adjusted laterally to highlight the preservation and coherency of high $\delta^{18}O$ summer values and low $\delta^{18}O$ winter values. Data from Kohn et al. (2002).

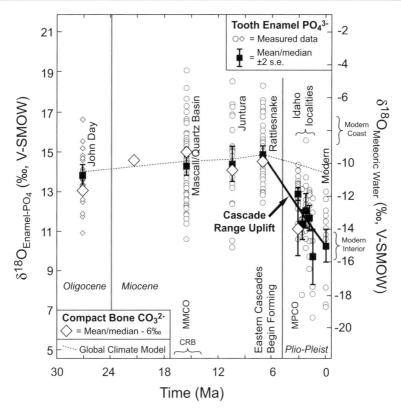

Figure 13. Summary of isotope data for teeth from southwestern Idaho, southeastern Oregon, and central Oregon. Ranges in composition reflect both intra-tooth compositional variability (due to preserved isotope seasonality), and compositional differences among teeth (see Fig. 12). Means and medians are expected to reflect yearly average compositions (i.e., average climate). Bone compositions are for compact bone only, and have been decreased by 6‰ to facilitate comparisons; bone data at 3.2 Ma are corrected for aridity (Kohn and Law 2006). Dotted line shows model prediction of isotope compositional shifts to tooth enamel due to global warming and cooling (Kohn et al. 2002; Kohn and Fremd 2007). The general inability of these models to reproduce the observed compositional trends since ~7 Ma implies strong tectonic control on isotope compositions. Note that water compositional scale implicitly assumes a fixed humidity, and that rhino rather than equid compositions are plotted at 27 Ma. MMCO and MPCO are the mid-Miocene and mid-Pliocene climatic optima. CRB indicates timing of eruption of the Columbia River Basalts. Data from Kohn et al. (2002), Kohn and Law (2006), and Kohn and Fremd (2007).

conversion factor is derived by dividing the apparent isotopic lapse rate (230 ± 30 m/‰, 2σ or -4.3 ± 0.55 ‰/km) by the slope of the correlation between equid composition and local water composition (0.83 ± 0.14‰/‰, 2σ; Kohn 1996; Kohn and Fremd 2007) or ~275 ± 60 (2σ) m per 1‰ shift to $\delta^{18}O$ values of tooth enamel. Because the isotope record as corrected for global climate remains essentially flat until ~7 Ma, then decreases sharply (Fig. 13), paleoelevations are calculated to have been static at ~800 m from ~30 Ma to ~7 Ma, followed by a sharp increase to 1600 m today. Because 800 m approximates the modern elevations of the basins from which the fossils were collected, paleoelevations cannot be directly estimated prior to 7 Ma. If the basins were always at 800 m elevation, then the Cascades could have been lower, but neither the Cascades nor the basins could have been higher than 800 m between 30 and 7 Ma.

As discussed by Kohn and Fremd (2007) and below, elevation estimates assume that isotopic lapse rates did not change substantially through time. Errors in that assumption in principle

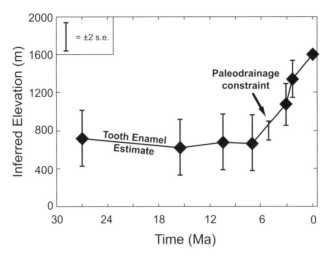

Figure 14. Model of paleoelevation of central Cascade Range, showing rapid increase since 7 Ma. Error bars include ± 2 s.e. uncertainties from composition data, isotopic lapse rate, and correlation between equid vs. meteoric water $\delta^{18}O$. Although the errors are relatively large (~40% of the total shift), quantitatively they are small because of the large numbers of analyses, the relatively large lapse rate of precipitation $\delta^{18}O$ for the central Cascades (-4.4 ± 0.6 ‰/km), and the extensive modern dataset for equid vs. meteoric water $\delta^{18}O$. The paleodrainage constraint is from Conrey et al. (2002).

contribute additional uncertainties to elevation estimates. Nonetheless, the high lapse rate for the modern Cascades implies that any former lapse rates were likely lower. This effect tends to decrease calculated elevations in the past, enhancing the elevation rise subsequent to 7 Ma (Kohn and Fremd 2007). More importantly, changes in the isotopic lapse rate would not affect the observation that the *timing* of initial rapid uplift must have occurred soon after 7 Ma.

The rapid increase in elevation after 7 Ma correlates with genesis and growth of the eastern Cascades and impingement of Basin and Range extension on the arc. Although detachment of lithospheric mantle could conceivably explain the recent elevation rise (Bird 1979), a more likely explanation appeals to lithospheric "erosion" of cooler metasomatized mantle and replacement by hotter MOR-type mantle on timescales of only a few million years. This interpretation implies that short timescale changes to lithospheric mantle can occur independently of delamination, and potentially provides alternative interpretations of rapid elevation rises in other arcs. Interestingly, Columbia River Basalt magmatism between 14 and 17 Ma did not express itself on arc elevation, presumably because it was too distant (~200 km) to influence the arc directly.

Late Cretaceous and Paleogene Rockies

The timing and rate(s) of uplift of the Colorado Plateau and Rocky Mountains are of great interest geodynamically, yet presently known only poorly (e.g., Pederson et al. 2002). One difficulty in applying stable isotopes to this tectonic problem stems from complex moisture sources, specifically large distances and multiple sources. A more general problem stems from the area's complicated tectonic history. Laramide deformation as old as late Cretaceous caused high-angle reverse faulting and probably created significant topography (e.g., Burchfiel et al. 1992), but other data suggest much later plateau uplift, as late as mid/late-Miocene to Pliocene (McMillan et al. 2002; Pederson et al. 2002; Sahagian et al. 2002). Of course early block faulting and formation of high topography may have been succeeded by erosion and later rejuvenation. Understanding these processes and their tectonic causes, especially plateau uplift, requires first characterizing the early Laramide history. Several studies have investigated late Cretaceous to

Eocene topographic relief by using stable isotope compositions of fossils in different basins and at different times to investigate the stable isotope hydrology of ancient river systems in the western interior.

In one such study, Dettman and Lohmann (2000) used the average $\delta^{18}O$ value of well-preserved unionid shells and the average growth temperature to calculate the $\delta^{18}O$ of river water in this region. These data were used to infer past elevations and regional topographic variation, and are re-plotted here (Fig. 15) using the relationship of Fig. 5 to calculate water $\delta^{18}O$ instead of the average growth temperature (21 °C). The samples spanned the interval from 76.5 to 51.8 Ma. The oxygen isotope ratio of water from the late Cretaceous to the end of the Paleocene was quite negative at times and the $\delta^{18}O$ values that are less than −15‰ V-SMOW imply precipitation that has been strongly distilled, either because of cold temperatures or because of long transport from moisture sources. In the modern IAEA database of isotopes in precipitation (http://isohis.iaea.org/GNIP.asp) few $\delta^{18}O$ values less than −14‰ occur at monthly temperatures above 2 °C. At a MAT of 0 °C or less in the contiguous USA the average annual precipitation $\delta^{18}O$ value is less than −13.5 ‰ (Fricke and Wing 2004; Dutton et al. 2005). Therefore we conclude that cold temperature precipitation is involved in waters with $\delta^{18}O$ values less than −15‰ (allowing for the more negative seawater of the Cretaceous and Paleogene). The beginning of the Eocene marks a time when these very negative $\delta^{18}O$ values disappear (Dettman and Lohmann 2000; Fig. 15).

Although temperatures throughout this period fluctuated significantly there are good indicators that the basin floors were relatively warm (Pocknall 1987; Wing et al. 1995; Fricke and Wing 2004). The presence of frost sensitive flora and fauna, e.g., alligators and palm trees, in the basins imply that freezing temperatures were rare (Sloan and Barron 1992; Greenwood and Wing 1995). The very negative $\delta^{18}O$ values are unusual in warm environments and they are more surprising considering the proximity of the Cretaceous and early Paleocene interior

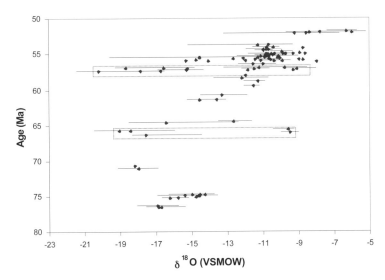

Figure 15. Water $\delta^{18}O$ calculated from the relationship of Figure 5 applied to bulk shell oxygen isotope ratios of fossils from southern Alberta, Montana, Wyoming, and northern Colorado. Diamonds represent the average of all shells from one stratigraphic horizon, lines are the range of $\delta^{18}O$ values when $n > 4$ for a shell bed. Calculated $\delta^{18}O$ values have changed very little relative to that of the original paper (Dettman et al. 2000); the most negative values move to slightly more positive $\delta^{18}O$ values, but changes are generally less than 1‰.

seaway. This suggests that the rivers carried non-local water from high elevation regions, where cold-temperature precipitation is not precluded by the paleontological data.

Strong isotopic contrasts in different rivers from the same region do not support the alternative explanation of the low $\delta^{18}O$ values as resulting from long transport from moisture sources. There are two intervals in this record when rivers with very negative and much more positive $\delta^{18}O$ values occur in close stratigraphic and spatial proximity (Fig. 15). This occurs in eastern Montana in the late Maastrichtian and in the Powder River Basin during the Paleocene. The simplest explanation is that the rivers were sourced by catchments at different elevations. One set, those with more positive $\delta^{18}O$ values, was derived from local basinal rainfall while the others were sourced by precipitation at high elevation and cold temperatures, likely snow and snowmelt. Both rivers were present on the basin floor and had a similar temperature history, but had large differences in $\delta^{18}O$.

If these two types of rivers represented high elevation vs. basinal recharge zones, the oxygen isotope difference between the two can be used to estimate the elevation difference between the floor of the Powder River Basin and the highlands surrounding the basin. At 57 Ma the spread in $\delta^{18}O$ values for river water in the Powder River Basin ranges from 9 to 11‰. One might be tempted to use an average lapse rate for meteoric precipitation of −2.9 ‰/km, implying that the Bighorn Mountains were over 3000 m higher than the basin floor. However, Drummond et al. (1993) concluded that the isotopic lapse rate in precipitation between latitudes of 38 and 48°N is −4.2 ‰/km based on data from the IAEA and Smith et al. (1979). Because the shells formed in river water at latitudes of ~45 °N, we assume a relatively large −4.2 ‰/km value, but additionally assign an uncertainty of ± 1 ‰/km acknowledging the inherent uncertainty in estimating a lapse rate for the Paleocene. Both the noisy nature of the river water $\delta^{18}O$ signal (up to ± 2‰ for isotopic differences) and the uncertainty of the isotopic lapse rate in this region during the Paleocene imply that the relief between the Bighorn Mountains and the Powder River Basin was on the order of 2400 ± 750 m. This is similar to or greater than the relief seen today, with the highest peaks 2800 m above the basin floor and much of the northern half of the Bighorns sitting 1500 m above the basin floor. The Paleocene sediments shed off the Bighorn Mountains form the Fort Union Formation and are a very thick foreland basin sediment package, which suggests that major deformation and uplift occurred during this initial mountain building event (Hoy and Ridgway 1997) and that the elevation difference between basin floor and mountain peaks may very well have exceeded that of today.

A similar calculation can be presented for the Hell Creek area, although there are no nearby mountain ranges to serve as the locus for the high elevation river source. It is most likely that the rivers with very negative $\delta^{18}O$ values of water in the Hell Creek area formed a paleo-Yellowstone River system, which carried high elevation runoff from the Rocky Mountains or perhaps from high elevation Laramide uplifts at the end of the Cretaceous. The difference between the more positive waters (−9.5‰) and the three more negative river water averages (−17.5 to −19‰) suggest an 8 to 9.5‰ contrast, which can be viewed as a ~2000 m elevation difference.

Seasonal cycles in oxygen isotope ratios in these shells support strong contrasts both in elevation and in river hydrology (Dettman et al. 2000). Among the shells with more positive $\delta^{18}O$ values there are some with small $\delta^{18}O$ cycles, for example −7 to −11‰ VPDB. This type of cycle is expected in a unionid living in a river with little seasonal variation in $\delta^{18}O$ and an average value between −8 and −10. Other shells show very large $\delta^{18}O$ amplitudes, for example −7 to −17‰ VPDB. This is consistent with a river that undergoes a large seasonal change to $\delta^{18}O$, from very negative values reflecting spring snow melt to much more positive values, perhaps established by late summer low-elevation rainfall. Finally, one shell has been measured which has very negative $\delta^{18}O$ values throughout the year, ranging from −19 to -23.3‰ VPDB. This shell grew in river water that remained close to −20 ‰VSMOW all year round. It is very likely that other shells with mean $\delta^{18}O$ values of $-20‰$ VPDB have similar cycles in $\delta^{18}O$. This last pattern suggests that high elevation recharge or permanent snowfields supported this river throughout the year. These patterns are identical to seasonal variability observed in modern river water collected in eastern Montana today (see discussion in Dettman and Lohmann 2000).

We note that Fricke (2003) agreed with the plausibility of Dettman and Lohmann's interpretations, but proposed a very different alternative explanation for the isotopic patterns in the region during the early Eocene. Based on three samples from Powder River Basin and three from the Bighorn Basin, Fricke (2003) concluded that the average river water $\delta^{18}O$ values were -7 and $-8‰$ (V-SMOW) respectively. The small $\delta^{18}O$ difference between the basins implied a small isotopic rain shadow and was used to infer low intervening elevation for the Bighorn Mountains: ~500 m or lower, compared to ~3000 m as estimated by Dettman and Lohmann (2000). The much lower $\delta^{18}O$ values measured in mollusks were proposed to reflect either seasonally cold conditions, even in warm climates and low elevation sites, or long transport distances from moisture sources. We disagree with the reinterpretation of the mollusk data for two reasons. Firstly, the large differences in *mean annual $\delta^{18}O$ values* for these rivers makes seasonal biases unlikely. Secondly, seasonal data from individual shells demonstrates that some rivers had very negative $\delta^{18}O$ values throughout the year. That is, although some rivers do appear to have been sourced at low elevation in a warm climate (Dettman and Lohmann 2000; Fricke 2003), and taken alone might be used to indicate low topography, others require seasonal and even perennial sources from high elevations, e.g., snow melt (Dettman and Lohmann 2000). The simple proximity of samples (the samples are within tens of kilometers of each other and quite close in geologic age) suggests that similar airmasses provided moisture and that the isotopic differences among and within shells resulted from precipitation originating at different local elevations. An approach similar to the mollusk study, using coeval fossil teeth with different average $\delta^{18}O$ values thought to represent different river systems, is now being used to infer late Eocene topography (Barton and Fricke 2006; Fricke 2006).

DISCUSSION

All studies to date that have endeavored to use carbonate or phosphate fossils to infer paleoelevations have focused on stable isotope trends. This principal reliance on isotopes presents potentially serious problems on two fronts—the presumed constancy of isotopic lapse rates, and the required models of the impact of global climate change on local climate, independent of local topography.

Lapse rates

Lapse rates have been discussed previously, so here we merely emphasize that the large variation in isotopic lapse rates and the failure to model isotopic enrichments accurately in the lees of mountain belts present probably the biggest problems in interpreting stable isotope records today. In the absence of direct measures of lapse rate (e.g., in southern Tibet) or isotopic depletion across a mountain belt (e.g., the western ranges of North America), errors in inferred changes to elevation can easily approach ± 50%. Until better predictive models are developed, decreasing uncertainties will depend on continued determination of lapse rates on a case-by-case basis. Consequently, younger orogens with analogous past and modern conditions are best suited for paleoaltimetry because we can measure modern lapse rates and extrapolate to the past with greater confidence. Older orogens with different boundary conditions will be investigated only with increased error. It is worth noting as well that we understand animal diets and physiologies better for younger fossils, again minimizing errors in isotope models (Kohn 1996).

Paleoclimate

A critical reader will recognize that in most cases we are measuring a climate proxy and assuming that local climate is dominated by mountain ranges and/or plateaux. Although perhaps true in many cases, the climate link presents a major complication in applying stable isotope methods because global climate change can also affect both the mean values and seasonal variations in precipitation and its oxygen isotope compositions. For example, cooling causes generally drier and more seasonal climates, and most paleoclimatically-based coefficients of precipitation $\delta^{18}O$ vs. temperature also imply that $\delta^{18}O$ values will decrease. Thus, during global cooling, $\delta^{18}O$ values may both decrease overall and increase their variation, potentially masquerading as an increase in elevation or an increase in plateau size respectively. In practice this means that one must model climate effects, if even semiquantitatively, to rule out global climate change as a driver of stable isotope trends. For example, interpretations that rely on isotope depletions at elevation or in leeward sides of ranges can be improved if a simultaneous measure of isotope compositions from low-elevation windward sites (i.e., nearer moisture sources) can be measured as a monitor of global climate change effects. This caution applies to *all* isotope records (not just fossils) that rely on water compositions to infer elevation.

Climate change drives isotopic changes, so retrieval of local tectonic effects necessitates subtracting influences from global climate change. Although simple models have been developed to predict and subtract effects of global climate change (Kohn et al. 2002, 2004, 2007; Kohn and Fremd 2007; see above), in reality the response of precipitation $\delta^{18}O$ to climatic changes is not well understood because it reflects not only temperature (as modeled), but also rainout histories and moisture sources. In most cases, global climate change appears to have minor influence, but without direct verification, this assumption remains problematic.

Potential improvements to paleoaltimetry

Besides predictive models, another means of improving paleoelevation estimates relies on obtaining additional climate proxies. In the case of paleosol carbonate, the advent of $\Delta47$ isotopic measurements retrieves both temperature and $\delta^{18}O$ (Ghosh et al. 2006; Quade et al. 2007), permitting the most accurate resolution of elevations through stable isotope methods. In principle, such analysis could be conducted using fossil shells or the CO_3^{2-} component of fossil bone, although for bone, prior calibration is required (and currently under investigation).

The possibility of simultaneous retrieval of mean annual temperatures (MATs) improves paleoelevation interpretations in two ways. Firstly, rather than relying on an *isotopic* lapse rate to infer elevation, the *temperature* lapse rate can also be used. Secondly, paleotemperatures help separate the influences of rain shadows, elevation, and aridity on $\delta^{18}O$ values. In most tectonically active areas, any particular site will be influenced by elevations both at the site and upwind. In some cases it may be difficult to determine which factor is most important. In rain shadows aridity may also play a significant role, driving oxygen isotope ratios to more positive values. For example, the paleoelevation history of Tibet plays a major role in determining how Tibetan lithosphere evolved thermomechanically and chemically, but the stable isotope story is complex. Did $\delta^{18}O$ values for carbonates as low as $-20\%o$ (V-PDB) in the interior of Tibet between 40 and 15 Ma (Currie et al. 2005; Rowley and Currie 2006) really result from high elevations in the Tibetan interior, or simply rainout over a high front, e.g., the Himalaya? Many mountain ranges today produce diminished $\delta^{18}O$ values in their lees, irrespective of leeward elevations. Conversely, did relatively high $\delta^{18}O$ values for middle Eocene carbonates ($\sim-11\%o$, V-PDB) in the Hoh Xil basin in the northern half of the Tibetan Plateau (~5000 m current elevation) really result from low elevations (Cyr et al. 2005), or simply from aridity and different moisture sources? Elevations across the modern plateau are nearly constant, yet summer precipitation and river water $\delta^{18}O$ values *increase* by $\sim10\%o$ from just north of the Himalayan front to the Hoh Xil basin (Tian et al. 2001) because of evaporative enrichment and changes in moisture sources. In both cases, the stable isotope record alone poorly distinguishes endmembers.

In contrast, temperature records distinguish well between low $\delta^{18}O$ values arising from truly high elevations vs. those arising from rainshadows. The temperature lapse rate is well established and less variable than isotopic lapse rates, and higher vs. lower elevation sites in a mountain range exhibit quite different MATs. In contrast, two sites at similar elevation (and latitude) on either side of a mountain range exhibit quite *similar* mean annual temperatures, although their isotope compositions and mean annual ranges of temperature (MARTs) may differ. To demonstrate this, we tabulated meteorological data for sites in the western United States, where precipitation travels mainly east-west, and is opposed by the north-south trending Coast Ranges, Sierra Nevada, and Cascade Range (Table 1). We recognize the complexity of weather systems that contribute moisture to the Great Basin, including primary moisture sources from the Gulf of California and the Gulf of Mexico, and fronts that skirt the highest elevations (Friedman et al. 2002a). But we believe our comparisons illustrate important principles, even if they are not completely comprehensive. Most importantly, despite radically different precipitation patterns and $\delta^{18}O$ values, similar elevation sites on either side of these ranges have MATs within 1-2 °C of each other. Thus, although $\delta^{18}O$ values alone could not demonstrate unequivocally whether an interior site is at high elevation or in a rainshadow, combination of these same $\delta^{18}O$ values with temperature information could show not only that a mountain range intervened, but also an estimate of the height of that range. We believe that combination of paleotemperatures with oxygen isotope compositions will ultimately improve interpretations of climate, paleoelevation, and tectonics.

Table 1. Meteorological data for sites at similar elevations but on different sides of a rainshadow.

Windward Site	Latitude (°N)	Elevation (m)	MAT (°C)	Precip. (cm)	MART (°C)	
Kern R., CA	35.47	825	16.6	33	19.6	
Yosemite, CA	37.45	1210	12.1	95	19.4	
Sierra City, CA	39.34	1280	11.1	162	17.8	
Cave Junction, OR	42.10	395	12.1	155	16.8	
Marion Forks, OR	44.36	755	8.0	178	17.4	
Gov't Camp & Three Lynx, OR	45.13	775	7.8	199	15.4	
Kid Valley, WA	46.22	210	9.7	154	14.7	
Carbon River, WA	47.00	527	7.5	179	15.3	
Diablo Dam, WA	48.43	271	9.2	191	17.9	

Leeward Site	Latitude (°N)	Elevation (m)	MAT (°C)	Precip. (cm)	MART (°C)	ΔMAT (°C)
Pahrump, NV	36.17	825	16.8	12	23.3	0.2
Bishop, CA	37.22	1255	13.3	13	21.9	1.2
Reno, NV	39.30	1340	10.3	19	21.0	−0.8
Medford, OR	42.20	395	11.9	50	18.4	−0.2
Metolius, OR	44.35	760	8.6	26	18.4	0.6
Friend, OR	45.21	745	8.4	41	19.6	0.6
Sunnyside, WA	46.19	230	11.4	18	22.7	1.7
Ellensburg, WA	47.02	525	8.5	23	25.3	1.0
Tonasket, WA	48.46	295	9.6	29	23.8	0.4

Note: ΔMAT is determined from the MAT(Leeward site) − MAT(Windward site). Elevations are rounded to nearest 5 m. MART is estimated from the maximum temperature difference in mean monthly temperatures (usually January and July). Data from *www.wrcc.dri.edu.*

ACKNOWLEDGMENTS

This material is based upon work supported the National Science Foundation under Grant Nos. EAR 9909568, EAR 0304181, and ATM 0400532 (to MJK) and EAR 9510033, EAR 9627766, and ATM 0354762 (to DLD). Ted Fremd is thanked for his longstanding support in providing vertebrate fossils, discussing paleoclimate and paleotopography, and facilitating collecting. Reviews by Henry Fricke and Bruce MacFadden helped focus discussion and improved the manuscript substantially.

REFERENCES

Araguas-Araguas L, Froelich K, Rozanski K (1998) Stable isotope composition of precipitation over southeast Asia. J Geophys Res 103:28721-28742
Awashi N, Prasad M (1989) Siwalik plant fossils from Surai Khola area, western Nepal. Paleobotanist 38:298-318
Ayliffe LK, Chivas AR (1990) Oxygen isotope composition of the bone phosphate of Australian kangaroos: potential as a palaeoenvironmental recorder. Geochim Cosmochim Acta 54:2603-2609
Ayliffe LK, Chivas AR, Leakey MG (1994) The retention of primary oxygen isotope compositions of fossil elephant skeletal phosphate. Geochim Cosmochim Acta 58:5291-5298
Ayliffe LK, Lister AM, Chivas AR (1992) The preservation of glacial-interglacial climatic signatures in the oxygen isotopes of elephant skeletal phosphate. Palaeogeog, Palaeoclim, Palaeoecol 99:179-191
Balasse M (2002) Reconstructing dietary and environmental history from enamel isotopic analysis: time resolution of intra-tooth sequential sampling. Int J Osteoarch 12:155-165
Bandel K (1990) Shell structure of the Gastropoda excluding Archaeogastropoda. *In:* Skeletal Biomineralization: Patterns, Processes and Evolutionary Trends, Vol. 1. Carter JG (ed) Van Nostrand Reinhold, New York, p 117-134
Barrick R (1998) Isotope paleobiology of the vertebrates: ecology physiology, and diagenesis. Paleontological Society Papers 4:101-137
Barrick RE, Showers WJ (1994) Thermophysiology of *Tyrannosaurus rex*: Evidence from oxygen isotopes. Science 265:222-224
Barrick RE, Showers WJ (1995) Oxygen isotope variability in juvenile dinosaurs (*Hypacrosaurus*); evidence for thermoregulation. Paleobiology 21:552-560
Barrick RE, Showers WJ (1999) Thermophysiology and biology of *Giganotosaurus*; comparison with *Tyrannosaurus*. Palaeontologia Electronica, *http://www.palaeo-electronica.org/1999_2/gigan/issue2_99.htm*
Barrick RE, Showers WJ, Fischer AG (1996) Comparison of thermoregulation of four ornithischian dinosaurs and a varanid lizard from the Cretaceous Two Medicine Formation; evidence from oxygen isotopes. Palaios 11:295-305
Barrick RE, Stoskopf MK, Marcot JD, Russell DA, Showers WJ (1998) The thermoregulatory functions of the *Triceratops* frill and horns; heat flow measured with oxygen isotopes. J Vert Paleo 18:746-750
Barton MA, Fricke HC (2006) Mapping topographic relief and elevation of the central Rocky Mountain region during the latest Eocene using stable isotope data from mammalian tooth enamel. Geol Soc Am Abstr Prog 38:202
Bird P (1979) Continental delamination and the Colorado Plateau. J Geophys Res 84:7561-7571
Blisniuk PJ, Stern LA, Chamberlain CP, Idleman B, Zeitler PK (2005) Climatic and ecologic changes during Miocene surface uplift in the Southern Patagonian Andes. Earth Planet Sci Lett 230:125-142
Blisniuk PM, Stern LA (2006) Stable isotope paleoaltimetry: a critical review. Am J Sci 305:1033-1074
Bocherens K, Koch PL, Mariotti A, Geraads D, Jaeger JJ (1996) Isotopic biogeochemistry (^{13}C, ^{18}O) of mammalian enamel from African Pleistocene hominid sites. Palaios 11:306-318
Botha J, Lee-Thorp J, Chinsamy A (2005) The paleoecology of the non-mammalian cynodonts Diademodon and Cynognathus from the Karoo Basin of South Africa, using stable light isotope analysis. Palaeogeog Palaeoclim Palaeoec 223:303-316
Bryant JD, Koch PL, Froelich PN, Showers WJ, Genna BJ (1996) Oxygen isotope partitioning between phosphate and carbonate in mammalian apatite. Geochim Cosmochim Acta 60:5145-5148
Burchfiel BC, Cowan DS, Davis GA (1992) Tectonic overview of the Cordilleran orogen in the western United States. *In:* The Cordilleran Orogen: Conterminous U.S. Burchfiel BC, Lipman PW, Zoback MC (eds) Geological Society of America, Boulder, CO, p 407-479
Carter JG (1980) Environmental and biological controls of bivalve shell mineralogy and microstructure. *In:* Skeletal Growth of Aquatic Organisms. Rhoads DC, Lutz RA (ed) Plenum, New York, pp 69-113

Chamberlain CP, Poage MA (2000) Reconstructing the paleotopography of mountain belts from the isotopic composition of authigenic minerals. Geology 28:115-118

Chivas AR, DeDeckker P, Wang SX, Cali JA (2002) Oxygen-isotope systematics of the nektic ostracod *Australocypris robusta. In:* The Ostracoda: Applications in Quaternary Research. Holmes JA, Chivas AR (ed) AGU Geophysical Monograph 131:301-313

Clementz MT, Koch PL (2001) Differentiating aquatic mammal habitat and foraging ecology with stable isotopes in tooth enamel. Oecologia 129:461-472

Conrey RM, Grunder AL, Schmidt M (2004) State of the Cascade Arc: stratocone persistence, mafic lava shields, and pyroclastic volcanism associated with intra-arc rift propagation, Oregon Department of Geology and Mineral Industries OFR O-04-04, Portland, OR, p 39

Conrey RM, Taylor EM, Donnelly-Nolan JM, Sherrod DR (2002) North-central Oregon Cascades: exploring petrologic and tectonic intimacy in a propagating intra-arc rift. *In:* Field Guide to Geologic Processes in Cascadia, vol 36. Moore GW (ed) Oregon Department of Geology and Mineral Industries, Portland, OR, p 47-90

Coplen TB, Kendall C (2000) Stable Hydrogen and Oxygen Isotope Ratios for Selected Sites of the U.S. Geological Survey's NASQAN and Benchmark Surface-water Networks. USGS Open-File Report 00-160

Craig H, Gordon LI (1965) Deuterium and oxygen-18 variations in ocean and the marine atmosphere. *In:* Proceedings of a conference on stable isotopes in oceanographic studies and paleotemperatures, Spoleto, Italy. Tongiorgi E (ed) p 9-130

Cummings ML, Evans JG, Ferns ML, Lees KR (2000) Stratigraphic and structural evolution of the middle Miocene synvolcanic Oregon-Idaho graben. Geol Soc Am Bull 112:668-682

Currie BS, Rowley DB, Tabor NJ (2005) Middle Miocene paleoaltimetry of southern Tibet: Implications for the role of mantle thickening and delamination in the Himalayan orogen. Geology 33:181-184

Curtis JH, Brenner M, Hodell DA, Balser RA, Islebe GA, Hooghiemstra H (1998) A multi-proxy study of Holocene environmental change in the Maya Lowlands of Peten, Guatemala. J Paleolimnology 19:139-159

Cyr AJ, Currie BS, Rowley DB (2005) Geochemical evaluation of Fenghuoshan Group lacustrine carbonates, north-central Tibet: implications for the paleoaltimetry of the Eocene Tibetan Plateau. J Geol 113:517-533

D'Angela D, Longinelli A (1990) Oxygen isotopes in living mammal's bone phosphate: Further results. Chem Geol 86:75-82

Dansgaard W (1964) Stable isotopes in precipitation. Tellus 16:436-468

De Deckker P (1981) Ostracods of athalassic saline lakes, a review. Hydrobiol 81:131-144

Delgado Huertas A, Iacumin P, Stenni B, Sanchez Chillon B, Longinelli A (1995) Oxygen isotope variations of phosphate in mammalian bone and tooth enamel. Geochim Cosmochim Acta 59:4299-4305

Dettman DL (2002) Refining the calibration of stable isotope paleo-elevation studies: Lessons and strategies from modern precipitation patterns. Geol Soc Am Abstr Prog 34:61

Dettman DL, Flessa KW, Roopnarine PD, Schöne BR, Goodwin DH (2004) The use of oxygen isotope variation in the shells of estuarine mollusks as a quantitative record of seasonal and annual Colorado River discharge. Geochim Cosmochim Acta 68:1253-1263

Dettman DL, Kohn MJ, Quade J, Ryerson FJ, Ojha TP, Hamidullah S (2001) Seasonal stable isotope evidence for a strong Asian monsoon throughout the past 10.7 m.y. Geology 29:31-34

Dettman DL, Lohmann KC (1995) Approaches to microsampling carbonates for stable isotope and minor element analysis: Physical separation of samples on a 20 micrometer scale. J Sedimentary Petrol A65:566-569

Dettman DL, Lohmann KC (2000) Oxygen isotope evidence for high altitude snow in the Laramide Rocky Mountains of North America during the late Cretaceous and Paleogene. Geology 28:243-246

Dettman DL, Palacios-Fest MR, Cohen AS (2002) Comment on G. Wansard and F. Mezquita, The response of ostracode shell chemistry to seasonal change in a Mediterranean freshwater spring environment. J Paleolimnology 27:487-491

Dettman DL, Palacios-Fest MR, Nkotagu H, Cohen AS (2005) Paleolimnological investigations of anthropogenic environmental change in Lake Tanganyika: VII, Carbonate isotope geochemistry as a record of riverine runoff. J Paleolimnology 34:93-105

Dettman DL, Reische AK, Lohmann KC (1999) Controls on the stable isotope composition of seasonal growth bands in aragonitic fresh-water bivalves (unionidae). Geochim Cosmochim Acta 63:1049-1057

Dettman DL, Smith AJ, Rea DK, Lohmann KC, Moore TC Jr (1995) Glacial melt-water in Lake Huron during early post-glacial times as inferred from single-valve analysis of oxygen isotopes in ostracodes. Quat Res 43:297-310

Dickinson JE, Land M, Faunt CC, Leake SA, Reichard EG, Fleming JB Pool DR (2006) Hydrogeologic Framework Refinement, Ground-Water Flow and Storage, Water Chemistry Analyses, and Water-Budget Components of the Yuma Area, Southwestern Arizona and Southeastern California. USGS Scientific Investigations Report 2006-5135

Driessens FCM, Verbeeck RMH (1990) Biominerals, CRC Press, Boca Raton

Drummond CN, Patterson WP, Walker JCG (1995) Climatic forcing of carbon-oxygen isotopic covariance in temperate-region marl lakes. Geology 23:1031-1034

Drummond CN, Wilkinson BH, Lohmann KC, Smith GR (1993) Effect of regional topography and hydrology on the lacustrine isotopic record of Miocene paleoclimate in the Rocky Mountains. Palaeogeog Palaeoclim Palaeoecol 101:67-79

Dutton A, Wilkinson BH, Welker JW, Bowen GJ, Lohmann KC (2005) Spatial distribution and seasonal variation in $^{18}O/^{16}O$ of modern precipitation and river water across the coterminous USA. Hydrol Proc 19:4121-4146

Easterling DR, Karl TR, Lawrimore JH, del Greco SA (1999) United States Historical Climatology Network daily temperature, precipitation, and snow data for 1871-1997. Oak Ridge National Laboratory, doc # ORNL/CDIAC-118, NDP-070

Elliott JC (2002) Calcium phosphate biominerals. Rev Mineral Geochem 48:427-453

England P, Molnar P (1990) Surface uplift, uplift of rocks, and exhumation of rocks. Geology 18:1173-1177

Estes RD (1991) The Behavior Guide to African Mammals, University of California Press, Berkeley

Fremd TJ, Bestland EA, Retallack GJ (1997) John Day Basin Paleontology, Northwest Interpretive Association, Seattle, WA, p 80

Fricke DC, Rogers RR (2000) Multiple taxon-multiple locality approach to providing oxygen isotope evidence for warm-blooded theropod dinosaurs. Geology 28:799-802

Fricke HC (2003) $\delta^{18}O$ of geographically widespread mammal remains as a means of studying water vapor transport and mountain building during the early Eocene. Geol Soc Am Bull 115:1088-1096

Fricke HC (2006) Looking for the edge of the Rocky Mountains during the Late Eocene: a stable isotope map from mammalian tooth enamel. EOS 87:T33C-0525

Fricke HC, Clyde WC, O'Neil JR (1998) Intra-tooth variations in $\delta^{18}O(PO_4)$ of mammalian tooth enamel as a record of seasonal variations in continental climate variables. Geochim Cosmochim Acta 62:1839-1850

Fricke HC, O'Neil JR (1996) Inter-and intra-tooth variation in the oxygen isotope composition of mammalian tooth enamel phosphate; implications for palaeoclimatological and palaeobiological research. Palaeogeog Palaeoclim Palaeoecol 126:91-99

Fricke HC, Wing SL (2004) Oxygen isotope and paleobotanical estimates of temperature and $\delta^{18}O$-latitude gradients over North America during the early Eocene. Am J Sci 304:612-635

Friedman GM (1977) Identification by staining methods of minerals in carbonate rocks. *In:* Subsurface Geology. LeRoy LW, Leroy DO (ed) Colorado School of Mines, Golden, CO, p 96-97

Friedman I, Harris JM, Smith GI, Johnson CA (2002a) Stable isotope composition of waters in the Great Basin, United States 1. Air-mass trajectories. J Geophys Res 107:ACL 14-11 - ACL 14-14

Friedman I, O'Neil JR (1977) Compilation of stable isotope fractionation factors of geochemical interest. *In:* Data of Geochemistry. Fleischer M (ed) USGS Prof. Paper 440-KK

Friedman I, Smith GI, Gleason JD, Warden A, Harris JM (1992) Stable isotope compositions of waters in southeastern California; I, Modern precipitation. J Geophys Res 97:5795-5812

Friedman I, Smith GI, Johnson CA, Moscati RJ (2002b) Stable isotope composition of waters in the Great Basin, United States 2. Modern Precipitation. J Geophys Res 107:ACL 15-11 - ACL 15-21

Garzione CN, Dettman DL, Quade J, DeCelles PG, Butler RF (2000) High times on the Tibetan Plateau; paleoelevation of the Thakkhola Graben, Nepal. Geology 28:339-342

Ghosh P, Garzione CN, Eiler JM (2006) Rapid uplift of the Altiplano revealed through $^{13}C-^{18}O$ bonds in paleosol carbonates. Science 311:511-515

Goodfriend GA (1992) The use of land snail shells in paleoenvironmental reconstruction. Quat Sci Rev 11:665-685

Greenwood DR, Wing SL (1995) Eocene continental climates and latitudinal temperature gradients. Geology 23:1044-1048

Grossman EL, Ku T-L (1986) Oxygen and carbon isotope fractionation in biogenic aragonite: temperature effects. Chem Geol 59:59-74

Harkness RE, Berkas WR, Norbeck SW, Robinson SM (2000) Water Resources Data, North Dakota, Water Year 1999, Vol. 1 Surface Water. USGS Water Resources of North Dakota ND-99-1

Harrison TM, Copeland P, Kidd WSF, Yin A (1992) Raising Tibet. Science 255:1663-1670

Holmes JA, Chivas AR (2002) Ostracod shell chemistry – Overview. *In:* The Ostracoda: Applications in Quaternary Research. Holmes JA, Chivas AR (ed) AGU Geophysical Monograph 131:185-204

Hoppe KA (2006) Correlation between the oxygen isotope ratio of North American bison teeth and local waters: Implications for paleoclimatic reconstructions. Earth Planet Sci Lett 244:408-417

Howard AD (1921) Experiments in the culture of fresh-water mussels. Bull U.S. Bureau Fisheries 38:63-90

Hoy RG, Ridgway KD (1997) Structural and sedimentological development of footwall growth synclines along an intraforeland uplift, east-central Bighorn Mountains, Wyoming. Geol Soc America Bull 109:915-935

Iacumin P, Bocherens H, Mariotti A, Longinelli A (1996) Oxygen isotope analyses of co-existing carbonate and phosphate in biogenic apatite; a way to monitor diagenetic alteration of bone phosphate? Earth Planet Sci Lett 142:1-6

IAEA/WMO (2004) Global network of isotopes in Precipitation. The GNIP database. *http://isohis.iaea.org*

Ingraham NL, Caldwell EA, Verhagen BT (1998) Arid Catchments. *In:* Isotope Tracers in Catchment Hydrology. Kendall C, McDonnell JJ (eds) Elsevier, Amsterdam, p 435-465

Ingraham NL, Taylor BE (1991) Light stable isotope systematics of large-scale hydrologic regimes in California and Nevada. Water Resources Res 27:77-90

Johnsen SJ, Dansgaard W, White JWC (1989) The origin of Arctic precipitation under present and glacial conditions. Tellus 41B:452-468

Kierdorf U, Kierdorf H (2000) Comparative analysis of dental fluorosis in roe deer (*Capreolus capreolus*) and red deer (*Cervus elaphus*): interdental variation and species differences. J Zool 250:87-93

Kim ST, O'Neil JR (1997) Equilibrium and nonequilibrium oxygen isotope effects in synthetic carbonates. Geochim Cosmochim Acta 61:3461-3475

Koch PL (1998) Isotopic reconstruction of past continental environments. Ann Rev Earth Planet Sci 26:573-613

Koch PL, Behrensmeyer AK, Tuross N, Fogel ML (1991) Isotopic fidelity during bone weathering and burial. Carnegie Inst Wash Yearbook:105-110

Koch PL, Fisher DC, Dettman D (1989) Oxygen isotope variation in the tusks of extinct proboscideans: A measure of season of death and seasonality. Geology 17:515-519

Koch PL, Tuross N, Fogel ML (1997) The effects of sample treatment and diagenesis on the isotopic integrity of carbonate in biogenic hydroxylapatite. J Arch Sci 24:417-429

Kohn M (2004) Comment: Tooth enamel mineralization in ungulates: Implications for recovering a primary isotopic time-series, by B. H. Passey and T. E. Cerling (2002). Geochim Cosmochim Acta 68:403-405

Kohn MJ (1996) Predicting animal $\delta^{18}O$: Accounting for diet and physiological adaptation. Geochim Cosmochim Acta 60:4811-4829

Kohn MJ (2006) REE and U zoning in fossil teeth. Geol Soc Am Abstr Prog 38:46

Kohn MJ, Cerling TE (2002) Stable isotope compositions of biological apatite. Rev Mineral Geochem 48:455-488

Kohn MJ, Fremd TJ (2007) Tectonic controls on isotope compositions and species diversification, John Day Basin, central Oregon. PaleoBios (in press)

Kohn MJ, Josef JA, Madden R, Kay R, Vucetich G, Carlini AA (2004) Climate stability across the Eocene-Oligocene transition, southern Argentina. Geology 32:621-624

Kohn MJ, Law JM (2006) Stable isotope chemistry of fossil bone as a new paleoclimate indicator. Geochim Cosmochim Acta 70:931-946

Kohn MJ, Miselis JL, Fremd TJ (2002) Oxygen isotope evidence for progressive uplift of the Cascade Range, Oregon. Earth Planet Sci Lett 204:151-165

Kohn MJ, Schoeninger MJ, Barker WW (1999) Altered states: effects of diagenesis on fossil tooth chemistry. Geochim Cosmochim Acta 18:2737-2747

Kohn MJ, Schoeninger MJ, Valley JW (1996) Herbivore tooth oxygen isotope compositions: Effects of diet and physiology. Geochim Cosmochim Acta 60:3889-3896

Kohn MJ, Schoeninger MJ, Valley JW (1998) Variability in herbivore tooth oxygen isotope compositions: reflections of seasonality or developmental physiology? Chem Geol 152:97-112

Kohn MJ, Zanazzi A, Josef JA (2007) Stable isotopes of fossil teeth and bones at Gran Barranca as a monitor of climate change and tectonics. *In:* The Paleontology of Gran Barranca: Evolution and Environmental Change through the Middle Cenozoic of Patagonia. Madden RH, Vucetich GM, Carlini AA, Kay R (eds) Cambridge University Press, Cambridge (in press)

Kouwenberg LLR, Kürschner WM, McElwain JC (2007) Stomatal frequency change over altitudinal gradients: prospects for paleoaltimetry. Rev Mineral Geochem 66:215-242

Kroopnick P, Craig H (1972) Atmospheric oxygen: isotopic composition and solubility fractionation. Science 175:54-55

LeColle P (1985) The oxygen isotope composition of landsnail shells as a climatic indicator: Applications to hydrogeology and paleoclimatology. Chem Geol 58:157-181.

Levin NE, Cerling TE, Passey BH, Harris JM, Ehleringer JR (2006) A stable isotope aridity index for terrestrial environments. Proc Natl Acad Sci 103:11201-11205

Riihimaki CA, Libarkin JC (2007) Terrestrial cosmogenic nuclides as paleoaltimetric proxies. Rev Mineral Geochem 66:269-278

Luz B, Cormie AB, Schwarcz HP (1990) Oxygen isotope variations in phosphate of deer bones. Geochim Cosmochim Acta 54:1723-1728

Luz B, Kolodny Y (1985) Oxygen isotope variations in phosphate of biogenic apatites, IV. Mammal teeth and bones. Earth Planet Sci Lett 75:29-36

Masson V, Vimeux F, Jouzel J, Morgan V, Delmotte M, Ciais P, Hammer C, Johnsen S, Lipenkov VY, Mosley-Thompson E, Petit JR, Steig EJ, Stievenard M, Vaikmae R (2000) Holocene climate variability in Antarctica based on 11 ice-core isotopic records. Quat Res 54:348-358

McCabe GJ, Palecki MA, Betancourt JL (2004) Pacific and Atlantic Ocean influences on multidecadal drought frequency in the United States. Proc Natl Acad Sci 101:4136-4141

McKenzie D (1978) Some remarks on the development of sedimentary basins. Earth Planet Sci Lett 40:25-32

McMillan MD, Angevine CL, Heller PL (2002) Postdepositional tilt of the Miocene-Pliocene Ogallala Group on the western Great Plains: Evidence of late Cenozoic uplift of the Rocky Mountains. Geology 30:63-66

McNab BK (2002) The Physiological Ecology of Vertebrates. A View from Energetics. Cornell University Press, Ithaca

Meyer HW (2007) A review of paleotemperature–lapse rate methods for estimating paleoelevation from fossil floras. Rev Mineral Geochem 66:155-171

Millard AR, Hedges REM (1996) A diffusion-adsorption model of uranium uptake by archaeological bone. Geochim Cosmochim Acta 60:2139-2152

Molnar P, England P, Martinod J (1993) Mantle dynamics, uplift of the Tibetan Plateau, and the Indian monsoon. Rev Geophys 31:357-396

Mulch A, Chamberlain CP (2007) Stable isotope paleoaltimetry in orogenic belts – the silicate record in surface and crustal geological archives. Rev Mineral Geochem 66:89-118

Negus CL (1966) A quantitative study of growth and production of unionid mussels in the River Thames at Reading. J Animal Ecol 35:513-532

Nickel E (1978) The present status of cathode luminescence as a tool in sedimentology. MineralsSci Eng 10:73-100

Passey BH, Cerling TE (2002) Tooth enamel mineralization in ungulates: implications for recovering a primary isotopic time-series. Geochim Cosmochim Acta 66:3225-3234

Pasteris JD, Wopenka B, Freeman JJ, Rogers K, Valsami-Jones E, van der Houwen JAM, Silva MJ (2004) Lack of OH in nanocrystalline apatite as a function of degree of atomic order: implications for bone and biomaterial. Biomatls 25:229-238

Pederson JL, Mackley RD, Eddleman JL (2002) Colorado Plateau uplift and erosion evaluated using GIS. GSA Today 12:4-10

Poage MA, Chamberlain CP (2001) Empirical relationships between elevation and the stable isotope composition of precipitation and surface waters: considerations for studies of paleoelevation change. Am J Sci 301:1-15

Pocknall DT (1987) Paleoenvironments and age of the Wasatch Formation (Eocene), Powder River Basin, Wyoming. Palaios 2:368-376

Quade J, Cater ML, Ojha TP, Adam J, Harrison TM (1995) Late Miocene environmental change in Nepal and the northern Indian subcontinent: Stable isotopic evidence from paleosols. Geol Soc Am Bull 107:1381-1397

Quade J, Cerling TE (1995) Expansion of C-4 grasses in the late Miocene of northern Pakistan - evidence from stable isotopes in paleosols, Palaeogr Palaeocl Palaeoecol 115:91-116

Quade J, Cerling TE, Barry JC, Morgan ME, Pilbeam DR, Chivas AR, Leethorp JA, Vandermerwe NJ (1992) A 16-Ma record of paleodiet using carbon and oxygen isotopes in fossil teeth from Pakistan. Chem Geol 94:183-192

Quade J, Cerling TE, Bowman JR (1989) Development of Asian monsoon revealed by marked ecological shift during the latest Miocene in northern Pakistan. Nature 342:163-166

Quade J, Garzione C, Eiler J (2007) Paleoelevation reconstruction using pedogenic carbonates. Rev Mineral Geochem 66:53-87

Ricken W, Steuber T, Freitag H, Hirschfeld M, Niedenzu B (2003) Recent and historical discharge of a large European river system – oxygen isotope composition of river water and skeletal aragonite of Unionidae in the Rhine. Palaeogeog Palaeoclim Palaeoecol 193:73-86

Roden JS, Ehleringer JR (1999) Observations of hydrogen and oxygen isotopes in leaf water confirm the Craig-Gordon model under wide-ranging environmental conditions. Plant Physiol 120:1165-1173

Rowley DB (2007) Stable isotope-based paleoaltimetry: theory and validation. Rev Mineral Geochem 66:23-52

Rowley DB, Currie BS (2006) Paleo-altimetry of the late Eocene to Miocene Lunpola basin, central Tibet. Nature 439:677-681

Rowley DB, Garzione CN (2007) Stable isotope-based paleoaltimetry. Ann Rev Earth Planet Sci 35:463-508

Rowley DB, Pierrehumbert RT, Currie BS (2001) A new approach to stable isotope-based paleoaltimetry; implications for paleoaltimetry and paleohypsometry of the High Himalaya since the late Miocene. Earth Planet Sci Lett 188:253-268

Royden L, Keen CE (1980) Rifting process and thermal evolution of the continental margin of eastern Canada determined from subsidence curves. Earth Planet Sci Lett 51:343-361

Rozanski K, Araguas-Araguas L, Gonfiantini R (1993) Isotopic patterns in modern global precipitation. *In*: Climate change in continental isotopic records. Swart PK, Lohmann KC, McKenzie JA, Savin S (eds) American Geophysical Union, Washington, p 1-36

Sahagian D, Proussevitch A (2007) Paleoelevation measurement on the basis of vesicular basalts. Rev Mineral Geochem 66:195-213Sarkar S (1989) Siwalik pollen succession from Surai Khola of western Nepal and its reflection on paleoecology. Paleobotanist 38:319-324

Schwalb A (2003) Lacustrine ostracodes as stable isotope recorders of late-glacial and Holocene environmental dynamics and climate. J Paleolimn 29:265-351

Schwalb A, Lister GS, Kelts K (1994) Ostracode carbonate $\delta^{18}O$ and $\delta^{13}C$ signatures of hydrological and climatic changes affecting Lake Neuchatel, Switzerland, since the latest Pleistocene. J Paleolimn 11:3-17

Shanahan TM, Pigati JS, Dettman DL, Quade J (2005) Isotopic variability in biogenic aragonite of freshwater gastropods living in near-constant temperature and $\delta^{18}O$ springs. Geochim Cosmochim Acta 69:3949-3966

Sheppard SMF, Nielsen RL, Taylor HP Jr (1969) Oxygen and hydrogen isotope ratios of clay minerals from porphyry copper deposits. Econ Geol 64:755-777

Shotwell JA (1968) Miocene Mammals of Southeast Oregon. University of Oregon, Eugene Oregon, p 67

Sloan LC, Barron EJ (1992) A comparison of Eocene climate model results to quantified paleoclimatic interpretations. Palaeogeog Palaeoclim Palaeoecol 93:183-202

Smith GA, Snee LW, Taylor EM (1987) Stratigraphic, sedimentologic, and petrologic record of late Miocene subsidence of the central Oregon High Cascades. Geology 15:389-392

Smith GI, Friedman I, Klieforth H, Hardcastle K (1979) Areal distribution of deuterium in eastern California precipitation. J Applied Meteorology 18:172-188

Stanton Thomas KJ, Carlson SJ (2004) Microscale $\delta^{18}O$ and $\delta^{13}C$ isotopic analysis of an ontogenetic series of the hadrosaurid dinosaur Edmontosaurus: implications for physiology and ecology. Palaeogeog Palaeoclim Palaeoec 206:257-287

Stern LA, Blisniuk PM (2002) Stable isotope composition of precipitation across the southern Patagonian Andes. J Geophys Res 107: doi:10.1029/2002JD002509

Sternberg LSL (1989) Oxygen and hydrogen isotope ratios in plant cellulose: mechanisms and applications. In: Rundel PW, Ehleringer JR, Nagy KA (eds) Stable Isotopes in Ecological Research, Ecological Studies Board, Los Angeles

Talbot MR (1990) A review of the palaeohydrological interpretation of carbon and oxygen isotopic ratios in primary lacustrine carbonates. Chem Geol 80:261-279

Tian L, Masson-Delmotte V, Stievenard M, Yao T, Jouzel J (2001) Tibetan Plateau summer monsoon northward extent revealed by measurements of water stable isotopes. J Geophys Res 106:28081-28088

Tian L, Yao T, Schuster PF, White JWC, Ichiyanagi K, Pendall E, Pu J, Yu W (2003) Oxygen-18 concentrations in recent precipitation and ice cores on the Tibetan Plateau. J Geophys Res 108: doi:10.1029/2002JD002173

Trueman CN, Tuross N (2002) Trace elements in recent and fossil bone apatite. Rev Mineral Geochem 48:489-521

Trueman CNG, Behrensmeyer AK, Tuross N, Weiner S (2004) Mineralogical and compositional changes in bones exposed on soil surfaces in Amboseli National Park, Kenya: diagenetic mechanisms and the role of sediment pore fluids. J Archaeol Sci 31:721-739

Tuross N, Behrensmeyer AK, Eanes ED (1989) Strontium increases and crystallinity changes in taphonomic and archaeological bone. J Archaeol Sci 16:661-672

Turpen JB, Angell RW (1971) Aspects of molting and calcification in the ostracod *Heterocypris*. Biol Bull 140:331-338

US Geol Surv (1982) Water Resources Data, Florida, Water Year 1981, Vol 2A, South Florida Surface Water. USGS/WRD/HD-82/020

Vennemann TW, Fricke HC, Blake RE, O'Neil JR, Colman A (2002) Oxygen isotope analysis of phosphates: a comparison of techniques for analysis of Ag_3PO_4. Chem Geol 185:321-336

Wang Y, Cerling TE (1994) A model for fossil tooth and bone diagenesis: implications for paleodiet reconstruction from stable isotopes. Palaeogeog Palaeoclim Palaeoecol 107:281-289

Wefer G, Berger WH (1991) Isotope palaeontology: growth and composition of extant calcareous species. Marine Geology 100:207-248

White RMP, Dennis PF, Atkinson TC (1999) Experimental calibration and field investigation bof the oxygen isotopic fractionation between biogenic aragonite and water. Rapid Comm Mass Spec 13:1242-1247

Wing SL, Alroy J, Hickey LJ (1995) Plant and mammal diversity in the Paleocene to early Eocene of the Bighorn Basin. Palaeogeog Palaeoclim Palaeoecol 115:117-155

Wright LD, Schwarcz HP (1996) Infrared and isotopic evidence for diagenesis of bone apatite at Dos Pilas, Guatemala: palaeodietary implications. J Archaeol Sci 23:933-944

Wurster CM, Patterson WP (2001) Late Holocene climate change for the eastern interior United States climate: evidence from high-resolution sagittal otolith stable isotope ratios of oxygen. Palaeogeog Palaeoclim Palaeoecol 170:81-100

Yonge CJ, Goldenberg L, Krouse HR (1989) An isotope study of water bodies along a traverse of southwestern Canada. Journal of Hydrology 106:245-255

Zanazzi A, Kohn MJ, MacFadden BJ, Terry DOJ (2007) Large temperature drop across the Eocene-Oligocene transition in central North America. Nature 445:639-642

Zazzo A, Lecuyer C, Sheppard SMF, Grandjean P, Mariotti A (2004) Diagenesis and the reconstruction of paleoenvironments: a method to restore original $\delta^{18}O$ values of carbonate and phosphate from fossil tooth enamel. Geochim Cosmochim Acta 68:2245-2258

Reviews in Mineralogy & Geochemistry
Vol. 66, pp. 155-171, 2007
Copyright © Mineralogical Society of America

6

A Review of Paleotemperature–Lapse Rate Methods for Estimating Paleoelevation from Fossil Floras

Herbert W. Meyer

National Park Service
P.O. Box 185
Florissant, Colorado, 80816, U.S.A.
Herb_Meyer@nps.gov

ABSTRACT

One widely-applied method for estimating paleoelevation from fossil floras uses the relation between vegetation and temperature to estimate mean annual temperature from two isochronous fossil assemblages (one at elevation and one at or near sea level), and the difference between these temperatures is multiplied by the reciprocal of a temperature lapse rate to estimate paleoelevation. Three fundamentally different approaches have been developed, and these vary both in the method by which paleotemperatures are estimated and in the way by which modern lapse rates can be measured. Paleotemperature methodologies include the use of plant physiognomy, or the use of nearest living relatives. Lapse rate methodologies use either a global terrestrial lapse rate as the world mean, a regional terrestrial lapse rate that represents a large portion of a continent, or a local terrestrial lapse rate for areas narrowly confined by latitude and longitude. Other variables and/or corrective factors that should be considered in these applications include the thermal effects of continentality and elevated land surfaces in the continental interior, and standardization to compensate for fluctuations or variability in sea level, global climate change, and paleolatitude. These different methodologies can be contrasted by their common application to the late Eocene Florissant flora of Colorado, with results ranging from 455 to 4133 meters.

INTRODUCTION

Fossil floras provide one of the most useful sources for obtaining data that can be used to estimate paleoelevation. The distribution of modern forests is clearly delineated largely in accordance with climate, which varies with both altitude and latitude. The correlation of modern vegetation with mean temperature parameters provides the basis for comparing Cenozoic fossil floras with the thermal distribution of these modern forest types to infer paleotemperatures. Such information is a valuable source for inferring climate fluctuations through time, and in combination with thermodynamic properties of the atmosphere, it also can be used to estimate paleoelevation.

Paleoelevation models based on fossil floras use three different approaches: 1) the use of floras to estimate temperature, which is used in combination with lapse rates to infer elevation; 2) the use of floras to estimate enthalpy, which is used with gravitational acceleration to estimate elevation; and 3) the use of stomatal frequency in leaves to indicate altitudinal changes in CO_2 partial pressure. This paper will focus on the first of these, the temperature–lapse rate method, which itself has three basically different approaches that differ in the way paleotemperatures can be estimated from fossil floras and in the methods by which lapse rates can be utilized in the calculations. The purpose of this paper is to provide a concise overview that summarizes the basic tenets of this methodology, comparing its three variations in application. This summary is intended for a paleoelevation workshop, in order that the temperature–lapse rate method can be compared with other methodologies described in this volume for inferring past elevations, including the other paleobotanical methods, chemical methods (e.g., stable isotopes of water, carbonates, and diagenetic clays/oxides), and low temperature thermochronology and cosmogenic radionuclide analysis of erosion and exposure.

REVIEW OF METHODOLOGIES FOR ESTIMATING PALEOELEVATION

Paleotemperature estimates from fossil floras

Estimates of paleotemperature from fossil floras provide the primary data on which paleoelevation estimates are based. Two fundamentally different approaches have been used in paleobotany to acquire these data. The first is based on the climatic distribution of nearest living relatives (the NLR method, also sometimes referred to as the floristic method), and the second has its basis in the correlation of plant physiognomic characters with climate.

The NLR method has been applied using both qualitative and quantitative approaches. Qualitative inferences indicate conditions such as "tropical," "subtropical," and "temperate," and quantitative estimates derive specific temperature parameters (Mosbrugger 1999). The quantitative approach analyzes the range of temperature tolerances for modern taxa that are presumed to be nearest living relatives of the various fossil species in order to determine the "coexistence" (i.e., overlap in mean annual temperature) indicated by the greatest number of taxa (Mosbrugger 1999). The NLR method assumes that morphological characters of a fossil leaf or fruit taxon indicate a phylogenetic relationship to a modern species and that the fossil had an environmental tolerance similar to that of the living form. There are several potential problems with the method, including: 1) identifications on which the comparisons are based usually must be made only from vegetative or fruit remains from the fossil record, rather than on the full suite of morphological and anatomical characters that define living taxa, 2) morphological characters, particularly those with a reproductive function such as pollen and seeds or fruits, may have evolved differently than the more climate-sensitive physiological tolerances of a taxon through time, and 3) the fossils may not be diagnostic to the generic level, may represent an extinct genus, or may be misidentified. The method is most reliable for very recent floras and confidence in the validity of the method progressively decreases for floras of greater antiquity. Futhermore, the method may have greater validity for some taxa than for others (Meyer 2001).

The plant physiognomic method is based on the correlation of particular plant morphological characters to climate. Modern forests from geographically separated but climatically similar regions of the world often have markedly different taxonomic compositions, yet they share the same physiognomic characteristics because these features have evolved similar adaptations to climate through convergent evolution. The application to paleobotany is based on the assumption that this correlation existed through time just as it does spatially today. Although such characters in modern forests can include large-scale features such as canopy characteristics and tree growth forms, in the fossil record it is necessary to limit the characters to those that are evident in fossil leaves. These include leaf size and the presence or absence of teeth. Most of the methods that currently are used to estimate temperature for paleoelevation calculations use this method rather than the NLR method.

Since the work of Bailey and Sinnott (1916), it has been recognized that leaf margins in modern floras are closely correlated with climate. The method became widely applied in paleobotany, particularly with the work of Wolfe (Wolfe and Hopkins 1967; Wolfe 1978, 1979), which showed the close correlation of the percentage of entire-margined species (i.e., those with leaves having smooth margins lacking teeth or lobes) in a flora to mean annual temperature

(MAT), and applied this to interpret climatic change through the Tertiary based on analysis of fossil floras.

The physiognomic method was further elaborated and quantifiably documented with the development of the Climate-Leaf Analysis Multivariate Program, referred to as CLAMP (Wolfe 1993, 1995; Wolfe and Spicer 1999). CLAMP is an extensive database that uses correspondence analysis to analyze numerous physiognomic characters of leaves from modern forests and their relation to climate variables. The data were obtained from leaf samples collected proximal to established climate stations, and a large suite of leaf characters was scored. Some of the primary conclusions of CLAMP were that it circumvents problems of accurate identification and changing evolutionary tolerances inherent to the NLR method, and that the climatic parameter that it most accurately estimates is MAT, with a standard error of 1 °C or less. Wilf (1997) argues that the MAT correlation in the CLAMP dataset is dominated by the leaf margin character, and that this character alone can be used in univariate analyses to obtain MAT estimates that are at least as accurate as the multivariate CLAMP method (Fig. 1); however, leaf margin analysis may only indirectly reflect MAT as a consequence of the closer correlation of leaf margins with winter temperature (Spicer et al. 2005). The reliability and precision of CLAMP has been criticized based on evidence suggesting the method's poor ability to predict MAT, as well as biases during fossilization, the allocation of specimens to taxa, and factors besides temperature that can affect physiognomy (Jordan 1997); however, this criticism was based on Austral floras for which the CLAMP database had not been not calibrated, and such biogeographic factors can degrade the capability of CLAMP's performance (Spicer et al. 2005). Taphonomic biases also can affect the climate interpretations from fossil floras (Greenwood 1992), particularly where they result in the loss of taxa and/ or the loss of characters (Spicer et al. 2005). CLAMP is based on woody dicotyledons and does not consider the climatic significance of conifers, which can be especially important in high elevation forests (Axelrod 1998b). Nevertheless, the method avoids some of the pitfalls of the NLR method, such as invalid taxonomic identification of fossil material and the potential for evolutionary change in the tolerances of taxa. CLAMP is reported to provide estimates for paleoclimate that are more repeatable and valid than the NLR approach (Wolfe

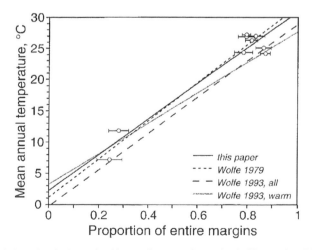

Figure 1. Correlation of entired-margined leaves (i.e., smooth margins lacking teeth or lobes) with MAT following the analyses of Wolfe (1979, 1993) and Wilf (1997, indicated as "this paper" in the figure). From Wilf (1997). [Reprinted from *Paleobiology*, v. 23, Wilf, P., When are leaves good thermometers? A new case for leaf margin analysis, p. 373-390, © 1997, The Paleontological Society, with permission from The Paleontological Society.]

1993, 1995). The CLAMP approach (to predict enthalpy and hence paleoelevation) has been independently verified against oxygen isotope data for Tibet, strengthening the argument that foliar physiognomy is not prone to systematic calibration changes over time due either to evolution or variations in CO_2 concentrations (Spicer et al. 2003; Currie et al. 2005; RA Spicer, written comm. in review 2007).

Another method recently being developed uses higher taxa as a means for estimating paleoclimate (Boyle et al., in press). This method uses cluster analysis based on genus and family level composition to determine the closest modern analog among modern forest plots that are documented by detailed ecological transect analyses. Paleotemperature estimates from this method are still being tested, but they hold promise as yet another tool for application to lapse rate paleoelevation modeling.

In the three different temperature–lapse rate methods for paleoaltimetry, Axelrod (e.g., Axelrod 1980, 1998a,b) relied on the NLR method for estimating temperature. Meyer (1986, 1992) estimated temperature primarily from characteristics of plant physiognomy (before the development of CLAMP) and secondarily from the climatic tolerances of particular genera. Wolfe (1994) and Gregory (1994; Gregory and Chase 1992; Gregory and McIntosh 1996) used CLAMP in their analyses. Besides these differences in the way that temperatures were estimated, each of these workers also used different types of lapse rates to estimate paleoelevation.

Lapse rate methodologies for estimating paleoelevation

Background. The rate of temperature change with elevation is referred to as the lapse rate. Meteorological lapse rates usually express this rate for the open atmosphere, yet lapse rates that are applicable to paleoaltimetry need to be measured near the ground surface, which is often heated by solar radiation during the day and may be warmer than adjacent parcels of the free atmosphere. These are referred to as terrestrial lapse rates. Terrestrial lapse rates have been measured in three different ways for application to paleoelevation modeling. The first uses the global mean terrestrial lapse rate for the world. The second measures local terrestrial lapse rates within areas confined by 1-2° latitude and 1-5° longitude. The third measures regional terrestrial lapse rates over long distances, specifically between the coast and the continental interior of a continent. These three methods of measuring lapse rates are very different ways of expressing temperature change as it relates to elevation, continentality, and the effect of broad uplifted land surfaces. The actual value of these lapse rates also varies widely, from 3.0 °C/km in the regional lapse rate method, to 5.5 °C/km in the global lapse rate method, to 3.64 to 8.11 °C/km in the local lapse rate method. Selection of the lapse rate has significant implications in calculating paleoelevation.

Global lapse rate method. Axelrod (1965, 1968, 1980; Axelrod and Bailey 1976) provided the first theoretical basis for systematically quantifying paleoelevation estimates using paleotemperatures and lapse rates. This method used a NLR approach for estimating MAT of sea level and upland fossil floras, and the difference between these was multiplied by the reciprocal (182 m/1 °C) of the mean terrestrial lapse rate for the world (5.5 °C/km) to derive paleoelevation. This can be expressed as a simple equation, $(MAT^{sl}-MAT^{u})(1000 \text{ m}/5.5)$, where MAT^{sl} is the mean annual temperature of a sea level flora, MAT^{u} is the mean annual temperature of the upland flora for which elevation is being calculated, and 1000/5.5 derives the reciprocal of the lapse rate. Axelrod's method has been criticized for its attempt to derive paleotemperature estimates, and for its assumption that a single global mean lapse rate can be applied validly to all geographical situations (Meyer 1986, 1992; Wolfe 1992a). Subsequent to Axelrod's work, two fundamentally different approaches to the lapse rate paleoelevation method were developed, herein referred to as the local lapse rate method and the regional lapse rate method.

Local lapse rate method. Meyer (1986, 1992) analyzed climatic data for 39 areas of the world to calculate lapse rates based on MAT as well as warm month mean temperature (WMMT) and cold month mean temperature (CMMT). These results show that local terrestrial lapse rates (i.e., those calculated from data within narrowly confined areas of latitude and longitude) based upon MAT varied regionally from 3.64 to 8.11 °C/km. Within the western United States (the area with the largest number of samples), progressive patterns could be demarcated by isopleths that showed a marked increase in lapse rates from 5.0 °C/km along the Pacific coast to 8.1 °C/km in the southwestern interior (Fig. 2; Meyer 1992). These data also show that the projected sea level MAT (i.e., a MAT value derived by projecting the MAT lapse rate to 0 m in regions where the lowest elevations are actually well above sea level) increases 2 to 10 °C from coastal to interior areas (Fig. 3), and mean annual range of temperature (MART; i.e., the difference between WMT and CMT) increases 18 to 21 °C (Fig. 4; Meyer 1992). Such variability in lapse rates and projected sea level temperatures can be explained by factors such as 1) continentality (i.e, distance from the cooling effects of a maritime influence); 2) the effects of elevated regional lowlands (i.e., the lowest land surfaces within a region are at moderate to high elevation), which are heated more intensely than what would be expected from the open atmosphere at comparable altitude; 3) mountain barrier effects; and 4) the location of the subtropical high pressure zone.

Meyer (1992) compensated for many of these factors in a revised paleoelevation methodology that incorporates variables such as the effects of continentality, elevated regional lowlands, extensive mountain massifs, geographical and seasonal variability in lapse rates, climatic changes, sea level fluctuations, and paleogeography (Meyer 2001). It is assumed that some of these effects were less pronounced during the Tertiary due to less intense pre-middle Miocene cold water upwelling along the Pacific coast, a more weakly developed subtropical high pressure zone, and a less pronounced effect of mountain barriers (Meyer 2001). Fundamental to this approach is correcting for the effects of continental climates and elevated regional lowlands by inferring a projected sea level MAT that essentially compensates for these factors by increasing MAT from 2-10 °C, as documented by the modern data between coastal and interior regions. Additionally, the method differs from the earlier Axelrod method by 1) using local terrestrial lapse rates selected for appropriate regions rather than a global mean terrestrial lapse rate; 2) considering climatic change during the Tertiary, particularly the significant cooling during the Eocene-Oligocene transition; 3) incorporating latitudinal corrections to standardize sea level and upland paleotemperatures; and 4) standardizing results to modern sea level to account for sea level fluctuations through time.

The primary criticisms of this method are that 1) it is conceptually and procedurally complicated because it introduces complexities such as calculating a hypothetical sea level temperature in the area of unknown altitude as a means for correcting for the effects of continentality and elevated land surfaces; 2) it makes assumptions with no meteorologically sound physical basis for assuring that particular lapse rates and climatic characteristics can be extended back in time; 3) the physiognomic basis for estimating paleotemperature was qualitative (i.e., pre-CLAMP); and 4) Tertiary climates were more equable and vegetation may have been zoned differently than at present (Gregory and Chase 1992; Wolfe 1992a; Wolfe et al. 1998; Axelrod 1998b).

Regional lapse rate method. Wolfe (1992a) took a different approach to calculating terrestrial lapse rates, and his analysis covered much of the same ground as Meyer (1986, 1992) but with significant differences. This analysis was confined to the western United States (whereas Meyer had also included many other regions throughout the world), and lapse rates were calculated over large regional areas relative to the Pacific Coast (whereas Meyer had used more localized lapse rates for particular areas). Specifically, Wolfe calculated regional lapse rates from two-point data comparisons over long distances using the difference between adjusted MAT for the Pacific Coast and actual MAT from a climate station at the same latitude in the interior. Because the modern lapse rates were calculated over long distances, some of

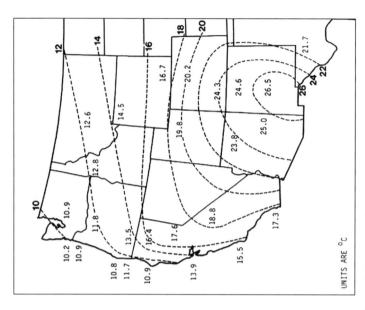

Figure 3. Isotherms of MAT at sea level. Coastal areas represent actual climate data, and interior areas are calculated from local lapse rates to 0 m elevation (referred to as projected sea level MAT). From Meyer (1992). [Reprinted from *Palaeogeography, Palaeoclimatology, Palaeoecology,* v. 99, Meyer, H.W., Lapse rates and other variables applied to estimating paleoaltitudes from fossil floras, p. 71-99, © 1992, Elsevier Science Publishers B.V., with permission from Elsevier.]

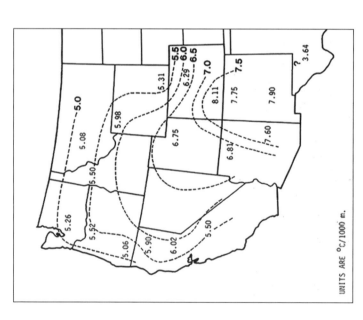

Figure 2. Local lapse rates (°C/km) calculated from MAT based on analysis of 18 areas in the western United States. From Meyer (1992). [Reprinted from *Palaeogeography, Palaeoclimatology, Palaeoecology,* v. 99, Meyer, H.W., Lapse rates and other variables applied to estimating paleoaltitudes from fossil floras, p. 71-99, © 1992, Elsevier Science Publishers B.V., with permission from Elsevier.]

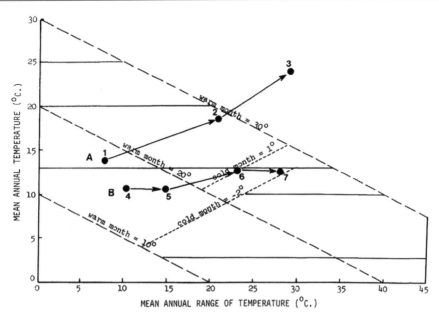

Figure 4. Projection of sea level temperature changes (MAT and MART) from coastal to interior areas for two transects in the western United States. Transect A: 1 = coastal California, 2 = southeastern California, 3 = northern Arizona; Transect B: 4 = coastal Washington, 5 = Washington, 6 = Idaho, 7 = central Montana. Points 1 and 4 represent actual values; all other values are calculated from local lapse rates to 0 m. From Meyer (1992). [Reprinted from *Palaeogeography, Palaeoclimatology, Palaeoecology*, v. 99, Meyer, H.W., Lapse rates and other variables applied to estimating paleoaltitudes from fossil floras, p. 71-99, © 1992, Elsevier Science Publishers B.V., with permission from Elsevier.]

the effects due to continentality and elevated regional lowlands are probably intrinsically incorporated into the results.

The regional lapse rates derived from this analysis are significantly lower than the global mean lapse rate of 5-6 °C/km and most local terrestrial lapse rates. Regional lapse rates for the western United States vary between physiographic provinces, but for specific provinces, Wolfe noted a tendency for lapse rates to decrease with decreasing latitude for a given altitude and to decrease with decreasing altitude for a given latitude, except for intramontane basins of the Rocky Mountains. Many of the regional lapse rates calculated for interior locations within 200-300 km of the west coast are negative. Wolfe (1992a) noted that basins leeward of mountain ranges are more strongly heated and can have lapse rates lower than on windward sides of the mountains, and that at constant altitude heating increases further into the continental interior. The calculation of lapse rates by this method thus to some extent reflects the effects of continentality and the steep temperature gradient inland from the Pacific coast (Meyer 2001). In the Rocky Mountains, lapse rate values are about 3 °C/km for intramontane basins and about 4 °C/km for adjacent mountains, yet values within even a few miles can vary significantly as calculated by this method (e.g., Wolfe 1992a, plate 1). Based on the observation that most fossil floras were deposited in intramontane basins, Wolfe used a lapse rate of 3 °C/km in his calculations of paleoelevation (e.g., Wolfe 1992b, 1994). This low lapse rate produces very high elevation estimates in most applications.

The primary criticisms of the method are that it derives modern lapse rates that reflect the modern magnitude of cold water upwelling along the Pacific Coast, and the presence today of near-coastal mountain barriers, both of which would have been different during the early and middle parts of the Tertiary (Meyer 2001). Meyer argues that intensification of cold water upwelling beginning during the Miocene may have had a cooling effect on coastal temperatures, hence lowering present-day regional lapse rates (as calculated by the Wolfe method) relative to the pre-Miocene and calling into question the validity of using these low lapse rates for earlier Tertiary floras. However, Wolfe (1992a) argued that upwelling probably occurred to some extent along the Pacific coast since the Cretaceous. Wolfe also argued that incursions of frigid arctic air affect modern terrestrial lapse rates and that in the absence of this condition, Tertiary regional lapse rates would have been even lower. Axelrod (1997, 1998b) noted that the lower values for regional lapse rates significantly overestimates the elevation of modern upland stations, although his test for this applied the regional lapse rates for the western United States to other areas of the world, and hence may not have been valid. Axelrod (1998a) also contended that such low lapse rates for the Tertiary would suggest global conditions of impossibly negligible atmospheric convection, although the validity of this criticism can be disputed by the fact that Wolfe's lapse rates were calculated by a methodology that did, in fact, use existing modern climate data.

Seasonal variations of temperature: Effects on vegetation distribution and lapse rates

Temperature parameters besides MAT, such as WMT and CMT, have been used to plot the distribution of modern vegetation types on various climagraphs, and these have also proven useful for determining the seasonality for fossil floras. Although MAT is the value that is most widely used for paleoaltimetry, WMT and CMT are often the ones that determine particular vegetation boundaries (Fig. 5; Wolfe 1979), and an understanding of the lapse rates for these seasonal temperatures is therefore also an important consideration in paleoelevation studies. For example, Meyer (1992) calculated WMT and CMT local lapse rates as well as

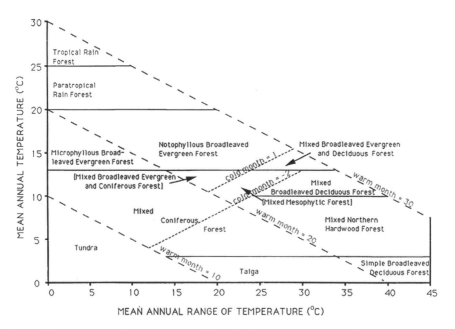

Figure 5. The distribution of forests in eastern Asia can be illustrated on a climagraph showing the relation of MAT, CMT, and WMT to vegetation boundaries. Diagram from Meyer (1992), based on Wolfe (1978). [Reprinted from *Palaeogeography, Palaeoclimatology, Palaeoecology,* v. 99, Meyer, H.W., Lapse rates and other variables applied to estimating paleoaltitudes from fossil floras, p. 71-99, © 1992 Elsevier Science Publishers B.V., with permission from Elsevier.]

the regression of MART and noted that within continental settings, the regression of MART showed a decrease with increasing elevation (Fig. 6); that is, WMT lapse rates are greater than CMT lapse rates in such settings. This pattern is reversed for island settings. For vegetation boundaries that are climatically determined by these seasonal means (e.g., the occurrence of treeline near 10 °C WMT), it is important to incorporate this seasonality of lapse rates into the model for estimating paleoelevation by inferring the regression of MART against elevation on a climagraph (Meyer 1992).

Climagraphs used by Axelrod incorporate additional, mathematically derived thermal parameters referred to as "warmth" and "temperateness." According to Axelrod (e.g., Axelrod and Bailey 1976), MART influences estimates of paleoelevation because the radiating lines of warmth diverge from one another as MART increases, and therefore there is increasing altitude separation between comparable lines of warmth with increasing MART (Fig. 7). Axelrod contends that vegetation zones converge under conditions of lower MART, and hence that during the Tertiary they were "more broadly zoned" (apparently implying that they did not cover as much range in elevation). Meyer (1986, 1992), however, notes that warmth and temperateness are not clearly coincident with vegetation boundaries and that they are values not widely used in climatology, thus questioning their usefulness in paleoaltimetry.

Other variables

Meyer (1986, 1992) considered a number of variables to be considered in the calculation of paleoelevation. As mentioned previously, these include the effects of continentality and elevated land surfaces in the continental interior, which produce high values for projected sea level temperatures. In addition, sea level fossil floras (ideally occurring in marine rocks) need

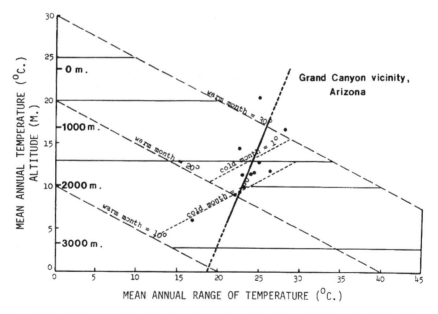

Figure 6. Results from analysis of modern temperature data (Meyer 1993) can be illustrated on a climagraph to show changes in mean temperatures with elevation, including the decrease in MART with increasing elevation. Solid line indicates the elevation range within which climate data occur; dashed lines are projections to higher and lower elevations; dots indicate mean temperatures from climate stations used in the analysis. From Meyer (1992). [Reprinted from *Palaeogeography, Palaeoclimatology, Palaeoecology*, v. 99, Meyer, H.W., Lapse rates and other variables applied to estimating paleoaltitudes from fossil floras, p. 71-99, © 1992, Elsevier Science Publishers B.V., with permission from Elsevier.]

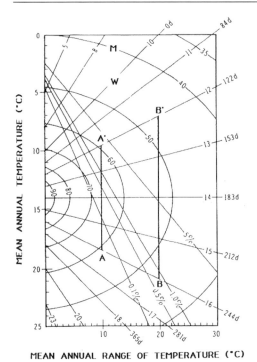

Figure 7. Thermal parameters purported to demarcate vegetation distribution according to Axelrod include radiating lines of warmth (W, expressed as number of days at or above specified temperature) and arcs representing an index of temperateness (M). According to the Axelrod model, MART influences estimates of paleoelevation because W diverges as MART increases; however, the relation of W to the distribution of vegetation zones is questionable. [Modification of Axelrod and Bailey (1976) from Meyer (1992). Original from *Paleobiology*, v. 2, Axelrod, D.I., Tertiary vegetation, climate, and altitude of the Rio Grande depression, New Mexico–Colorado, p. 235-254, © 1976, The Paleontological Society, used with permission from The Paleontological Society. Modified version reprinted from *Palaeogeography, Palaeoclimatology, Palaeoecology*, v. 99, Meyer, H.W., Lapse rates and other variables applied to estimating paleoaltitudes from fossil floras, p. 71-99, © 1992, Elsevier Science Publishers B.V., with permission from Elsevier.]

to be adjusted in elevation to account for changes in sea level through time, in order to calibrate the elevation results to the datum of modern sea level. Comparison of temperatures from sea level and interior floras also requires that the two be at the same latitude; in some cases, it may be necessary to calibrate temperature estimates to the same latitude using a paleolatitudinal temperature gradient. Because climate has fluctuated through geologic time, including abrupt, significant changes in temperature, it is important that two such floras be demonstrably contemporaneous. Another consideration is the effect of cold air drainage within closed basins, which can lower mean temperatures; many fossil floras are, of course, preserved in closed basins such as calderas. Other changes in paleogeography, such as development of mountain barriers and their effects on producing different windward vs. leeward slope temperatures and lapse rates, can affect the extrapolation of modern lapse rates into the past. Considering the combined complexity of all of these estimated variables, Meyer (1992) emphasized that the calculation of paleoelevation becomes problematic, comprising the potential for accuracy.

Gregory-Wodzicki (2000) incorporated many of these variables into the following equation for calculating elevation:

$$Z = Z_m - \frac{MAT_i + \Delta MAT_{gc} + \Delta MAT_{cd} + \Delta MAT_{pg} - MAT_m}{\gamma} + S$$

where Z = paleoelevation; Z_m = modern elevation; MAT_i = MAT from the fossil flora; ΔMAT_{gc}, ΔMAT_{cd}, ΔMAT_{pg} = the change in MAT since deposition of the fossil flora due to global climate change (gc), latitudinal continental drift (cd), and changes in paleogeography, respectively; MAT_m = modern MAT, γ = terrestrial lapse rate; and S = ancient sea level relative to modern sea level.

APPLICATIONS

Late Eocene Florissant flora

A comparison of the various applications of lapse rates to paleoelevation modeling is best exemplified by the late Eocene Florissant flora of central Colorado because of the large number of studies that have been done there (Table 1). Indeed, Florissant has become a good testing ground for any new methodology in paleoelevation, and it has been used by various workers to cross-check similar methodologies. The data used in these previous estimates and the resulting paleoelevations are summarized by Meyer (2001).

MacGinitie (1953, p. 53) made the first estimate for Florissant, noting that "the plant association indicates a region of moderate elevation" to derive an estimate of 300 to 900 m. For decades, this estimate was widely used as a benchmark for the late Eocene erosion surface and was cited as evidence for significant late Tertiary uplift of the region (e.g., Epis and Chapin 1975), strongly influencing models of the tectonic history of the southern Rocky Mountains.

Meyer (1986, 1992) was the first to apply a quantitative methodology to Florissant using paleotemperatures and lapse rates, providing an estimate of 2450 m (1900 to 3200 m range). This was derived using both physiognomic and floristic criteria for estimating paleotemperature, and using the local lapse rate method with corrections for paleolatitude, continentality, elevated interior lowland surfaces, and calibration to modern sea level (Fig. 8). These corrective factors specifically included 1) adding 3 °C to MAT of the sea level flora to calibrate for 8-10° higher paleolatitude than Florissant; 2) adjusting MAT by an additional +3 °C to derive a hypothetical projected interior sea level temperature in accordance with modern climate data (see Fig. 4), but assuming that this pattern was less pronounced during the Tertiary; and 3) adding 200 m to adjust for the higher late Eocene sea level.

Wolfe (1992b) used regional lapse rates and CLAMP, giving an estimate of 2700 m (2900 m calibrated to modern sea level). Later, Wolfe (1994) used both a regional lapse rate (3.0 °C/km) and a mean global lapse rate for comparison (5.5 °C/km) to derive estimates of 4133 m and 2255 m, respectively, with no corrections for other factors.

Both local and regional lapse rates were used in calculations by Gregory and Chase (1992), Gregory (1994), and Gregory and McIntosh (1996), providing a means for comparison of the same methodologies used by Meyer (local lapse rate method) and Wolfe (regional lapse rate method). Their application of the local lapse rate method used a lower lapse rate value than Meyer (1992), however, and did not incorporate corrective factors to MAT to compensate for the effects of continentality and elevated regional lowlands.

Following the development of the local and regional lapse rate methods and their application to Florissant, Axelrod (1998b) used his original global lapse rate method to make his first paleoelevation estimate for Florissant. This method derived a low estimate of 455 m, which is within the range of the original estimate by MacGinitie (1953).

Besides the lapse-rate based methodologies, other methods have also been applied to estimate paleoelevation for Florissant. Another approach based on paleobotany has used enthalpy (Forest et al. 1995; Wolfe et al 1998), based on principles of atmospheric energy conservation, and results from this method have produced high elevation estimates (2900-3800 m) comparable to those of the local and regional lapse rate methods. Analysis of stable isotopes from mammalian tooth enamel provides preliminary results that indicate low relief and low elevation (Barton and Fricke 2006).

Western Tertiary floras

The various lapse rate methods have been applied to other Tertiary floras, although in most cases these floras have been analyzed by only one of the methods and do not provide the

TABLE 1. Comparison of paleoelevation methods applied to the late Eocene Florissant flora.

Method	Paleoelevation (meters)	Lapse Rate (°C/km)	MAT for Florissant	MAT Sea Level	References	Comments
Qualitative	300-900	n/a	18		MacGinitie 1953	Non-quantitative approach based on modern plant distributions
Global Lapse Rate	455 (Axelrod) 2255 (Wolfe)	5.5	10.8 (Wolfe) 15.5 (Axelrod)	18 (Axelrod) 23.2 (Wolfe)	Axelrod 1998b Wolfe 1994	
Local Lapse Rates with regional temperature correction	2450	6.7	14	26	Meyer 1986, 1992	Estimates are corrected for paleolatitude and/or modern sea level. Application uses correction factors for regional affects of continentality and elevated base level (see text)
Local Lapse Rates without regional temperature correction	1900-2300	5.9	10.7-12.8	22.7-22.9	Gregory and Chase 1992 Gregory 1994 Gregory and McIntosh 1996	Some estimates are corrected for paleolatitude and/or modern sea level.
Regional Lapse Rates	2700-4133	3.0	10.7-12.8	20-23.2	Wolfe 1992, 1994 Gregory and Chase 1992 Gregory 1994 Gregory and McIntosh 1996	Some estimates are corrected for paleolatitude and/or modern sea level
Enthalpy	2900-3800	n/a	n/a	n/a	Forest et al. 1995 Gregory and McIntosh 1996 Wolfe et al. 1998	Included for comparison with above estimates. Method is not based on lapse rates

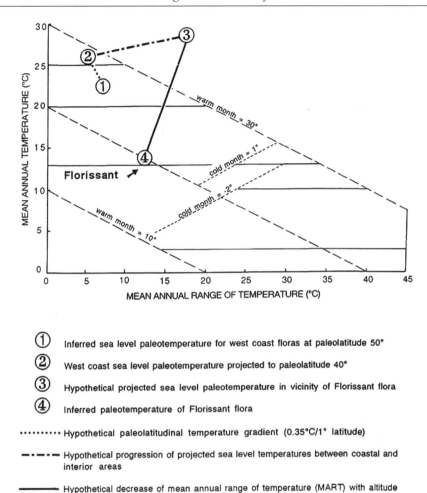

① Inferred sea level paleotemperature for west coast floras at paleolatitude 50°

② West coast sea level paleotemperature projected to paleolatitude 40°

③ Hypothetical projected sea level paleotemperature in vicinity of Florissant flora

④ Inferred paleotemperature of Florissant flora

·········· Hypothetical paleolatitudinal temperature gradient (0.35°C/1° latitude)

—·—·— Hypothetical progression of projected sea level temperatures between coastal and interior areas

————— Hypothetical decrease of mean annual range of temperature (MART) with altitude

Figure 8. Calculation of paleoelevation for the late Eocene Florissant flora, Colorado, based on Meyer (1986, 1992). The difference in MAT (15 °C between points 3 and 4) is multiplied by the reciprocal of a mean terrestrial lapse rate (150m/-1 °C) to give a value of 2250 m. Calibration to present sea level (+200 m) gives a paleoelevation estimate of 2450 m. From Meyer (1992). [Reprinted from *Palaeogeography, Palaeoclimatology, Palaeoecology*, v. 99, Meyer, H.W., Lapse rates and other variables applied to estimating paleoaltitudes from fossil floras, p. 71-99, © 1992, Elsevier Science Publishers B.V., with permission from Elsevier.]

same multi-method comparison as Florissant. Notable examples of these works include:

1)   Global Lapse Rate Method: Axelrod and Bailey (1976) estimated elevations ranging from 1200 to 2600 m for various Eocene and Oligocene floras of New Mexico and Colorado; Axelrod (1980) estimated 610-760 m for the Miocene Mt. Reba flora in the Sierra Nevada of California, suggesting significant subsequent uplift; and Axelrod (1998a) estimated 1730 m for the Eocene Thunder Mountain flora of Idaho and 1000 m for the Oligocene Haynes Creek flora of Idaho. For comparison with his regional lapse rate method, Wolfe (1994) calculated estimates based on global lapse rates as well, although these were generally about half of the values estimated from regional lapse rates, and he considered that they were minimal estimates.

2) Local Lapse Rate Method: Meyer (1986) estimated high elevations ranging from 1800 to 4100 m for various Eocene and Oligocene floras of New Mexico and Colorado, including 1800 m for the late Oligocene Creede and Platoro floras of Colorado, and 3000 m for the early Oligocene Hermosa flora of New Mexico.

3) Regional Lapse Rate Method: Wolfe (1994) calculated high paleoelevations for several classic Eocene floral localities, including Salmon (4000 m), Copper Basin (2533 m), Green River (2867 m), Republic (1667 m), and Kissinger Lakes (1000 m).

DISCUSSION

Comparison between the three methods for estimating paleoelevation from temperatures and lapse rates is not simple. Discrepancies between the various methods are due to complex differences in the way that paleotemperatures are estimated for fossil floras, and the means by which climate data are modeled and utilized. Each method utilizes a unique combination of 1) methodology for estimating MAT; 2) type of lapse rate (i.e., global, local, or regional lapse rate); 3) type of climagraph that describes the thermal distribution of vegetation with temperature; and 4) consideration of other variables and the use of corrective factors to compensate for these. Each method emphasizes the importance of selecting an "appropriate" terrestrial lapse rate, although equally important is considering whether a particular methodological approach adequately incorporates the thermal effects of continentality, elevated land surfaces in the continental interior, and windward vs. leeward slopes of mountain barriers. Selection of a lapse rate in itself has a major impact on paleoelevation estimates; for example, a regional lapse rate of 3.0 °C/km (333m/1 °C) produces estimates nearly double those of a global lapse of 5.5 °C/km (182 m/°C/km) (Wolfe 1994; Axelrod 1997, 1998b). These variations produce widely ranging results, as evident by their application to the Florissant flora (Table 1).

The most simplistic and therefore easily understood of the various approaches is the global lapse rate method (e.g., Axelrod 1965, 1968, 1980, 1987, 1998a,b), although its simplicity may be misleading. It only minimally attempts to unravel some of the complexities of climate patterns within particular regions, and these can be important considerations if accurate results are to be obtained. However, when the local lapse rate method (Meyer 1986, 1992, 2001) is applied in conjunction with corrective factors to compensate for the effects of regional climate patterns and elevated land surfaces, it is difficult to determine the exact extent to which modern patterns would have prevailed during the past. Application of numerical general circulation models may provide a basis for overcoming some of these difficulties (Spicer et al. 2003). Nevertheless, one of the greatest strengths of the local lapse rate method is that it probably best illustrates the complexity and problems in applying modern climate data to paleoelevation modeling. The regional lapse rate method (Wolfe 1992a, 1994) intrinsically incorporates some of the same large-scale climate and geographic variations, although the value of these lapse rates in part may be reflective of the modern magnitude of cold water upwelling along the Pacific coast and the presence of near coastal mountain barriers, and hence early to middle Tertiary regional lapse rates may have been higher and paleoelevations lower than those calculated (Meyer 2001).

Besides estimating paleoelevation relative to sea level floras, some workers have instead made estimates of elevation *uplift* (surface height change) within a region by using the difference between modern temperatures near the fossil site and estimated paleotemperature from the fossil assemblage, multiplied by the reciprocal of the lapse rate. For example, Axelrod (1987) estimated 1500-1750 m of uplift for the Oligocene Creede flora (Colorado) by using the difference in MAT between Creede today and that inferred from the fossil flora. Similarly, Axelrod and Bailey (1976) estimated a minimum uplift of 1300 m for the Miocene Skull Ridge assemblage (New Mexico) using the difference between its paleotemperature and the present temperature at Santa Fe. Such estimates are highly suspect because they do not account for global climate changes (Wolfe 1978) that occurred subsequent to the deposition of the fossil floras; that is, temperature at a given elevation did not remain constant through time. Because of climate change through time, it is essential that the floras and temperatures being compared are isochronous.

The use of lapse rate methodologies has been restricted to Cenozoic floras. This is largely because CLAMP and NLR methods for estimating paleotemperature are based on predominantly angiospermous floras. CLAMP can be applied to late Cretaceous floras (e.g., Herman and Spicer 1996), although its reliability is less certain for mid-Cretaceous floras because this was a time during the early radiation of angiosperms (Huber 1998). Older, pre-angiospermous floras would require another method to estimate paleotemperatures. Application of modern lapse rates to earlier periods of time is also more problematical than it is for the Cenozoic due to greater geographic and climatic dissimilarities from today.

CONCLUSIONS

The paleobotanically-based paleotemperature–lapse rate method provides one of several important proxies for estimating paleoelevation. Its primary strengths are that 1) it is internally consistent based on comparison to the quantifiable relation between modern temperatures, elevation, and geography; 2) it provides an independent proxy that is not derived solely from geologic evidence; 3) it has fundamental variations in the methodological approach that allow cross-testing of results, and 4) because plants grow on the Earth's surface, they provide data for actual surface height information rather than the relative height of rocks that geologic models derive. The primary weaknesses are that 1) it utilizes a strongly uniformitarian approach to the application of climate data, when in fact the factors that can influence such data (e.g., mountain barriers, cold-water upwelling along the coast, intensity of the subtropical high pressure zone, and atmospheric circulation patterns) may have changed through time (although these aspects of uniformitarianism potentially could be reduced by incorporating a climate modeling approach); and 2) modern data demonstrate significant regional variability in lapse rates, making the selection of an appropriate lapse rate problematical. Nevertheless, the fact that two fundamentally different lapse-rate-based approaches (local lapse rates and regional lapse rates) applied by various workers, as well as the enthalpy-based methodology, have all indicated high to very high elevation for the late Eocene Florissant flora provides encouragement that these methods are capable of providing a quantitative basis for making solid qualitative statements as to whether paleoelevation is low, moderate, high, or very high, even though the prospect for obtaining demonstrably accurate, precise quantitative results may never be completely achievable in *any* of the various paleobotanical or other proxies for paleoelevation modeling.

ACKNOWLEDGMENTS

I thank Charles Chapin and Robert Spicer for reviews of the manuscript, and Matt Kohn for inviting me to contribute to this volume.

REFERENCES

Axelrod DI (1965) A method for determining the altitudes of Tertiary floras. Paleobotanist 14:144-171
Axelrod DI (1968) Tertirary floras and topographic history of the Snake River basin, Idaho. Geol Soc Am Bull 79:713-734
Axelrod DI (1980) Contributions to the Neogene paleobotany of central California. Univ California Publ Geol Sci 121:1-212
Axelrod DI (1987) The Late Oligocene Creede flora, Colorado. Univ California Publ Geol Sci 130:1-236
Axelrod DI (1997) Paleoelevation estimated from Tertiary floras. Int'l Geology Rev 39:1124-1133

Axelrod DI (1998a) The Eocene Thunder Mountain flora of central Idaho. Univ California Publ Geol Sci 142:1-61, pl 1-15

Axelrod DI (1998b) The Oligocene Haynes Creek flora of eastern Idaho. Univ California Publ Geol Sci 143:1-99, pl 1-23

Axelrod DI, Bailey HP (1976) Tertiary vegetation, climate, and altitude of the Rio Grande depression, New Mexico—Colorado. Paleobiology 2:235-254

Bailey IW, Sinnott EW (1916) The climatic distribution of certain types of Angiosperm leaves. Am J Bot 3:24-39

Barton MA, Fricke HC (2006) Mapping topographic relief and elevation of the central Rocky Mountain region druing the latest Eocene using stable isotope data from mammalian tooth enamel. Geol Soc Amer Abstr Prog 38:202

Boyle B, Meyer H, Enquist B, Salas S (in press) Higher taxa as paleoecological and paleoclimatic indicators: A search for the modern analog of the Florissant fossil flora. In: Paleontology of the upper Eocene Florissant Formation, Colorado. Meyer, HM and Smith DM (eds) Geol Soc Amer Spec Pap 435

Currie BS, Rowley DB, Tabor NJ (2005) Middle Miocene paleoaltimetry of southern Tibet: Implications for the role of mantle thickening and delamination in the Himalayan orogen. Geology 33:181-184.

Epis RC, Chapin CE (1975) Geomorphic and tectonic implications of the post-Laramide, late Eocene erosion surface in the southern Rocky Mountains. In: Cenozoic History of the Southern Rocky Mountains. Curtis BF (ed) Geol Soc Am Mem 144:45-74

Forest CE, Molnar P, Emanuel KA (1995) Paleoaltimetry from energy conservation principles. Nature 374:347-350

Greenwood DR (1992) Taphonomic constraints on foliar physiognomic interpretations of late Cretaceous and Tertiary paleoclimates. Rev Palaeobot Palynology 71:149-190

Gregory KM (1994) Paleoclimate and paleoelevation of the 35 Ma Florissant flora, Front Range, Colorado. Palaeoclimates 1:23-57

Gregory KM, Chase CG (1992) Tectonic significance of paleobotanically estimated climate and altitude of the late Eocene erosion surface, Colorado. Geology 20:581-585

Gregory KM, McIntosh WC (1996) Paleoclimate and paleoelevation of the Oligocene Pitch-Pinnacle flora, Sawatch Range, Colorado. Geol Soc Am Bull 108:545-561

Gregory-Wodzicki KM (2000) Uplift history of the central and northern Andes. A review. Geol Soc Am Bull 112:1091-1105

Herman AB, Spicer RA (1996) Paleobotanical evidence for a warm Cretaceous Arctic Ocean. Nature 380:330-333

Huber BT (1998) Tropical paradise at the Cretaceous poles? Science 282:2199-2200

Jordan GJ (1997) Uncertainty in paleoclimatic reconstructions based on leaf physiognomy. Austr Jour Botany 45: 527-547

MacGinitie HD (1953) Fossil Plants of the Florissant Beds, Colorado. Carnegie Inst Wash Publ 599: p 1-198, pl 1-75

Meyer HW (1986) An evaluation of the methods for estimating paleoaltitudes using Tertiary floras from the Rio Grande rift vicinity, New Mexico and Colorado. PhD dissertation, Univ Calif Berkeley, p. v-vii, 1-217

Meyer HW (1992) Lapse rates and other variables applied to estimating paleoaltitudes from fossil floras. Palaeogeog Palaeoclim Palaeoecol 99:71-99

Meyer HW (2001) A review of the paleoelevation estimates from the Florissant flora, Colorado. In: Fossil Flora and Stratigraphy of the Florissant Formation, Colorado. Evanoff E, Gregory-Wodzicki KM, Johnson KR (eds) Proc Denver Museum Nature Sci ser 4 no 1: p 205-216

Mosbrugger V (1999) The nearest living relative method. In: Fossil Plants and Spores: Modern Techniques. Jones TP, Rowe NP (eds) Geol Soc London, p 233-239

Spicer RA, Harris NBW, Widdowson M, Herman AB, Guo S, Valdes PJ, Wolfe JA, Kelley SP (2003) Constant elevation of southern Tibet over the past 15 million years. Nature 421:622-624

Spicer RA, Herman AB, Kennedy EM (2005) The sensitivity of CLAMP to taphonomic loss of foliar physiognomic characters. Palaios 20:429-438

Wilf P (1997) When are leaves good thermometers? A new case for leaf margin analysis. Paleobiology 23:373-390

Wolfe JA (1978) A paleobotanical interpretation of Tertiary climates in the Northern Hemisphere. Am Sci 66:694-703

Wolfe JA (1979) Temperature parameters of humid to mesic forests of eastern Asia and relation to forests of other regions of the northern hemisphere and Australasia. US Geol Surv Prof Paper 1106, p 1-37, pl 1-3

Wolfe JA (1992a) An analysis of present-day terrestrial lapse rates in the western conterminous United States and their significance to paleoaltitudinal estimates. US Geol Surv Bull 1964, p 1-35, pl 1-2

Wolfe JA (1992b) Climatic, floristic, and vegetational changes near the Eocene/Oligocene boundary in North America. In: Eocene-Oligocene Climatic and Biotic Evolution. Prothero DR, Berggren WA (eds) Princeton University Press, Princeton, NJ, p 421-436

Wolfe JA (1993) A method of obtaining climatic parameters from fossil leaf assemblages. US Geol Surv Bull 2040, p 1-71, pl 1-5

Wolfe JA (1994) Tertiary climatic changes at middle latitudes of western North America. Palaeogeog Palaeoclim Palaeoecol 108:195-205

Wolfe JA (1995) Paleoclimatic estimates from Tertiary leaf assemblages. Ann Rev Earth Planet Sci 23:119-142

Wolfe JA, Forest CE, Molnar P (1998) Paleobotanical evidence of Eocene and Oligocene paleoaltitudes in midlatitude western North America. Geol Soc Amer Bull 110:664-678

Wolfe JA, Hopkins DM (1967) Climatic changes recorded by Tertiary land floras in northwestern North America. *In*: Tertiary Correlations and Climatic Changes in the Pacific. Hatai H (ed) Sasaki, Sendai, Japan, p 67-76

Wolfe JA, Spicer RA (1999) Fossil leaf character states: multivariate analyses. *In*: Fossil plants and spores: modern techniques. Jones TP, Rowe NP (eds) Geol Soc London, p 233-239

Reviews in Mineralogy & Geochemistry
Vol. 66, pp. 173-193, 2007
Copyright © Mineralogical Society of America

7

Paleoaltimetry:
A Review of Thermodynamic Methods

Chris E. Forest

Center for Global Change Science
Massachusetts Institute of Technology
Cambridge, Massachusetts, 02139, U.S.A.
ceforest@MIT.EDU

ABSTRACT

This review presents the use of conserved thermodynamic variables to estimate paleoaltitudes. The method based on conservation of moist static energy (the combined internal, latent heat, and gravitational potential energy of moist air) is discussed. This method exploits the physical relation between of the wind fields and the spatial and vertical distributions of both temperature and humidity. Given the climatological distributions of these three atmospheric fields, the method identifies moist enthalpy (the combined internal and latent heat energies of air) as a thermodynamic variable that varies with height in the atmosphere in a predictable fashion. To use this method, the major requirements are: (1) *a priori* knowledge of the spatial distribution of moist static energy for the paleoclimate and (2) the ability to estimate moist-enthalpy in the paleo-environment for two isochronous locations: one at sea level, the other at some unknown elevation. As presented here, the method incorporates basic physical principles of atmospheric science and inferences of paleoclimates from plant leaf physiognomy. Assuming that expected errors estimated from present-day relationships between physiognomy and enthalpy apply to ancient climates and fossil leaves, an uncertainty estimate of ± 910 m in the paleoaltitude difference between two isochronous fossil assemblage locations can be assessed.

INTRODUCTION

The fields of paleobotany, paleoclimate, and geophysics require accurate paleoaltitude estimates. Within a given field, progress requires significant advances in all fields to interpret key data. Changes in paleoflora can be interpreted as either changes in paleoclimate or paleoaltitudes. Changes in paleoclimate can be either at global or regional scales in response to multiple causes, not least of which would be a change in altitude. Likewise, temporal changes in altitudes of large regions will inform basic theory of uplift and mountain building while also providing insights for paleoclimate and paleobotanical work. The motivation for such progress is exemplified by the present volume. Furthermore, this demonstrates a renewed interest in providing robust estimates for paleo-altitudes from whatever means are available for the common goal of exploring paleo-earth. In this context, this is a review of the thermodynamic methods for estimating the paleo-elevations that have been used in the literature and it will attempt to discuss their strengths and weaknesses. The complementary chapters of this volume will provide similar discussions so that a more complete picture of paleo-altimetry research can be assessed.

At the outset, it is worthwhile to consider some simple questions:

- Why do thermodynamic methods work well for paleo-altimetry?

- Why are paleo-botanical estimates of paleoclimate well suited for this?

1529-6466/07/0066-0007$05.00

DOI: 10.2138/rmg.2007.66.7

- What are the requirements for estimating paleo-altitudes with such techniques?

- What are the limitations?

The introduction addresses these in short while the majority of the paper reviews the method from Forest et al. (1999).

Paleobotanical finds offer the most direct indications of paleoclimatic changes in continental regions, but clearly many ecological conditions vary similarly with latitude and altitude. (i.e., plants indicate the conditions of the local surface environment) Correspondingly, to exploit all climate information from a fossil assemblage, the component of climate relating to local elevation must be determined. Thus, the paleobotanical record on both local and global scales, climate changes of both local and global scales, and the evolution of mean elevations of large tracts of land should be closely linked to one another.

Calculating elevation from climate variables requires measuring an atmospheric quantity that varies with altitude. In other chapters, paleo-pressure are inferred by techniques such as basalt vesicularity (Sahagian and Maus 1994) or measuring cosmogenic nuclide concentrations in exposed rocks. For example, Brown et al. (1991) and Brook et al. (1995) used concentrations of ^{10}Be and ^{26}Al to place constraints on the uplift rate and duration of exposure of rocks in the Transantarctic Mountains, Antarctica. Additional methods and further discussions are available in the present volume.

Here we use paleoenthalpy (a thermodynamic variable combining the sensible and latent heat of moist air) as the quantity that varies with altitude. Paleo-temperature has been used by itself however implicit in the paleotemperature methods are the dependence of surface-temperature lapse rates on the regional moisture content. When more moisture is in the air, the surface temperature will decrease more slowly with elevation changes. In atmospheric physics, this phenomenon is explained by the conversion of latent heat energy to the sensible heat energy of air as water vapor condenses into precipitation. Given that precipitation is typically a local or regional event, one needs to understand how this relates to large-scale atmospheric circulations and remains a key question.

Estimating paleotemperatures has been used successfully for estimating paleoaltitudes. Mean temperatures show correlations with leaf morphology of living plant assemblages with uncertainties in mean annual temperature perhaps as small as 1 °C (Wolfe 1979, 1993). Hence, assuming that foliar characteristics of leaf fossil assemblages obey the same relationships, we should be able to infer correspondingly accurate paleotemperatures. Because temperature varies with latitude, longitude, and height, and because temperature has varied over geologic time, the calculation of paleoelevation from paleotemperature involves comparing the surface temperature of a high-altitude location to that of a low-altitude location at the same latitude and of the same age (Axelrod 1966). Hence, the use of paleobotanical data for paleoaltimetry includes two required steps: first, estimating a climatic parameter, like mean annual temperature, from fossil plants, and second, using differences in that parameter from separate sites to estimate elevation differences, given *a priori* knowledge of the variation of the parameter with altitude.

Two approaches to inferring paleoclimates from fossil plants have distinct philosophical bases. (Axelrod 1966; Axelrod and Bailey 1969) suggested that taxonomic similarities between floral fossil assemblages and present-day forests could be used to infer paleoclimates. Each fossil's taxon, defined at the species level when possible, is assigned a nearest living relative. Then, the climatic parameter of a present day forest containing as many nearest living relatives of the floral assemblage as possible is assigned to the paleoclimate of the fossil locality. Criticisms of this assumption are summarized in Forest et al. (1999). This assumption can be avoided if physiognomic characteristics of plants are used to infer local climates. Much work has shown that relationships exist between foliar characteristics and various climate parameters (Bailey and Sinnott 1915, 1916; Wolfe 1979, 1993; Gregory and Chase 1992; Gregory 1994) indicating that this latter approach is more useful and appropriate.

Estimating elevation from differences in climatic parameters has been accomplished via two methods. In the first method, differences in surface temperatures are used to infer paleo-altitudes. This method requires using empirical relationships between altitude and temperature derived from the present climate. Such relationships (a.k.a. surface temperature lapse rates) lack a firm theoretical basis, and variations in present-day surface temperatures with altitude in the western United States show large spatial variations (Meyer 1986, 1992; Wolfe 1992). Moreover, we have no reason to expect that such laterally varying empirical relationships should hold in the different climates that have prevailed in geologic time. We therefore seek an inferrable thermodynamic quantity whose distribution with altitude and longitude is constrained well by both theory and observation. The use of a quantity derived from fundamental thermodynamic laws is more reliable than one fit empirically to data spanning a fraction of the twentieth century—a small fraction of geologic history. The next section addresses such a quantity and develops its use as a second method for estimating paleoaltitudes. This follows the description in Forest et al. (1999).

USING MOIST STATIC ENERGY

Two conservative thermodynamic variables commonly used in atmospheric physics are moist static energy and equivalent potential temperature, each derived from the first law of thermodynamics (e.g., Wallace and Hobbs 1977; Emanuel 1994). Moist static energy (per unit mass), the sum of moist enthalpy and gravitational potential energy per unit mass, is the total specific energy content of air, excluding kinetic energy, which is very small ($<<1\%$) compared with the other terms (Peixoto and Oort 1992). Moist static energy, h, is written

$$h = c'_p T + L_v q + gZ = H + gZ \qquad (1)$$

where c'_p is the specific heat capacity at constant pressure of moist air, T is temperature (in K), L_v is the latent heat of vaporization for water, q is specific humidity, g is the gravitational acceleration, Z is altitude, and H is moist enthalpy. We use the specific heat capacity of moist air, $c'_p = c_{pd}(1 - q) + c_w q$, to account for compositional changes of the air, where c_{pd} and c_w are the specific heat capacities of dry air and liquid water respectively. For consistency, we must also account for the temperature dependence of the latent heat of vaporization, $L_v = L_{vo} + (T - 273)(c_{pv} - c_w)$, where c_{pv} is the specific heat capacity at constant pressure of water vapor and L_{vo} is the latent heat of vaporization at 0 °C. These second order effects (in both L_v and c'_p) are most important in warm and moist climates, where they contribute variations of ~5–10 kJ/kg, which, if ignored, would introduce errors in elevation of 500-1000 m.

Changed only by radiative heating and surface fluxes of latent and sensible heat, moist static energy, like equivalent potential temperature, is virtually conserved following air parcels. We consider only moist static energy because the relationship between altitude and equivalent potential temperature is less simple and direct. Considering the conservative properties of h, we note that typical variations in its components are roughly of equal magnitudes (Table 1).

Table 1. Variations in moist static energy components.

Parameter	Range	Energy Component	Range
Temperature, T	0-30 °C	Specific Heat, $c_p T$	0-30 kJ/kg
Specific Humidity, q	0-20 g/kg	Latent Heat, $L_v q$	0-50 kJ/kg
Elevation, Z	0-4000 m	Gravitational Potential, gZ	0-40 kJ/kg

Two properties of moist static energy make it a desirable candidate for inferring paleoelevations. First, as mentioned above, it is nearly conserved following air parcels and, hence, is approximately constant along trajectories. Second, the value of h in the boundary layer is usually strongly constrained by convection to be nearly equal to the value of h in the upper troposphere (see "Introduction" section). Owing to the Earth's rotation, air in the middle-latitude upper-troposphere flows nearly from west to east, so that contours of h there should also run approximately west-east. This implies that h at the surface should be nearly invariant with longitude. We will test this inference of longitudinal invariance in the "Spatial Distribution of Moist Static Energy" section.

Assuming that h is invariant with longitude along the Earth's surface, if we can estimate enthalpy at sea level ($H_{sea\ level}$) for a particular latitude, it follows from (1) that the altitude of another location at the same latitude is given by

$$Z = \frac{H_{sea\ level} - H_{high}}{g} \qquad (2)$$

where H_{high} is the enthalpy at the high altitude location (see Fig. 1).

In summary, the necessary assumptions to estimate paleoaltitude are that the surface moist static energy is invariant with longitude and that the moist enthalpy is a measurable quantity for the regional paleoclimate. The remainder of this paper tests these two assumptions and quantifies each error source. Given that previous work uses paleobotanical data, the second requirement implies that moist enthalpy must be accurately estimated from extant or newly discovered paleobotanical collections.

SPATIAL DISTRIBUTION OF MOIST STATIC ENERGY

Theoretical constraints

Before testing the assumption of longitudinal invariance, we discuss several meteorological constraints on the distribution of h.[1] Particularly, we consider three processes, each affecting h differently: (1) boundary layer convection, (2) free atmosphere convection and (3) large-scale horizontal motions. The interaction of these atmospheric dynamic processes constrains moist static energy to be longitudinally invariant. First, we consider convection in an unsaturated atmospheric boundary layer, approximately the lowest 1 km of the atmosphere. Boundary layer convection maintains a dry adiabatic vertical temperature gradient (lapse rate) which is $g/c_{pd} = 9.8$ K/km. Hence, $c_{pd}T + gZ$ is essentially constant with altitude in the boundary layer. Moreover, the concentration of tracers is well mixed in such layers implying that q should be invariant with altitude in this layer (see Fig. 2). Thus h should be nearly constant with altitude, a result confirmed by observations (Stull 1989). This allows us to write

$$h_{surface} \cong h_{topBL} \qquad (3)$$

where $h_{surface}$ is the value of h at the Earth's surface and h_{topBL} is the value of h at the top of the boundary layer.

---

[1]   These concepts highlight the basic dependence of climate on latitude (not longitude) and mechanisms for supporting them. Fundamental to these are that the pole to equator temperature gradient sets up a zonal jet (a region of enhanced west-to-east winds in the upper-troposphere, see Peixoto and Oort 1992). With no topography or continents, this jet structure would be the dominant feature of the climate system (e.g., consider the banded image of Jupiter's atmosphere). For mid-latitudes on Earth (poleward of ~30°), this is approximately the appropriate framework although both continents and topographic effects alter this slightly. Were certain features of Earth to be fundamentally different (rotation rate, equator-to-pole temperature difference, radiative balance, etc.), these features would change and necessarily must be reconsidered. Assessing these fundamental constants for Earth in the past must also be considered from the paleo-record.

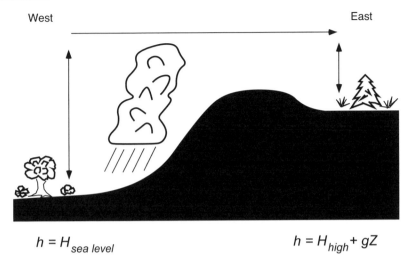

West   East

$$h = H_{sea\ level}$$

$$h = H_{high} + gZ$$

Figure 1. Surface air conserves moist static energy as it traverses a mountainous region by converting between internal heat, latent heat, and potential energy. The potential energy for the high elevation site is the difference in moist enthalpies, $H_{sea\ level} - H_{high}$, which yields the elevation estimate. The vertical arrows indicate the maintenance of the moist static energy profile by convective transports. The horizontal arrow indicates the typical west to east flow in the upper troposphere. [Used by permission of Geological Society of America, from Forest et al. (1999), Geol. Soc. Am. Bull., Vol. 111, Fig. 1, p. 499.]

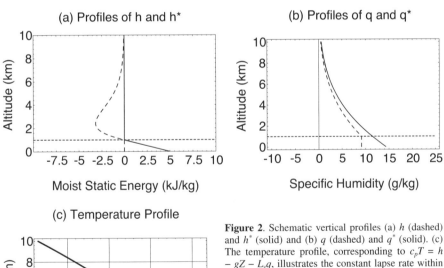

Figure 2. Schematic vertical profiles (a) h (dashed) and h^* (solid) and (b) q (dashed) and q^* (solid). (c) The temperature profile, corresponding to $c_p T = h - gZ - L_v q$, illustrates the constant lapse rate within the boundary layer and the reduced lapse rate above the boundary layer. The boundary level (1 km) is indicated by the horizontal dashed line in each panel. These profiles illustrate typical climatic values that are determined by moist convective adjustment in the free atmosphere and dry adiabatic convection in the boundary layer. [Used by permission of Geological Society of America, from Forest et al. (1999), Geol. Soc. Am. Bull., Vol. 111, Fig. 2, p. 500.]

Above the atmospheric boundary layer, the vertical temperature gradient is often close to its moist adiabatic value (Betts 1982; Xu and Emanuel 1989), implying the near invariance of the saturation moist static energy:

$$h^* = c'_p T + L_v q^* + gZ \qquad (4)$$

where q^* is the saturation specific humidity. Unlike q, q^* is a state variable (i.e., a function of temperature and pressure alone). Now suppose a sample of air is displaced upward from the boundary layer far enough that the cooling makes it saturated; its value of h will also be its value of h^*. Comparing the air sample's value of h to the value of h^* in the immediate environment is equivalent to comparing its temperature with that of its environment. If we neglect the small dependence of density on water substance in the sample, the buoyancy of the sample is a function of T, which directly determines h^* at a given altitude. Thus, the vertical profile of h^* indicates the buoyancy of vertically displaced air parcels. The frequently observed neutral state for moist convection is thus characterized by $\partial h^*/\partial Z \approx 0$. In addition, the climatic value of h at the bottom of the free atmosphere is constrained to be the climatic value of h^*. If $h_{topBL} > h^*_{topBL}$, the atmosphere would be unstable to convection. If $h_{topBL} < h^*_{topBL}$, surface heating and boundary layer convection would raise h_{topBL} until convection occurs throughout the troposphere. For climatic time-scales, the equilibrium between free atmosphere convection and surface heating results in the statement, $h_{topBL} \approx h^*_{topBL}$. Because the vertical profiles are continuous, we can also write $h_{topBL} \approx h^*_{botFA}$ where h^*_{botFA} is the value of the saturation moist static energy at the bottom of the free atmosphere. We recognize that for mid-latitudes, this approximation holds along surfaces of constant angular momentum, rather than strictly along the vertical direction, but these surfaces are nearly vertical. Finally, in the high troposphere the absolute temperature is small and $q \approx 0$. Therefore, at these altitudes, $h \approx h^*$. These constraints allow us to write

$$h_{surface} \cong h_{top} \qquad (5)$$

where h_{top} is the value in the upper troposphere (Fig. 2).

Aside from convective constraints on h, we expect h to be nearly longitudinally invariant in the upper troposphere. This results from the conservative property of h along trajectories and from the generally west-to-east winds at these high altitudes. Dynamical constraints require that basic flow properties vary only over horizontal scales greater than about 1000 km, the Rossby radius of deformation in the middle latitude troposphere. We assume that radiative cooling will not affect the distribution of h because the time-scale for radiative transfer is much longer than that associated with transport across the continent. In summary, since h should not vary rapidly with longitude in the upper troposphere and h^* is nearly equivalent to h there, and since h^* is constrained to equal h in the boundary layer by the condition of moist convective neutrality, h within the boundary layer (and thus at the surface) should be approximately invariant with longitude.

This theoretical discussion is based on a conceptual model involving the interaction of the parcel theory for convection and the large-scale circulation of the atmosphere. It is important to note that other theoretical approaches exist to explain the global-scale climatological patterns of atmospheric temperatures and wind distributions. Although their explanations may differ from the continental-scale approach here, the implications for exploring paleoclimate should be considered in the future. As an example of where one can begin, Schneider (2007) provides a recent review of the maintenance of the thermal structure in mid-latitude troposphere and provides verification for the alignment of angular momentum surfaces and moist static energy via slantwise convection (see Fig. 3.5 in Schneider 2007). In brief, this results from the combination of constraints by both baroclinic adjustment (see Stone 1978; Stone and Carlson 1979) and moist convective adjustment (Xu and Emanuel 1989) in which the large-scale mixing occurs owing to baroclinic instability but with effects of moisture included. The resulting mixing by baroclinic eddies (length-scales ~1000 km) transports heat, moisture, and momentum from high values in the low-latitude extratropics towards low values in the higher latitude regions. As discussed above and in Schneider (2007), the effects of moisture add an additional constraint on the atmosphere's thermal structure (i.e., the lapse rate is moist adiabatic along angular momentum surfaces within the baroclinic eddies) and likewise provide a constraint on the atmospheric circulations. The dynamics of these interactions is a broad field and beyond the scope of this paper although the reader is encouraged to explore these further. Two recent papers (Korty and Schneider 2007; Korty and Emanuel 2007) have explored both the assumption of zonal invariance and the implications of these constraints in more equable climates. With these discussions, they provide additional support for the assumption of the zonal invariance of h in the paleoaltimetry context.

Surface observations of moist static energy

To test the assumption of longitudinal invariance of moist static energy at least in the present climate, we use the observed distribution of h across the North American continent. Forest et al. (1999) present a thorough discussion of climatic parameters related to h and important choices regarding annual and seasonal averages. Several conclusions were drawn in Forest et al. (1999). First, the zonal invariance of h is least for autumn averages followed by annual averages. Further, given the seasonality of plant life, one does not expect the autumn climatology to influence strongly the foliar characteristics (or other botanical data.) Thus, annual averages are the best choice for estimating paleoenthalpy for use in paleoaltimetry. Second, a comparison with the equivalent method for temperature variations indicated similar errors for constant surface lapse rates but thse final error estimates are altitude dependent.

Forest et al. (1999) examined the spatial distribution of h (Fig. 3). The winter and summer means of h (not shown, see Forest et al. 1999) deviate from longitudinal invariance much more than spring, autumn, and annual means, but all show similarities to large scale circulation patterns (Oort 1983) as expected. A method to estimate the circulation patterns for paleoclimates would provide a first order correction to the assumption of zonal invariance. However, this requires predicting stationary waves for the paleoclimate setting which is a current topic of active research. One such method is based on simulations with general circulation models (GCMs) using paleoclimate boundary conditions for the relevant eras. The sensitivity to surface elevation itself must be considered in addition to geographic distributions of continents and ocean bathymetry. Further discussion on land-surface, ice-distribution, greenhouse gas concentrations, natural dust aerosols, solar constant, rotation rate, and other factors must also be considered.

Forest et al. (1999) quantified the longitudinal variability by statistically fitting h to monotonic functions of latitude, φ. Because invariance is being tested, the standard deviation from the function is more important than the actual function. The mean values of h were fit to a cubic function of φ using a standard least squares technique (Table 2) (see Fig. 3). As expected, winter and summer means show the largest deviations. Autumn exhibits the smallest deviations. We do not expect plant characteristics to correlate with mean autumn enthalpy, when photosynthesis and plant growth are minimal. Since foliar physiognomic characteristics correlate with mean annual temperature, we expect them to correlate with mean annual enthalpy. If one season revealed markedly smaller zonal variation, the correlations of climatic parameter in that season with foliar physiognomy could have been investigated, but from Table 2, we note that no seasonal deviation appears significantly better than the mean annual case. For the mean annual

Table 2. Zonal variability of moist static energy (in kJ/kg).

Period →	winter	spring	summer	autumn	annual
h_{mean}	6.80	5.15	5.44	3.79	4.48

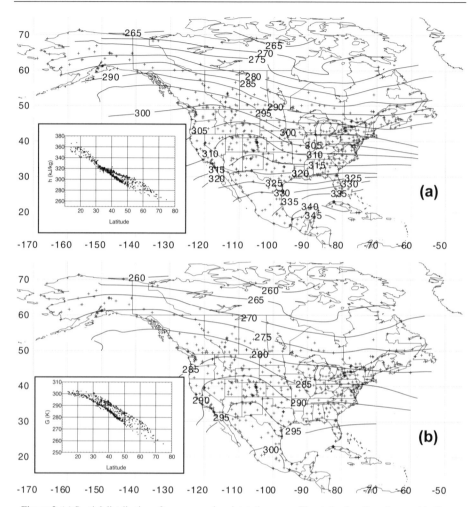

Figure 3. (a) Spatial distribution of mean annual moist static energy. The station locations (crosses) indicate the spatial coverage of the data set. (b) Spatial distribution of the function $G = T + \gamma_t Z$ where T is the mean annual temperature, Z is the station elevation, and $\gamma_t = 5.9$ K/km. The distributions of each variable as functions of latitude are shown in the insets. [Used by permission of Geological Society of America, from Forest et al. (1999), Geol. Soc. Am. Bull., Vol. 111, Fig. 3, p. 502.]

values, the standard deviation from zonal invariance of h is 4.5 kJ/kg (see Table 2). Dividing 4.5 kJ/kg by g yields a minimum estimate of the error in altitude of 460 m.

Moist static energy versus mean annual temperature

The focus of the previous section was to estimate the expected error from assuming the zonal invariance of mean values of moist static energy. This expected error contributes to the total expected error of a paleoaltitude estimate. Before proceeding to the next section (inferring paleoclimate from plant fossils), we examine the zonal invariance assumption as applied in the mean annual temperature approach to paleoaltimetry. Based on the initial method of Axelrod (1966), paleoaltitudes can be estimated by comparing mean annual temperature differences using the formula

$$Z = \frac{T_{sea\ level} - T_{high}}{\gamma_t} \tag{6}$$

where γ_t defines an empirical coefficient relating surface temperature linearly to elevation and is usually chosen to agree with the local climate (Axelrod 1966; Wolfe and Schorn 1989; Gregory and Chase 1992). Meyer (1992) calculated local γ_t for 39 areas of the world with surface topography greater than 750 meters and found $\gamma_t = 5.9 \pm 1.1$ K/km, but with a range from 3.64-8.11 K/km. Wolfe (1992) showed that average values of γ_t calculated from mean annual temperatures at high altitudes and at sea level at the same latitudes vary over a similar range.

In using mean annual temperatures, one implicitly assumes the longitudinal invariance of the surface distribution of a quantity $G = T + \gamma_t Z$ (bottom of Fig. 3). For $\gamma_t = 5.9$ K/km, we calculated deviations from longitudinal invariance of G to be 3.2K for the present climate. Hence, a minimum standard error, σ_Z, for estimating altitude can be computed from these results, using

$$\sigma_Z^2 = \frac{\sigma_G^2 + \sigma_T^2}{\gamma_t^2} + \frac{(\Delta T)^2 \sigma_{\gamma_t}^2}{\gamma_t^4} \tag{7}$$

where σ_G and σ_{γ_t} are the standard errors of G and γ_t respectively. With the inclusion of an uncertainty in γ_t of ± 1.1 K/km (Meyer 1992), minimum uncertainties in the estimated altitudes increase with altitude: standard errors of 540 m, 660 m, and 920 m for altitudes of 0 m, 2000 m, and 4000 m, respectively. Hence, even ignoring the unpredictable temporal changes in γ_t over geologic time, these estimates of minimum uncertainties exceed that of 460 m from assuming longitudinal invariance in h. We note that Wolfe and others typically use lower values of γ_t for the warm periods such as the Eocene. This implies that the error due simply to an uncertainty in γ_t, $[(\Delta T)\sigma_{\gamma_t}]/\gamma_t^2$, will reach 1500 m with $\gamma_t = 3$ K/km for a 4 km high mountain, and the overall altitude error, σ_Z, will increase for higher estimates of altitude.

We suspect that the variations in γ_t from 3 to 9 K/km, determined by Meyer (1986, 1992) and Wolfe (1992), are consistent with approximately longitudinal invariance of h and large variations in q. Differentiating (1) with respect to Z while holding h constant yields

$$-\frac{\partial T}{\partial Z} = (g + L_v \frac{\partial q}{\partial Z}) \frac{1}{c'_p} \tag{8}$$

Because air cools adiabatically as it rises, and because q depends strongly on T (at saturation) (see Forest et al. 1999), q will not be conserved as an air mass rises over high terrain. Condensation and precipitation reduce q, but conservation of h requires that at a constant pressure, T increases when condensation occurs. Thus, in such regions like the Sierra Nevada, vertical gradients of specific humidity directly determine the vertical temperature gradients. Indeed, in that region, temperatures decrease slowly with altitude, at only 3 K/km (Wolfe 1992). Regardless of whether or not conservation of h can account for other large variations in q, clearly H, instead of temperature, should be used to estimate elevation changes, at least insofar as mean annual enthalpy can be inferred well from paleobotanical material.

RELEVANT PALEOBOTANICAL WORK

A major goal of paleobotanists is to quantitatively estimate the climatic environment of a paleoflora. To achieve this goal, paleobotanists first distinguish the various taxa represented by the fossil collection/assemblage that defines the paleoflora. The research presented here does not address problems associated with this task and assumes the taxonomy is done properly. Secondly, paleobotanists identify features of the paleoflora that are related to the local climate.

Their goal is to characterize the plant life in such a manner to distinguish ecological niches determined by climate. The method chosen here describes the average characteristics of sizes and shapes of leaves for the flora in order to make the distinctions and to estimate the mean climate. Before proceeding to quantify the relationship between leaf characteristics and climate, we recognize that from a paleoaltimetric view , one can be agnostic about the method used to determine the paleo-enthalpy of the local region provided the estimated errors have sufficiently small contribution to the total error. Were a better method available, it could complement the paleobotanical approach given here and lead to more robust paleo-altitude estimates. One recent example would be Green (2006) in which estimates of paleoenvironmental variables are done with a nearest-neighbor approach to relating foliar physiognomy to paleoclimate. As others methods are developed, they will be quite useful for obtaining robust paleoaltitude estimates.

Previous physiognomic methods for estimating paleoclimate

As summarized in Forest et al. (1999), previous work has shown (Bailey and Sinnott 1915, 1916; Wolfe 1971, 1979; Wolfe and Hopkins 1967) that foliar physiognomy correlates with climatic parameters, such as mean annual temperature or growing season precipitation, suggesting that we might expect to predict mean annual enthalpy through similar techniques. Because of other correlations between foliar shape and living environment (Givnish 1987), Wolfe (1993) expanded the list of characteristics, from the single characteristic of the fraction of species with smooth margins of leaves, to include other foliar shape parameters: size, apex and base shape, lobedness, compoundness, and overall shape. With these additions, he developed the Climate-Leaf Analysis Multivariate Program (CLAMP) utilizing correspondence analysis (Wolfe 1993) to determine the relationship between climate and leaf shapes and sizes. Others have also employed multivariate regression techniques using the CLAMP data to infer paleoclimate parameters (Wolfe 1990; Gregory and Chase 1992; Gregory 1994; Forest et al. 1995; Greenwood and Wing 1995). We note that several critiques (Wilf 1997; Royer et al. 2005; Spicer et al. 2005) of the method used here have appeared and should be considered as they relate to the robustness of estimates for paleoclimate based on foliar physiognomy. These highlight the need for careful analysis of these methods.

ESTIMATING PALEOCLIMATE

The task remains to determine the mean annual enthalpy from plant physiognomy. An analysis is presented relating foliar physiognomic characters to mean annual values of enthalpy, temperature, specific humidity, and relative humidity that exploits the method and data in the Climate-Leaf Analysis Multivariate Program (Wolfe 1993). From present-day plant data collected from North America, Puerto Rico, and Japan, the leaf parameters are searched for linear combinations of the foliar characteristics that covary with the local climates. By doing so, the foliar characteristics can be determined that covary with one another and which best correlate with climate parameters.

Data

To extract climate information (e.g., mean annual temperature, precipitation, specific humidity or moist enthalpy) from fossil flora, one must first determine which typical foliar characteristics of extant plants show relations to present-day local climate. For forests in well characterized climates, Wolfe (1993) measured average characteristics of foliar physiognomy, weighting each taxon equally (otherwise, taxa are ignored). For each taxon, he identified and recorded the characteristics of its representative leaves. The characteristics can be separated into seven categories (see Tables 3 and 4 and Fig. 4). For each vegetation site, he totaled the number of displayed characteristics for all represented species and calculated the percentage of species exhibiting a given characteristic. He called this percentage the score of the leaf

Table 3. Foliar characteristics (see Fig. 4 for examples).

Characteristic	No. of Categories	Description
lobed	1	
margin shape	6	no teeth, regular, close round, acute, compound
leaf size	9	nanophyll;leptophyll I,II; microphyll I,II,III; mesophyll I,II,III
apex shape	4	emarginate, round acute, attenuate
base shape	3	cordate, round, acute
length to width ratio	5	<1:1, 1-2:1, 2-3:1, 3-4:1, >4:1
leaf shape	3	obovate, elliptic, ovate

Table 4. Specification of foliar size categories.

Size Category	Leaf Area (mm²)
Nanophyll	< 5
Leptophyll I	5–25
Leptophyll II	25–80
Microphyll I	80–400
Microphyll II	400–1400
Microphyll III	1400–3600
Mesophyll I	3600–9000
Mesophyll II	9000–15000
Mesophyll III	>15000

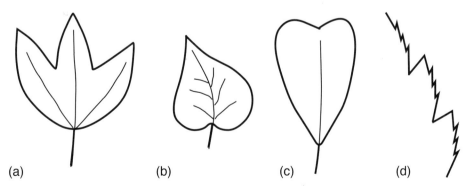

(a)   (b)   (c)   (d)

Figure 4. Sketches of representative leaves showing various leaf characteristics. (a) illustrates a lobed leaf with an acute apex, a round base, and no teeth (like (b) and (c)). (b) illustrates an ovate shaped leaf with an attenuated apex (drip-tip), and a cordate base. (c) illustrates an obovate shaped leaf with a round and emarginate apex, and an acute base. (d) illustrates a leaf margin with teeth that are compound, acute, closely spaced, and regularly spaced. (see Wolfe (1995)) [Used by permission of Geological Society of America, from Forest et al. (1999), Geol. Soc. Am. Bull., Vol. 111, Fig. 6, p. 505.]

characteristic for a given site. In all, Wolfe (1993) scored leaf samples for 31 leaf characteristics, using 123 vegetation sites distributed across the northern hemisphere, of which 112 are in North America, 9 are in Puerto Rico, and 2 are in Japan. Wolfe chose these sites to be near meteorological stations to obtain measurements of temperature and precipitation, but most stations did not record humidity.

To obtain an accurate score, a sufficient number of species for a given location are required. Wolfe estimated that at least 20 species are necessary to infer climate parameters reliably. This is supported by Povey et al. (1994) and by considering possible sources of error in the score data and comparing the error to some measure of information content in the data.

Present-day environmental data are also required for the leaf collection sites. Because humidity data are not available for each vegetation site, we could not directly calculate mean enthalpy for all sites. Instead, we used data from the meteorological stations (Fig. 3) and interpolated moist static energy to the vegetation sites. For the same reasons that we expect h to be longitudinally invariant, we expect h to be a smoothly varying field. Hence, the local variability of h should be small over dimensions of mountains and valleys. Having a non-gridded dataset, we followed meteorological methods (Barnes 1964) and interpolated from nearby stations using a distance-weighted mean,

$$h_{plantsite} = \frac{\sum_i (h_i w_i)}{\sum_i w_i} \tag{9}$$

where $w_i = exp[(r_i/d)^2]$, r_i = radial distance from plant site to the meteorological station i, and d is the average distance to the nearest station, 90.8 km. Knowledge of the elevation, Z, and mean annual temperature, T, at the vegetation sites allows the calculation of mean moist enthalpy, $H = h - gZ$, and specific humidity, $q = (H - c'_p T)/L$, from the interpolated value of h. Using data from 123 sites, we sought correlations of the 31 leaf characteristics with moist enthalpy and also, separately, with mean annual values of temperature, specific humidity, and relative humidity.

Data analysis

The relation between leaf physiognomy and climate parameters has been analyzed by Canonical Correspondence Analysis (CANOCO) (ter Braak 1986; ter Braak and Prentice 1988), a form of direct gradient analysis. Similar to ecological research regarding the relation between species and environment, predicting climate parameters from foliar physiognomic characteristics follows an analogous technique (ter Braak and Prentice 1988). Fundamentally, two problems exist in this type of research: (1) the removal of redundant information in the foliar physiognomic data and (2) estimating a robust relation between the climate parameter and physiognomic data. Traditionally, the first problem requires removing redundant information (Fig. 5) by identifying combinations of foliar data that vary together. Following the ecological community's terminology, this is the problem of ordination. The second problem has been termed gradient analysis and is a form of standard regression. Two forms of gradient analysis are common. First, indirect gradient analysis involves the use of some ordination technique before the regression relationship is found and does not make use of the climate information to constrain the linear combination of physiognomic variables. The alternative, direct gradient analysis, provides such a constraint and is implemented in CANOCO. The distinction of CANOCO from other gradient analysis techniques (e.g., canonical correlation analysis (CCA) or linear regression using principal component analysis (LR/PCA)) lies with the method of ordination. Rather than assuming a linear relation, Canonical Correspondence Analysis, like its counterpart Correspondence Analysis, assumes that the distribution of foliar characteristics along the environmental gradient has a Gaussian shape, as observed in the CLAMP data.

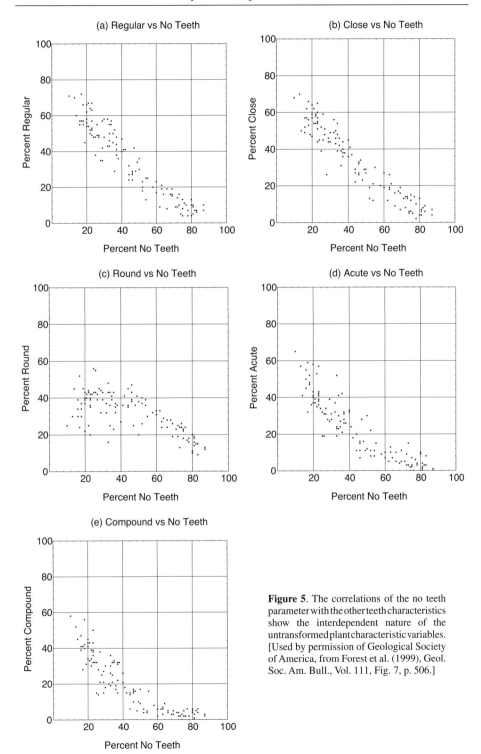

Figure 5. The correlations of the no teeth parameter with the other teeth characteristics show the interdependent nature of the untransformed plant characteristic variables. [Used by permission of Geological Society of America, from Forest et al. (1999), Geol. Soc. Am. Bull., Vol. 111, Fig. 7, p. 506.]

The CANOCO output includes physiognomic and environmental axes and the eigenvalues of the respective axes. Each axis represents a weighted linear combination of the physiognomic scores or environmental parameters that explains a certain amount of variance in the given data. The method chooses the first axis by determining the direction in data space that maximizes the variance explained by the first axis. Subsequent axes are chosen to maximize the variance explained in the remainder of the data. Thus, the first environmental axis represents the weighted linear combination of climate parameters that maximizes the variance explained in the physiognomic data. Similarly, the first physiognomic axis represents the weighted linear combination of physiognomic scores that explains the most variance in the environmental data. (This is only true for the first axes of each data set, however, because the subsequent axes are only constrained to be orthogonal to the previously determined axes.) Because of these relations, the environmental and physiognomic data sets are transformed onto the principal axes and the scores are combined into a single plot.

The estimate of a climate parameter is obtained by projecting the physiognomic scores of a given sample onto the vector obtained for the climate parameter. The value of the climate parameter is plotted against the value of the projection onto the climate vector in the physiognomic space. A least squares fit is obtained and provides a predictive formula for the given climate parameter. A complete discussion of this technique is given in Wolfe (1995).

Results of physiognomy/climate analysis

From the CLAMP data and associated mean annual climate data, Forest et al. (1999) obtained estimates of enthalpy, temperature, relative humidity, and specific humidity (Fig. 6). The data set has been reduced by removing the outliers as indicated by scores along the third and fourth axes (see Wolfe 1995 for a description). The axis eigenvalues from CANOCO indicate that significant information is contained in the first 6 axes and implies that the use of the axes three and four as an outlier indicator should be robust. The estimates of the climate data indicate that mean annual enthalpy can be predicted from fossil leaf physiognomy with an uncertainty of $\sigma_H = \pm 5.5$ kJ/kg. Additionally, the standard errors for the estimates of temperature, specific humidity, and relative humidity are respectively, $\sigma_T = 1.8$ °C, $\sigma_q = 1.7$ g/kg, and $\sigma_{RH} = 13\%$.

The interpretation of the projected scores along the given axes is simplified because CANOCO transforms both the environmental and physiognomic data onto the same axes (see Fig. 7). The scores of enthalpy along the first and second axes indicate that enthalpy is related to the temperature and moisture stress axes (one and two, respectively) with about equal importance. As anticipated, the enthalpy score plots roughly equally along the first two axes. The scores of the relative humidity and specific humidity on the first and second axes indicate that relative humidity aligns more strongly with the moisture stress axis than does specific humidity. Intuitively, relative humidity should align with the moisture stress axis if we consider that relative humidity is a measure of the departure from saturation conditions and nearly independent of temperature. Alternatively, the specific humidity strongly depends on the temperature as well as on the availability of moisture which implies the specific humidity score should have a component along the temperature axis. The relative humidity vector being nearly orthogonal to the temperature vector only highlights this more clearly.

The relative lengths of the environmental parameter vectors on the first and second axes (Fig. 7) can be used a measure of importance for the various climate parameters at constraining the directions of the axes. The mean annual temperature has the longest length, followed by enthalpy, specific humidity, and relative humidity. From these scores, we can identify the first axis with mean annual temperature while the second axis aligns with the mean specific or relative humidity. These associations also allow us to infer which character states are most important for estimating the climate parameters as discussed next.

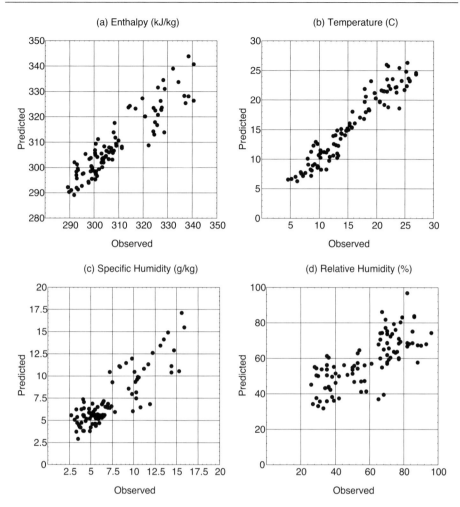

Figure 6. The predictions of the climate parameters from the plant characteristic variables as given by CANOCO are plotted versus the observations for the plant collection sites for (a) mean annual enthalpy (kJ/kg), (b) temperature (°C), (c) specific humidity (g/kg), and (d) relative humidity. [Used by permission of Geological Society of America, from Forest et al. (1999), Geol. Soc. Am. Bull., Vol. 111, Fig. 8, p. 507.]

The projections of the physiognomic characteristics onto a given axis represent the relative importance of the characteristics for explaining the environmental variations along that axis. The first axis (Fig. 8), which is strongly related to mean annual temperature, has significant contributions from the entire margin (i.e., no teeth), small leaf sizes, and emarginate apex character states. In contrast to this, the second axis, related to moisture stress, has contributions from the large leaf sizes, attenuate apices, and long narrow leaves. The projections onto the third through fifth axes are shown as well.

Before estimating the total expected error in paleoaltitude, we estimate the contribution, 560 m, from the uncertainty, $\sigma_H = 5.5$ kJ/kg, in predicting mean annual enthalpy. We estimate a comparable error, 390 m with $\gamma_t = 5.9$ K/km, for the mean annual temperature approach. Clearly, this latter error is an underestimate, because it is dependent on the choice of γ_t whereas the former error will remain constant.

Forest

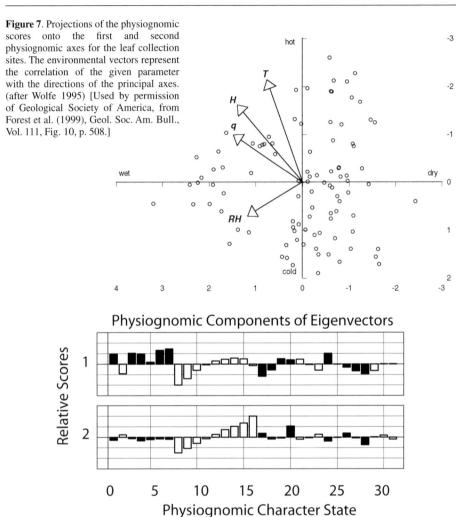

Figure 7. Projections of the physiognomic scores onto the first and second physiognomic axes for the leaf collection sites. The environmental vectors represent the correlation of the given parameter with the directions of the principal axes. (after Wolfe 1995) [Used by permission of Geological Society of America, from Forest et al. (1999), Geol. Soc. Am. Bull., Vol. 111, Fig. 10, p. 508.]

Figure 8. Normalized components of the first two eigenvectors of the Canonical Correspondence Analysis, showing the relative contributions from the 31 leaf characteristics. The order of the characteristics corresponds to that in Tables 4 and 5. Within each eigenvector, the components are normalized by the maximum value which are respectively, 1.08 and 0.88. [Used by permission of Geological Society of America, from Forest et al. (1999), Geol. Soc. Am. Bull., Vol. 111, Fig. 11, p. 509.]

TOTAL EXPECTED ERROR IN PALEOALTITUDE

We obtain an expected error for the paleoaltitude by combining the expected errors from the zonal asymmetry, $\sigma_h = 4.5$ kJ/kg, and from the botanical inferences of enthalpy, $\sigma_H = 5.5$ kJ/kg, at each location.

$$\sigma_z = \sqrt{\frac{2\sigma_H^2 + \sigma_h^2}{g^2}} = 910\,\text{m} \tag{10}$$

where the errors in enthalpy from sites at sea level and inland have been combined to yield the factor of two. Other quantifiable sources of error could be included in a similar manner.

We can also estimate the standard error in altitude by predicting the altitude of the present-day plant collection sites. We assume the sea level enthalpy follows a linear function of latitude based on the latitudinal distribution of moist static energy as described in the "Using Moist Static Energy" section. Restricting our data to latitudes south of 55°N , where our assumption of zonal symmetry is valid, the standard deviation of the predicted altitude is 620 m (see Fig. 9). We believe that this lower error estimate results partially from the use of altitude to estimate enthalpy at plant sites for which we have no humidity data. For such sites, we relied on meteorological estimates of *h* and heights of sites to infer *H* causing an unavoidable dependence of the value of *H* on altitude. The lower error estimate of 620 m implies that the error estimate of 910 m calculated from expected errors in the components is robust.

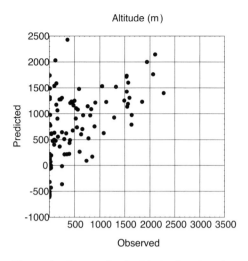

Figure 9. The predicted altitude for the plant collection sites vs. the true altitude of the sites. [Used by permission of Geological Society of America, from Forest et al. (1999), Geol. Soc. Am. Bull., Vol. 111, Fig. 12, p. 509.]

Forest et al. (1999) discussed some potential sources of error related to taphonomic, microclimatic, and fossil dating considerations, which are difficult to quantify for a paleoaltitude estimate. The reader is encouraged to review these as they represent specific issues for improving the interpretation of the fossil record. In total, only the fossil dating uncertainty can be quantified in a straightforward fashion and included as an additional source of error in Equation (10). Forest et al. (1999) estimated an additional error of 5 kJ/kg resulting from uncertainty in unforced variability in the climate system on time-scales shorter than the temporal resolution of the paleo-record. When this is included, the total expected error increases to 1080 m. In any case, the potential errors associated with transport of leaves, fossil dating, and local variations in climate are reminders that we should expect modifications and improvements in the approach taken here.

In addition to these errors, it is worth considering that the errors for the zonal invariance of *h* are the expected random errors and would not include any systematic biases. As discussed in the "Spatial distribution of Moist Static Energy" section, the assumptions were derived for mid-latitudes (i.e., north of ~30°N) and tested only in the northern hemisphere. There are at least two cases where the moist convective neutrality assumptions can break down and these are offered as additional caveats to those looking to apply this method or as areas to consider for theoretical improvements. The first case is in regions where moist convection is not dominating the thermal stratification of the troposphere. The descending branches of the Hadley or Walker circulations (see Peixoto and Oort 1992) are such places where the atmospheric radiative cooling is balancing the large-scale descent. When applying this in the tropical regions, the possible locations could be in one of these regions and a bias would be present. Because the thermodynamic paleoaltimetry method requires estimates of moist enthalpy from two locations, the application would require an alternative assumption for assuming the invariance of moist static energy.

A second situation is where the boundary layer may contain an additional term in its energy budget from surface fluxes as the air moves along its surface trajectory. This leads to a decoupling of the moist static energy in the boundary layer from the values in the upper troposphere and can be found in the coastal regions where colder deep-ocean waters are upwelling and affecting surface temperatures (M. Huber, pers. comm.). This would then lead to a bias in the estimated enthalpy at the sea-level location and need to be considered. In the present-day climate, this affects temperatures along the west coast of North America. The extent that this can be quantified for the present-day and incorporated into this method would be worthwhile. In both cases, the underlying assumptions of invariance of moist static energy would not be valid and so the application of this method in the appropriate paleoclimate setting would need to be considered. These issues highlight the need for estimating a complete picture of the paleoclimatic conditions that includes the oceanic and atmospheric circulations to the best extent possible. Whether theoretical conceptual models (as done here for mid-latitudes) or numerical models (e.g., Huber and Sloan 2001; Huber et al. 2003; Spicer et al. 2003) are employed, climate models of some sort are required to supply such information. Advances in these areas will be critical to improving the thermodynamic methods of paleoaltimetry.

APPLICATIONS OF THERMODYNAMIC METHODS

The following papers make reference to thermodynamic methods based on paleobotanical estimates of paleoclimate: Gregory (1994), Forest et al. (1995), Wolfe et al. (1997), Wolfe et al. (1998), Chase et al. (1998), Gregory-Wodzicki (2000), Spicer et al. (2003), and Graham et al. (2001). Though this is probably not a complete list, it indicates that the use of these methods is expanding where data are available. Three of these results are briefly discussed here with the estimated paleo-elevations and their relevance for paleoclimate or tectonics.

Wolfe et al. (1997) analyzed twelve mid-Miocene floras from western Nevada to explore the paleoaltitudes of this region and time. The analyses indicated that this part of the Basin and Range Province stood ~3 km above sea level at 15 to 16 Ma ago, which is 1 to 1.5 km higher than its present altitude. This suggests a significant collapse of this region to its present-day altitude by about 13 Ma ago.

Wolfe et al. (1998) explored the paleoclimates from Eocene and Oligocene fossil leaf assemblages from middle latitudes in western North America. These results indicated paleoaltitudes were comparable or higher than present-day altitudes. High altitudes during Eocene and Oligocene time in western North America appear to have been normal, even in areas such as the Green River basin, and therefore cast doubt on the commonly inferred late Cenozoic uplift of that region.

Spicer et al. (2003) present paleo-altitude estimates from fossil leaf assemblages from the Namling basin in southern Tibet and dated to ~15 Ma ago. One unique aspect of this work is that they use a numerical general circulation model to estimate moist static energy at the location of the fossil leaves as opposed to estimating the sea-level moist enthalpy directly from fossil leaf assemblages. Despite the uncertainties (as discussed earlier) in using general circulation models for this purpose, they estimate the elevation of the Namling basin to be about 4.7±0.9 km, which is comparable to the present-day altitude of 4.6 km. This suggests that the elevation of the southern Tibetan plateau probably has remained unchanged for the past 15 Ma. We note that this raises a question regarding the shift in monsoon climate of the region estimated to occur near 6-9 Ma ago (Molnar et al. 1993) and has given rise to a recent debate on the uplift history of Tibet (Molnar et al. 2006; Rowley and Currie 2006).

SUMMARY AND CONCLUSIONS

The methods for estimating paleoaltitudes based on thermodynamic methods have been summarized. The basic algorithm requires comparing the local paleoclimate from some elevated region to that at sea level and at approximately the same latitude. It is argued that comparing mean annual temperatures alone has inherent uncertainties for estimating paleoaltitudes and should be replaced with the method in which moist enthalpies are compared. This is equivalent to assuming the invariance of moist static energy, h, between locations. Supported by basic physics, the moist enthalpy approach incorporates the effect of moisture on the temperature distribution. To estimate a paleoaltitude, it is required that h be invariant between fossil sites and that the moist enthalpy, H, be inferrable from fossil plant leaves. These requirements were examined using the present-day climatic and foliar data to quantify the expected errors for the method.

The surface distribution for mean annual h results from two properties of atmospheric flow: conservation of h following the large-scale flow and the maintenance of the vertical profile of h by convective processes. These features of the climate system allow one to quantify the expected errors for assuming that mean annual h is invariant with longitude and altitude for the present-day distribution. Forest et al. (1999) examined the distribution and calculated the expected error from assuming zonal invariance to be 4.5 kJ/kg for the mean annual climate. This error translates to an altitude error of 460 m and is compared with an equivalent error of 540 m from the mean annual temperature approach. Moreover, the uncertainty of the terrestrial lapse rate, γ_t, increases the expected error in elevation as elevations increase, particularly when small lapse rates are assumed.

We also reviewed the method for estimating paleo-moist-enthalpy. To estimate paleoenthalpy from plant fossils, Forest et al. (1999) quantified a relationship between leaf physiognomy and enthalpy from present-day plants and their local climate. Using Canonical Correlation Analysis, mean annual moist enthalpy can be estimated with an uncertainty of 5.5 kJ/kg. The contribution to the uncertainty in altitude is 560 m and is comparable to using temperature alone. Other statistical techniques that improve the ability to estimate enthalpy could replace the current method.

To apply these methods to a paleoclimate problem, one tacitly assumes that the errors as estimated from today's climate and plants are similar to those of the past. The deposition of plant fossils appears not to affect our climate estimates. Combining the estimated uncertainties indicates an altitude error of 910 m and compares favorably with altitude errors of 1100-1400 m from differences in paleopressure (Sahagian and Maus 1994) and 800-1500 m from differences in paleotemperature.

These methods have been used effectively to explore the tectonic and paleo-climate for regions in western North America, the Andes of South America, and the Tibetan plateau. The results provide additional evidence that must be included in debates about the paleobotany, paleoclimate, and tectonic evolution of these regions. The range of these results provides additional support for the need for such evidence in interpreting Earth's history and the need for continuing research in this area.

ACKNOWLEDGMENTS

The manuscript was significantly improved with two reviews, one anonymous and one by M. Huber. CEF also thanks M. Kohn for his patience and encouragement to include this manuscript in this volume.

REFERENCES

Axelrod DI (1966) A method for determining the altitudes of Tertiary floras. Paleobotanist 14:144–171
Axelrod DI, Bailey HP (1969) Paleotemperature Analysis of Tertiary Floras. Palaeogeog Palaeoclim Palaeoecol 6:164–195

Bailey IW, Sinnott EW (1915) A botanical index of Cretaceous and Tertiary climates. Science 41:831–834

Bailey IW, Sinnott EW (1916) The climatic distribution of certain types of angiosperm leaves. Am J Bot 3:24–39

Barnes SL (1964) An objective scheme for interpolation of data. J Appl Meteor 3:396–409

Betts AK (1982) Saturation point analysis of moist convective overturning. J Atmos Sci 39:1484–1505

Brook EJ, Brown ET, Kurz MD, Ackert Jr RP, Raisbeck GM, Yiou F (1995) Constraints on age, erosion, and uplift of Neogene glacial deposits in the Transantarctic Mountains determined from *in situ* cosmogenic ^{10}Be and ^{26}Al. Geology 23:1063–1066

Brown ET, Edmond JM, Yiou GMRF, Kurz MD, et al. (1991) Examination of surface exposure ages of Antarctic moraines using *in situ* ^{10}Be and ^{26}Al. Geochim Cosmochim Acta 55:2269–83

Chase CG, Gregory-Wodzicki KM, Parrish-Jones JT, DeCelles P (1998) Topographic history of the western cordillera of North America and controls on climate *In:* Tectonic boundary conditions for climate model simulations. Oxford Monographs on Geology and Geophysics 39. Crowley TJ, Burke K (eds) Oxford University Press, New York, 73–99

Emanuel KA (1994) Atmospheric Convection. Oxford University Press, New York

Forest CE, Molnar P, Emanuel KE (1995) Palaeoaltimetry from energy conservation principles. Nature 343:249–253

Forest CE, Wolfe JA, Molnar P, Emanuel KE (1999) Paleoaltimetry incorporating atmospheric physics and botanical estimates of paleoclimate. Geol Soc Am Bull 111:497–511

Givnish TJ (1987) Comparative studies of leaf form: assessing the relative roles of selective pressures and phylogenetic constraints. New Phytologist 106:131–160

Graham A, Gregory-Wodzicki KM, Wright KL (2001) Studies in Neotropical Paleobotany. XV. A Mio-Pliocene palynoflora from the Eastern Cordillera, Bolivia: implications for the uplift history of the Central Andes. Am J Botany 88:1545–1557

Green WA (2006) Loosening the CLAMP: An Exploratory Graphical Approach to the Climate Leaf Analysis Multivariate Program. Palaeontologia Electronica 9:1–17 <*http://palaeo-electronica.org/paleo/2006_2/clamp/index.html*>.

Greenwood DR, Wing SL (1995) Eocene continental climates and latitudinal temperature gradients. Geology 23:1044–1048

Gregory KM (1994) Palaeoclimate and Palaeoelevation of the 35 Ma Florissant Flora, Front Range, Colorado. Paleoclimates 1:23–57

Gregory KM, Chase CG (1992) Tectonic significance of paleobotanically estimated climate and altitude of the late Eocene erosion surface, Colorado. Geology 20:581–585

Gregory-Wodzicki KM (2000) Uplift history of the Central and Northern Andes: A review. Geol Soc Am Bull 112:1091–1105

Huber M, Sloan L, Shellito C (2003) Early Paleogene oceans and climate: A fully coupled modeling approach using the National Center for Atmospheric Research Community Climate System Model. *In:* Causes and Consequences of Globally Warm Climates in the Early Paleogene. Wing SL, Gingerich PD, Schmitz B, Thomas E (eds) Geological Society of America Special Paper 369:25–57

Huber M, Sloan LC (2001) Heat transport, deep waters, and thermal gradients: Coupled simulation of an Eocene "greenhouse" climate. Geophys Res Lett 28:3481–3484

Korty RL, Emanuel KA (2007) The dynamic response of the winter stratosphere to an equable climate surface temperature gradient. J Climate (in press)

Korty RL, Schneider T (2007) A climatology of the tropospheric thermal stratification using saturation potential vorticity. J Climate (in press)

Meyer HW (1986) An Evaluation of the Methods for Estimating Paleoaltitudes Using Tertiary Floras from the Rio Grande Rift Vicinity, New Mexico and Colorado. Ph.D. dissertation, Univ. Calif. Berkeley

Meyer HW (1992) Lapse rates and other variables applied to estimating paleoaltitudes from fossil floras. Palaeogeog Palaeoclim Palaeoecol 99:71–99

Molnar P, England P, Martinod J (1993) Mantle dynamics, uplift of the Tibetan Plateau, and the Indian monsoon. Rev Geophys 31:357–396

Molnar P, Houseman GA, England PC (2006) Palaeo-altimetry of Tibet. Nature 444:E4

NCDC (National Climatic Data Center) (1989) International Station Meteorological Climate Summary — CD-ROM v. 1 United States Government

Oort A (1983) Global atmospheric circulation statistics, 1958–1973. NOAA Prof Paper 14

Peixoto JP, Oort AH (1992) The Physics of Climate. American Institute of Physics

Povey DAR, Spicer RA, England P (1994) Palaeobotanical investigation of early Tertiary palaeoelevations in northeastern Nevada: initial results. Rev Palaeobot Palynol 81:1–10

Rowley DB, Currie BS (2006) Reply to: Palaeo-altimetry of Tibet by Molnar et al. (2006), Nature, 444, doi:10.1038/nature05368. Nature 444:E4–E5

Royer DL, Wilf P, Janesko DA, Kowalski EA, Dilcher DL (2005) Correlations of climate and plant ecology to leaf size and shape: Potential proxies for the fossil record. Am J Botany 92:1141–1151

Sahagian DL, Maus JE (1994) Basalt vesicularity as a measure of atmospheric pressure and palaeoelevation. Nature 372:449–451

Schneider T (2007) The thermal stratification of the extratropical troposphere. *In:* The Global Circulation of the Atmosphere. Schneider T, Sobel AH (eds) Princeton University Press, Princeton, NJ 47–77

Spicer RA, Harris NBW, Widdowson M, Herman AB, Guo S, Valdes PJ, Wolfe JA, Kelley SP (2003) Constant elevation of southern Tibet over the past 15 million years. Nature 421:622–624

Spicer RA, Herman A, Kennedy EM (2005) The sensitivity of CLAMP to taphonomic loss of foliar physiognomic characters. Palaios 20:429–438

Stone PH (1978) Baroclinic adjustment. J Atmos Sci 35:561–571

Stone PH, Carlson JH (1979) Atmospheric lapse rate regimes and their parameterization. J Atmos Sci 36:415–423

Stull R (1989) An Introduction to Boundary Layer Meteorology. Kluwer Academic Publ.

ter Braak CJF (1986) Canonical correspondence analysis: A new eigenvector technique for multivariate direct gradient analysis. Ecology 67:1167–1179

ter Braak CJF, Prentice IC (1988) A theory of gradient analysis. Adv Ecol Res 18:271–317

Wallace JM, Hobbs PV (1977) Atmospheric Science: An Introductory Survey. Academic Press, New York

Wilf P (1997) When are leaves good thermometers? A new case for Leaf Margin Analysis. Paleobiology 23:373–390

Wolfe JA (1971) Tertiary climatic fluctuations and methods of analysis of Tertiary floras. Palaeogeog Palaeoclim Palaeoecol 9:27–57

Wolfe JA (1979) Temperature parameters of the humid to mesic forests of eastern Asia and their relation to forests of other regions of the Northern Hemisphere and Australasia. US Geol Surv Prof Pap 1106:1–37

Wolfe JA (1990) Palaeobotanical evidence for a marked temperature increase following the Cretaceous/Tertiary boundary. Nature 343:153–156

Wolfe JA (1992) An analysis of present-day terrestrial lapse rates in the western conterminous United States and their significance to paleoaltitudinal estimates. US Geol Surv Bull 1964:1–35

Wolfe JA (1993) A method of obtaining climatic parameters from leaf assemblages. US Geol Surv Bull 2040:1–71

Wolfe JA (1995) Paleoclimatic estimates from Tertiary leaf assemblages. Ann Rev Earth Planet Sci 23:119–142

Wolfe JA, Forest CE, Molnar P (1998) Paleobotanical evidence on Eocene and Oligocene paleoaltitudes in mid-latitude western North America. Geol Soc Am Bull 110:664–678

Wolfe JA, Hopkins HE (1967) Climatic changes recorded by Tertiary land floras in northwestern North America. *In:* Tertiary Correlations and Cimatic Changes in the Pacific. Hatai K (ed) Symp. Pacific Sci. Congr., 11th, Tokyo, Aug. - Sept. 1966 Sasaki, Sendai, Japan vol. 25 67–76

Wolfe JA, Schorn HE (1989) Paleoecologic, paleoclimatic, and evolutionary significance of the Oligocene Creede flora, Colorado. Paleobiology 15:180–198

Wolfe JA, Schorn HE, Forest CE, Molnar P (1997) Palaeobotanical evidence for high altitudes in Nevada during the Miocene. Science 276:1672–1675

Xu KM, Emanuel KA (1989) Is the tropical atmosphere conditionally unstable? Mon Wea Rev 117:1471–1479

Reviews in Mineralogy & Geochemistry
Vol. 66, pp. 195-213, 2007
Copyright © Mineralogical Society of America

8

Paleoelevation Measurement on the Basis of Vesicular Basalts

Dork Sahagian

Environmental Initiative and Dept. of Earth & Environmental Sciences
Lehigh University
Bethlehem, Pennsylvania, 18015, U.S.A.
dork.sahagain@lehigh.edu

Alex Proussevitch

Complex Systems Research Center
Institute for the Study of Earth, Oceans and Space
University of New Hampshire
Durham, New Hampshire, 03824, U.S.A.
alex.proussevitch@unh.edu

ABSTRACT

The vesicles (bubbles) in basaltic lava flows can be used to determine paleoelevation at the time of eruption. In the repertoire of paleoelevation proxies presently available to the research community, it represents one of very few direct proxies of elevation. The technique is based on the sizes of vesicles at the tops and bottoms of lava flows. We assume that bubbles do not know a priori that they will reside in one part of the flow or another when they are erupted from a volcanic vent. As such, the mass of gas is evenly distributed throughout the flow. The volume of the bubbles will therefore depend on pressure, which at the top of the flow is just atmospheric pressure, and at the bottom is atmospheric plus hydrostatic pressure from lava overburden. Since lava thickness can be measured in the field, and bubble size distributions (most notably the modal size) can be measured in the lab, a simple relation can be solved for atmospheric pressure, and using the standard atmosphere, elevation can be determined.

The key parameters that must be determined are lava flow thickness at the time of emplacement and cooling of the upper and lower parts of the flow, and the sizes of the vesicles. Each of these requires some analysis. Measured flow thickness in the field is not necessarily the same as that during solidification of the upper and lower several cm of the flow, as inflation and deflation of the flow could have occurred, thus changing the thickness from that which provided lava overburden to determine vesicles sizes in the lower part of the flow. Lava flows must be examined in cross-section in the field to ensure that the indications of inflation or deflation are absent. Either would alter the proper vesicularity profile as a function of stratigraphic position within the flow for a given flow thickness. Only flows displaying the correct vesicularity profile can be meaningfully sampled. Vesicle sizes can be measured in a number of ways, but the most accurate is Computed X-Ray Tomography, which provides 3D information regarding size, shape, distribution, and connectivity of vesicles in a sample. We have developed numerical routines for extracting quantitative size distributions from tomographic data.

The basalt paleoaltimeter can be applied to any tectonic region in which basalts are present. We tested the method by measuring the elevation of Mauna Loa, Hawaii, and then as a first application to a tectonic issue, placed constraints on the timing of uplift of the Colorado Plateau. While further testing of this tool for paleoelevation analysis would always be helpful,

 DOI: 10.2138/rmg.2007.66.8

the technique is sufficiently refined for general use by the geologic community, provided appropriate samples are collected from lavas demonstrating simple emplacement history, and vesicle sizes are measured with sufficient quantitative accuracy to resolve elevation to within a few hundred meters. As for any proxy, the best results can be obtained with multiple sampling, and when possible, multiple independent proxies for the most reliable determination of paleoelevation history.

INTRODUCTION

Numerous techniques have been developed for measuring paleoelevations. While paleo-elevations near sea level are relatively simple to establish through the stratigraphy of deposited sediments (Sahagian 1987, 1988), at higher elevations erosional regimes do not generally allow this. Over the past decade, new direct and indirect methods for quantifying ancient topography have been developed and validated. Direct methods record changes in atmospheric character-istics that scale with the elevation of the Earth's surface. Indirect methods include a variety of newly-developed thermochronometers, which together record the timing of cooling across a range of temperatures, in addition to other methods including interpretation of late Cenozoic landforms such as fluvial terraces, erosion calculations based on topographic reconstructions and deconvolution of basin sediments, and paleogradient calculations (Sahagian 2005). In each proxy for paleoelevation, it is helpful to identify other sources of variation such as climate (e.g., temperature, precipitation, wind direction) and other factors that may confound the proxy.

We have developed a direct technique for measuring paleoelevation via paleoatmospheric pressure on the basis of the vesicularity of basaltic lava flows (Gregory 1994; Sahagian and Maus 1994; Sahagian et al. 2002b). It is based on the difference in internal pressure between bubbles at the top and base of lava flows. At the top, the pressure is simply atmospheric pres-sure, while at the base there is an additional $\rho g H$ "lavastatic" pressure. Vesicularity can measure changes in sea level pressure over Earth history, and would involve analysis of ancient basalts (such as the Keweenawan Basalts) emplaced at or near sea level. Conversely, because most researchers agree that sea level pressure has not changed significantly in the Cenozoic (Abe and Matsui 1986; Azbel and Tolstikhin 1990; Williams and Pan 1992; Azbel and Tolstikhin 1993; Tajika and Matsui 1993), basalt vesicularity can be used to determine paleoelevation for Cenozoic (and probably earlier) events. An overview of the principles involved in the vesicular basalt paleoaltimeter is illustrated in Figure 1.

We applied this technique to make an initial estimate of the timing and extent of Cenozoic uplift of the Colorado Plateau (Sahagian et al. 2002a). Because the technique "measures" paleo-atmospheric pressure, it is not subject to uncertainties stemming from the use of proxies that depend on environmental factors other than elevation alone. Vesicular lavas preserve a record of paleopressure at the time and place of lava emplacement because the difference in internal pressure in bubbles at the base and top of a lava flow depends on atmospheric pressure and lava flow thickness. The modal size of the vesicle (bubble) population is larger at the top than at the bottom. This leads directly to paleoatmospheric pressure and thus elevation because the thickness of the flow can easily be measured in the field, and the vesicle sizes can be measured in the lab. All proxies have their limitations and hence are not applicable in all places and at all times. Vesicular basalts are no exception. For a lava flow to record atmospheric pressure, the flow thickness must remain constant between the time the upper and lower crusts (10-20 cm) form, and the time of complete solidification of the flow.

The accuracy of the vesicular lava method depends on our ability to determine two properties of lava flows: size distributions of vesicles (Sahagian 1985; Sahagian et al. 1989) and flow emplacement history. The most accurate method to determine the size distribution of vesicles in lava samples uses High-Resolution X-ray Computed Tomography (HRXCT) (Proussevitch

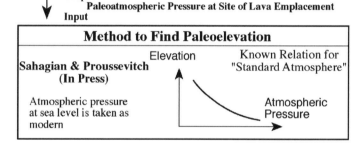

Figure 1. Diagram of methodology for determining paleoelevation from vesicular basalts (vesicular basalt paleoaltimeter) (Sahagian et al. 2002b).

and Sahagian 2001; Song et al. 2001). We have previously used plastic casts (Sahagian and Maus 1994; Shin et al. 2005) that involved error bars of over a kilometer of elevation. We also developed more robust ways to use 2D information from cross-sections on the basis of stereological analysis (Sahagian and Proussevitch 1998), but stereology has some strict limitations and involves important assumptions regarding vesicle morphology that are rarely satisfied and thus lead to errors unacceptable for the paleoaltimeter application. Our HRXCT techniques have reduced this uncertainty to about ± 400 m, enabling quantitative analysis of the amount as well as timing of epeirogeny. In addition, it allows the rapid analysis of a large number of samples, reducing uncertainties due to local variations in basalt vesicularity by multiple redundant sampling.

The second aspect of lava flows that must be determined is flow emplacement history. Because the calculation of ambient pressure and thus paleoelevation is based on flow thickness at the time the bubbles at the top and base of the flow were trapped in quenched glass, it is critical to be sure that flow thickness did not vary before the entire flow solidified. Inflation and deflation of lava flows is a common process during emplacement, so it is critical in the field to examine the criteria for the identification of inflated, deflated, and simply emplaced flows. Only flows with simple emplacement can be used in paleoelevation analysis.

CALCULATING PALEOATMOSPHERIC PRESSURE

When lava is emplaced, bubbles containing equal amounts of gas at the base and top of a flow are subject to different total pressures due to difference in overburden. At the top of the flow, there is atmospheric pressure only, while at the base there is an additional hydrostatic overburden of lava. The atmospheric pressure-dependence of vesicle size can be expressed by the ratio of vesicle size modes at the top and bottom of a flow:

$$\frac{V_t}{V_b} = \frac{P + \rho g H}{P} \tag{1}$$

where V_t and V_b are the volumes of the modal bubble sizes at the quenched top and bottom of the flow, respectively (measurable), ρ is lava density (2650 kg/m^3 for basalts), g is gravity (known), H is flow thickness (measured in the field), and P is atmospheric pressure at emplacement. Here the density of bubble free melt should be used since there is no melt drag by bubbles in basaltic melts due to their low viscosity and relatively low vesicularity (Amon and Denson 1986). Paleopressure can thus be determined because all other variables are known. This is predicated on the thickness measured in the field being the same as the thickness at the time of emplacement and solidification of the upper and lower parts of the flow. Using Equation (1) for paleoelevation measurement is also predicated on invariant sea level pressure since the time of emplacement.

In assessing the size distribution of a population, a few approaches can be taken, including mean size, mode, shape of distribution, bulk vesicularity, etc. Of these, we have found that the mode is the most useful, as it is not influenced by measurement resolution that may miss the smaller end of the population, or sample size that may miss some large bubbles in the upper end. (In fact, we use the entire size distribution to determine the modal ratio for greatest accuracy.) Likewise, simple bulk rock density measurement (for bulk vesicularity) would be subject to large errors particularly if vesicles are filled, or if the sample includes large bubbles that may make significant contributions to the sample void fraction. In theory, lava chemical composition and viscosity do not enter into the analysis because the same conditions affect bubbles at the top and bottom of the flow. The method has been developed for and is applicable to low viscosity pahoehoe basaltic lava flows that satisfy requirements of flow thickness limitations (discussed below) and have large enough modal bubble sizes to be measurable by currently available technologies (HRXCT). Any shear during transport would not affect the ratio of modal bubble sizes (top/bottom) because significant lateral variations in top or bottom vesicularity (modal sizes) would not be expected, nor have they been observed in previous studies (Sahagian et al. 1989, 2002b; Sahagian and Maus 1994). Any bubble nucleation during post-eruptive crystallization would produce a new mode within the smallest vesicles in the observed population (usually bimodal), and would thus not affect the mode of pre-eruptive bubbles (typically 1 mm modal diameter for basalts). This is another reason to use the modes rather than statistical mean.

NECESSARY CONDITIONS AND ASSUMPTIONS

The application of Equation (1) to natural lavas for the purpose of determining atmospheric pressure requires certain conditions to be satisfied:

(1) Lava was extruded with a well mixed population of bubbles so that their initial mass distribution is not a function of vertical position in the flow. This is typically the case in effusive eruptions that produce fluid lava (rather than ash).

(2) The lava flow experienced a simple emplacement history, meaning that the flow was erupted from a volcanic vent, traveled to its final resting place, and solidified *in situ*

without further disturbance. Examples of disturbance include inflation from within additional lava after top and base had solidified, deflation by drainage of interior lava, and additional lava emplacement over a partially solidified flow causing the solidified top to shear away from the emplacement site.

(3) There were no sources outside of the flow such as soil moisture or burning vegetation to introduce bubbles into the base of the flow before its solidification.

Fortunately, each of these conditions can be assessed in the field, and appropriate sites can be sampled on the basis of simple emplacement history. Of these conditions, it is the second that requires careful scrutiny. (The first is clearly satisfied, while the third is recognized by pipe vesicles, large vesicles at the base of a flow, and complex lower vesicular zone geometry.)

For the purpose of paleoelevation measurement, it is critical to sample only lavas exhibiting clear evidence of simple emplacement histories. Vesicularity profiles can clearly be observed in the field, and compared to "ideal" vesicularity described below (Sahagian 1985; Sahagian et al. 1989). The reason this is important is that we must be sure that the upper and lower 10 cm (or so) of the flow solidified after the flow reached its final thickness (thickness to be measured in the field). If the lava is emplaced without late-stage inflation or any other complicating factors (after top and base have crystallized to a significant distance into the flow), the vesicularity profile will include vesicular zones in the upper and lower parts of the flow, with recognizable characteristics as described below. These flows are good candidates for paleoelevation sampling. The problem of inflation/deflation of pahoehoe lava flows in application to the paleoaltimeter has been discussed in the literature (Bondre 2003) as being common in basaltic flows thus limiting the availability of sampling sites. This highlights the importance of visibly checking the vesicularity profiles in the field for selection of suitable flows (Sahagian et al. 2003).

To avoid potential complexities in size distribution, and to obtain maximum pressure resolution, it is best to collect samples from near the top and bottom of the flow to capture the initially erupted modal sizes before bubbles rise and coalesce and, thus, had a chance to alter the size distribution. As the theory indicates, samples should be taken from the very top and very base of each flow. However, there is often a glassy rind on the top, with evidence of small-scale deformation during pahoehoe emplacement causing some bubble escape. Likewise, the very base is often emplaced on an underlying pahoehoe surface so that it is deformed. While these small-scale (about a centimeter) deformations do not affect the emplacement history and internal pressure regime, they do introduce unwanted and unnecessary complications at the surface. Consequently, it is more productive to obtain samples from slightly (1-3 cm) below the top and above the bottom (taking account of this in the $\rho g H$ calculation, of course) (Figs. 2 and 3). In our sampling, the deep interior of the flow (30-50 cm from top or bottom) was avoided, where coalescence can alter the bubble size distributions and produce additional modes. Although bubble rise and coalescence can be accounted for (Sahagian 1985; Sahagian et al. 1989), it is better not to introduce an unnecessary potential source of error.

The nature of the lower vesicular zone is not particularly dependent on flow thickness beyond size compression due to lava overburden. As bubbles rise to escape the rising lower crystallization front, the size of the largest bubble "caught" depends on the velocity of the front, and once the velocity (slowing with the square-root of time like a cooling half space) is reduced below the Stokes velocity of the smallest bubbles in the distribution, all can escape and the lower boundary of the "massive zone" (Sahagian et al. 1989) is defined at that point. This is true of any flow thickness, so that the only factor that controls the nature of the lower vesicular zone (relative to that of the upper vesicular zone, which is much more complex) is the overlying pressure of the lava. A thicker flow would result in proportionally smaller size mode, which is the basis of the entire analysis for paleoelevation.

The nature of the upper vesicular zone is determined not by hydrostatic pressure of the overlying flow thickness, but rather by the bubble rise and coalescence processes within the flow and their interaction with the descending upper crystallization front. The upper crystallization front descends at a reducing rate (like cooling half space, but with a free radiative boundary condition rather than an insulated, conducting one; Sahagian 1985). As it does so, it captures rising bubbles. Initially, near the top of the flow, it descends quickly, capturing all bubbles before they have a chance to rise and escape, coalesce, or alter the size distribution of the observable population. This is the region in which samples should be collected for paleoelevation analysis. As time goes on and the upper vesicular front descends, it slows, so that bubbles from lower down in the flow have time to rise through the lava to meet the front. This leads to an increase in vesicularity farther down from the flow top.

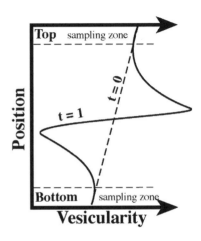

As the front descends and slows further, bubbles that started lower in the flow have time to reach the front. These bubbles may have escaped the lower crystallization front and had time on their way upward to interact and coalesce; the larger bubbles catching and joining with smaller ones with lower terminal Stokes velocities. These coalesced bubbles reach the upper crystallization front and lead to the formation of a highly vesicular zone that includes the largest bubbles in the flow. Below that zone, only the smaller

Figure 2. Typical vesicularity profile before and after vesicle rise during lava solidification in a 3 m flow (Sahagian et al. 2002b).

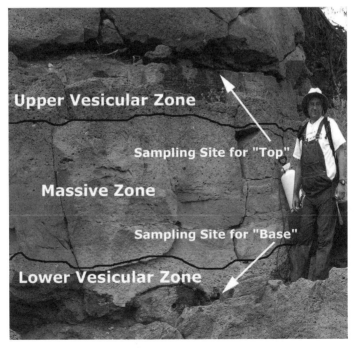

Figure 3. Typical outcrop of vesicular basalt. Note upper and lower vesicular zones, and massive vesicle free region in the interior of this 2.5 m flow.

bubbles that barely escaped the lower vesicular zone remain, and vesicularity tapers off rapidly into the massive zone.

The process described above is universal, but the time available for it to evolve the size distribution as a function of stratigraphic position in the flow depends on flow thickness and magma properties such as melt viscosity, crystallinity, etc. In very thin flows, the upper and lower crystallization fronts meet so quickly that there is insufficient time for bubbles to rise, coalesce or alter their size distributions. This would seem like an ideal circumstance for sample collection, but thin flows are not thick enough to accurately record atmospheric pressure. The reason for this is that in order for Equation (1) to faithfully reflect atmospheric pressure, the ρgH lava overburden should be roughly equivalent to the atmospheric pressure to be determined so that pressure is one atmosphere at the top and two atmospheres at the bottom of the flow. For typical lavas at sea level, the ideal thickness is about 3 m. At this thickness, typical basalts (with typical bubbles) have a lower vesicular zone, a very small massive zone, and an upper vesicular zone. Thinner flows (2 m or less) do not have a massive zone, Rather, vesicularity decreases from bottom to top as bubbles get smaller and less numerous, and there is a sudden increase in vesicularity and bubble sizes at the bottom of the upper vesicular zone where the large bubbles that escaped the lower crystallization front are captured. Vesicularity then decreases upward to the top, where the initial vesicularity and size distribution are frozen in by the initially rapidly descending upper crystallization front.

In relatively thick flows (5 m or more), the lower vesicular zone resembles, for most part, that of any thinner flow (except that the size mode is smaller due to the lava overburden). However, thick flows have a very well developed massive zone, and because all the bubbles from the thick massive zone are to be found in the upper vesicular zone, the latter is thick, highly vesicular, and includes very large bubbles that resulted from extensive bubble coalescence during an extended time of bubble rise and evolution.

Thus, the ideal flow thickness yields one bar at the top and 2 bars at the bottom of a flow (about 3 m) but we have found that flows anywhere from 1 to 5 m are quite well suited for analysis. A 10 m flow would be too thick as the lava overburden would overwhelm the atmospheric pressure in Equation (1), while a 50 cm flow would be so thin that atmospheric pressure would dominate, thus reducing sensitivity below the threshold necessary for analysis based on current ability to measure size distributions. This is one of the reasons why the method works better on basaltic rather than on more silicic and, thus, viscous and thick flows.

In the field, it is critical to assess the vesicularity profile as well as size distribution as a function of stratigraphic position within the flow so that the expected profile can be identified. If the expected profile is observed, it can be reliably inferred whether the final, observed flow thickness was the same as that which controlled the modal size of the bubble distribution at the base of the flow (up to about 10 cm up from the bottom). This is the key to determining that there were no post-emplacement complications in flow thickness or other processes within the flow that would confound the analysis for paleoelevation. In addition, extensive lensing, rip-up clasts, or other complexities in the profile of a flow should disqualify it from paleoelevation analysis.

Complications would be introduced by inflation or deflation after the top and bottom of the flow have solidified. During flow cooling and solidification, bubbles are frozen in at the top and bottom of the flow, and this provides the sizes to be used in Equation (1). If additional lava were to surge into the space between the solidified top and bottom after significant movement of bubbles and crystallization fronts, but before the two crystallization fronts meet for final solidification of the flow, the standard vesicularity profile described would be altered (as would, of course, the thickness, thus confounding the analysis). The newly introduced lava would bring with it its own (unevolved) bubble distribution which would then reside in the center of the older, evolved material. The bubbles within the new lava would then begin to rise, making a new vesicularity pattern within the older pattern. Again, before crystallization fronts merge, additional inflation could occur with new lava introducing new bubbles. This process can occur any number of times resulting in a series of nested units with their own vesicular zones. Inflation after partial solidification is thus readily identified by multiple layers of small vesicles in a relatively thick flow (e.g., >3 m) that may not contain a massive zone.

Alternatively, after initial emplacement and crystallization at the top and bottom of a flow, deflation could occur, and lava could drain away from the flow interior, resulting in a thinning of the flow after the vesicles at the top and bottom "recorded" the ambient atmospheric pressure. This would cause the upper vesicular and lower vesicular zones to be brought in closer contact with each other at an early stage of solidification, without the time for the crystallization fronts to intercept the evolving bubble distribution within the flow interior. This would prevent the development of a highly vesicular upper vesicular zone if deflation occurred early, and would remove a massive zone if deflation occurred late. Deflation can thus be identified by a continuous and relatively uniform vesicularity profile, lacking in a vesicle-free zone in the middle of the flow, and/or by large, coalesced vesicles in a thin flow with no corresponding vesicle-free zone. Since the deflation makes the massive zone thinner, it is less obvious in field observations than is inflation with its extra vesiculation zones (see above).

MEASURING VESICLE SIZE DISTRIBUTIONS

Calculation of paleopressure and thus the success of this approach to determining paleoelevation depend upon our ability to accurately measure the size distribution of vesicles in hand samples of basalt (Fig. 4). In the course of our research, we have developed three techniques for measurement. The first involved injecting hand samples with plastic monomer which subsequently polymerized, and then dissolving away the basalt to leave a large number of plastic casts of vesicles for measurement and counting. The plastic casts were resistant to the effects of HF acid, so they reliably represented the actual vesicle sizes and shapes but were difficult to measure accurately and very tedious to work with (Sahagian et al. 1989). This was a laborious and inefficient technique that should be abandoned in favor of faster and more accurate methods.

A more efficient technique is to determine vesicle size distributions from 2D cross-sections of hand samples (Dullien and Dhawan 1973; Russ 1986; Marsh 1988; Mangan 1990; Toramaru 1990). Sample cross-sections can be optically scanned and the data numerically processed to

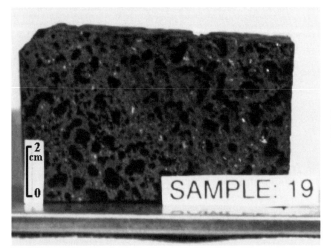

Figure 4. Vesicular basalt sample (Sahagian et al. 2002b).

produce 2D size distributions, which can be converted to 3D using Stereology (Saltikov 1967). To treat non-spherical particles such as vesicles, we refined stereological methods and developed appropriate conversion coefficients for 2D to 3D conversion (Sahagian and Proussevitch 1998) without a priori assumed size distribution (e.g., log normal, unimodal, etc.) (Higgins 1994; Peterson 1996).

The most accurate method for measuring bubble size distributions is High Resolution X-Ray Computed Tomography (HRXCT) (Carlson 2006) that can produce highly detailed 2D (Fig. 5) and 3D images (Fig. 6). HRXCT measures the transmission of X-rays through an object; contrast in the images is produced by differences in the linear attenuation coefficient for X-rays in different materials within the object's interior. The attenuation of X-rays of a given energy is a function primarily of mass density (and secondarily of atomic number) in the material being irradiated, so images closely approximate the density structure of objects. For a more detailed description of HRXCT imaging of geological materials, consult Ketcham and Carlson (2001). For quantitative analysis of HRXCT images, it was necessary to develop a numerical formulation to directly compute size distributions from the CT scan data (Proussevitch and Sahagian 2001).

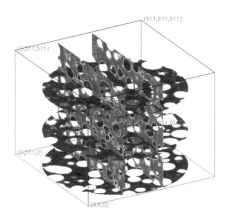
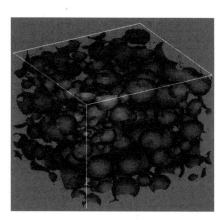

Figure 5. Image slices through vesicular basalt sample processed by HRXCT (Sahagian et al. 2002b).

Figure 6. Reconstructed 3D image of vesicles in a vesicular basalt processed by HRXCT (Sahagian et al. 2002b).

TESTING THE TECHNIQUE

In an initial test of the technique, we "measured" the elevation of the base and summit of Mauna Loa (Sahagian et al. 2002a) as illustrated in Figure 7. The proper vesicularity profile is so easy to recognize when one is standing in front of a cross-section of a lava flow, that the most difficult part of the fieldwork is invariably finding localities in which the base of the flow is exposed while the top of the flow is preserved. Potential sampling sites are ubiquitous at the summit of Mauna Loa, along many cracks and fissures near the crater rim (Fig. 8). Many recent flows have been well mapped, and there are small lobes available for sampling. These exhibit vesicularity profiles that reflect simple emplacement history, apparently removed from the main channels of the lava flows.

Sampling sites are also available at low elevation in Hilo and elsewhere (Fig. 8). Road cuts, stream channels, natural cliffs, and fissures provide many sampling sites near the terminus of the flows, or in breakout lobes. With the large number of potential sampling sites available at the base and summit, the only constraint on the number of samples was funds for analysis.

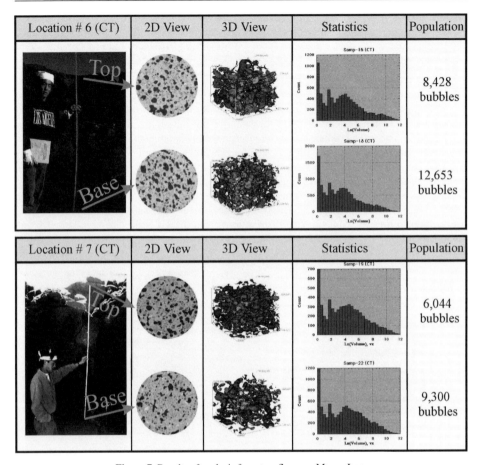

Figure 7. Results of analysis from two flows on Mauna Loa.

However, sampling sites are much more limited along the flanks of Mauna Loa. Road cuts and natural fissures are generally absent, so fully exposed cross sections are generally unavailable. Nevertheless, we found a site on the SE flank where a road cut exposed a flow with vesicularity reflecting simple emplacement history.

The accuracy of the technique can be seen in Figure 9, a plot of "measured" elevation using our technique vs. actual elevation on Mauna Loa. Perfect results (0 error) would lie along the diagonal line, and deviation from that line is a measure of error in the analysis assuming there was no elevation change of the tested flows since emplacement during the 20th century.

SENSITIVITY ANALYSIS AND POTENTIAL SOURCES OF ERROR

As in any measurement, there are error sources whose effects must be quantified. There are two ways to assess error in our analysis. The first, a simple comparison of analytical results with actual elevations (Fig. 9), results in a standard deviation (between measured and actual) of $\sigma = 372$ m, which is small relative to the elevation changes we consider in major tectonic events. This simple "empirical" approach to the error depends on the number of samples analyzed and is thus not intrinsic to the technique. The error can be reduced simply by taking more samples

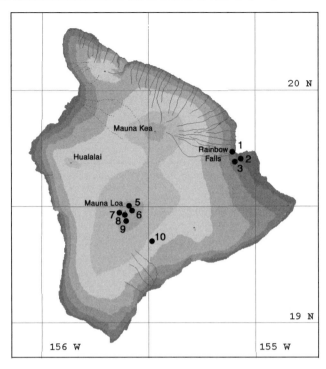

Figure 8. Location map of sampled lava flows on Mauna Loa. Numbers on the map correspond to flow numbers in Table 1 (Sahagian et al. 2002b).

Table 1. Elevation calculations for Hawaiian basalt flows.
Error bars in measured elevation are ± 200 m (see text).

Flow #	Flow Thickness (m)	Sample Location	Analyzed Vesicle Population	Modal Size (mm³)	Measured Pressure (atm)	*Measured Elevation (m)*	*Actual Elevation (m)*
1	1.52	Top	4,946	1.266	0.972	**307**	**46**
		Base	13,961	0.896			
2	1.52	Top	19,656	1.333	0.996	**45**	**46**
		Base	9,871	0.950			
3	1.78	Top	6,704	1.118	0.929	**778**	**335**
		Base	18,946	0.742			
10	1.24	Top	5958	0.955	0.783	**1951**	**2388**
		Base	8271	0.673			
5	1.02	Top	6,390	2.363	0.676	**3559**	**3978**
		Base	5,222	1.689			
6	1.22	Top	8,428	2.362	0.628	**4093**	**3978**
		Base	12,653	1.560			
7	1.65	Top	6,044	1.823	0.704	**3254**	**3932**
		Base	9,300	1.125			
8	1.27	Top	1,974	2.488	0.604	**4361**	**3932**
		Base	8,096	1.598			
9	1.68	Top	1,558	1.130	0.648	**3867**	**3972**
		Base	5,231	0.670			

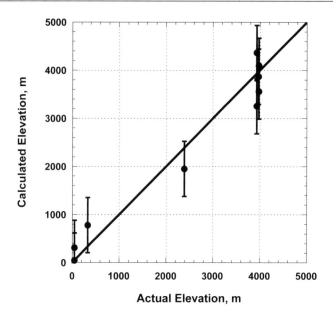

Figure 9. Results of analysis of using vesicular basalt as a measure of paleoelevation.
Perfect results would lie along the diagonal line (Sahagian et al. 2002b).

(to within budgetary constraints and availability of sampling locations). These results suggest
that even with a small number of measurements, the accuracy of the technique is sufficient to
resolve paleoelevations for application to various geologic problems.

The second approach to error assessment is a factor analysis of the various error sources
within the procedure. These sources can be ascribed to the three major parameters in Equation
(1) above: vesicle size, hydrostatic pressure, and sea level atmospheric pressure.

Measurement of vesicle sizes. The discretization of vesicles into voxelized representations
using X-ray tomography introduces a specific source of error in the vesicle volume. This error
can be calculated on the basis of the relative size of vesicles and the voxels with which they
are constructed. Based on a vesicle diameter of 1 mm, with a linear measurement resolution of
47 μm, error arises from the uncertainty of including or not including voxels from the surface
of a vesicle, where the boundary between rock and air cuts through a voxel. This uncertainty
regarding the nature of the boundary voxel makes a maximum measurement deviation of one
complete voxel length, or $\sigma = 47$ μm. If we were to measure a single 1 mm diameter bubble,
and in so doing leave the entire exterior shell of voxels either in or out of the vesicle radius, R,
the volumetric error, σ_v, would be $(R+2\sigma)^3/R^3 = (500+94)^3/500^3 = 1.67$ or $\sigma_v=67\%$. However,
we make the bubble interface measurement for every voxel in the exterior shell. The number
of voxels included in that exterior shell is theoretically $4\pi(\text{radius in voxels})^2 = 4\pi(1000/47/2)^2$
$= 1422$ voxels. This leads to a total measurement error of the volume of the vesicle as $\sigma_v/$
$(N_{population})^{1/2} = 67\%/(1422)^{1/2} = 1.77\%$. This analytical formulation for the surface area of the
vesicle does not account for the lattice orientation of cubic voxels, but gives an approximate
analytical treatment of the problem. The actual number of voxels is easily counted, and for one
specific example is 1480, leading to a slightly smaller error of 1.74%. This leads to an error in
paleopressure of ± 17 mb, corresponding to ± 190 m in elevation uncertainty.

Hydrostatic pressure at the base of lava flow. Densities of lavas are known to within about
± 1% (Ochs and Lange 1999). For basaltic flows, the density of the fluid lava is used because

at the time the top and bottom solidify, the flows do not include any highly vesicular zones of foams in which interaction between bubbles is strong enough to support the overlying fluid and thus reduce the pressure at the base of the flow to below hydrostatic (Amon and Denson 1986). Gravity is known "perfectly." The thickness of a flow can be measured in the field, but we will account for potential error resulting from unrecognized minor inflation or deflation after solidification of upper and lower parts of the flow. Each centimeter of measurement error or inflation/deflation would lead to an error of ± 33 m in elevation. Although inflation/deflation is readily identified in the field, we include an error term to account for 10 cm of inflation/ deflation, providing an elevation uncertainty of ± 330 m for a typical 3 m flow. However, with judicious choice of sampling sites, this source of error should be reduced.

Sea level pressure at the time of eruption. There are normal variations in barometric pressure due to changing weather conditions. Typically, the time scale of these variations is no longer than several days for synoptic systems. (Seasonally averaged variations are smaller.) Taking a conservative approach, however, variations in barometric pressure (due to weather) at the time of eruption of up to 30 mb lead to an uncertainty of about ± 150 m for a given flow. It is unlikely that a hurricane or other very low pressure system will have been present throughout the cooling of a lava flow, and very likely that barometric variations over several days will have averaged out in the slowly cooling lava, but a conservative estimate is ± 150 m.

Using the larger errors above, the total error from the various contributions is thus estimated as $\sqrt{(190)^2 + (330)^2 + (150)^2}$, or about ± 410 m. The different approaches to error assessment agree (372 vs. 410 m), indicating that in cases when the actual elevations of samples are not known a priori, the proposed analysis can be used to reliably determine paleoelevations to within about 400 m of uncertainty (or better with multiple sampling).

Further testing is certainly warranted to determine the limits of the technique with larger number of samples and different volcanic environments, but this initial test appeared sufficiently successful to warrant some initial applications to tectonic problems.

COMPARISON WITH OTHER PALEOELEVATION PROXIES

The basalt vesicularity proxy for paleoelevation measurement is limited to areas in which suitable basalts have been deposited, preserved, and exposed. As such, there are only certain geographic areas that can be explored in this way. Even in these areas (and in all areas) it is best to apply as many proxies as possible to get the most robust estimate of paleoelevation in any tectonic environment.

Other direct paleoaltimeters include oxygen isotopic techniques, paleoflora, and cosmogenic nuclides, and additional indirect proxies such as stratigraphy in destination basins, fission tracks and unroofing, and geomorphology make it possible to apply as much information as available to the elevation history of a specific locality. There has been particularly intense effort in oxygen isotopes (e.g., Chamberlain and Poage 2000; Garzione et al. 2000b; Rowley 2006), and the resolution of paleoelevation so measured is generally comparable to that of vesicular basalts (± 400 m), and can in some instances be better depending on isotopic lapse rate. Oxygen isotope paleoaltimetry has been recently used for first-order tectonic reconstructions in the Puna-Altiplano (Garzione et al. 2006; Ghosh et al. 2006).

APPLICATION TO THE UPLIFT HISTORY OF THE COLORADO PLATEAU

Once the technique was tested on Mauna Loa, it was possible to apply it to a tectonic province in which the paleoelevation was not known a priori. Plateau uplift has been the subject of considerable discussion and controversy in recent years because the timing and extent of uplift have been difficult to constrain except within very broad limits (McQuarrie and Chase 2000; Abbot et al. 1997). Cenozoic epeirogeny has affected the Colorado Plateau and Great Plains, the Sierra Nevada, the Great Basin, the Altiplano, and the Tibetan Plateau to name a few (Allmendinger and Gubbels 1996; Ruddiman 1998; Garzione et al. 2000a). The exact timing has been controversial in each case. Although epeirogenic activity is simple to constrain in lowlands, the sedimentary and faunal information that make this possible are generally absent from highlands.

The timing and extent of the Colorado Plateau uplift bears on the possible mechanisms that caused it. Various hypotheses have been suggested including influence of a mantle plume (Parsons et al. 1994), lithospheric thinning and/or delamination (Spencer 1996; Lastowka et al. 2001), and subduction of the Farallon plate margin (East Pacific Rise) (Dickinson and Snyder 1979). With a definitive uplift history, it will be possible to place constraints on the mechanisms of uplift and thereby shed light on mantle processes and lithospheric evolution.

We collected and analyzed suites of samples from around the plateau and conducted a robust size distribution analysis (Proussevitch et al. 2007). The analysis of the basalts resulted in a variety of uni- and polymodal distribution types that have been grouped and speculatively interpreted from the standpoint of alternative scenarios of vesiculation processes such as single nucleation and bubble growth event, multiple nucleation events, and coalescence. In order to use vesicular basalts for paleoelevation, it is necessary to separate pre-eruptive from post-emplacement nucleated bubbles so that only modes of pre-eruptive distributions are used. We found distinct distribution types in the samples we analyzed. Within each, they can be grouped into categories with respect to types and values of distribution functions. Each category appears to represent a progressive transition in the shape of the number density graphs, which may reflect different stages within the vesiculation process. It is likely that nucleation and bubble growth are the primary vesiculation processes operating in basaltic magmas and formed most of the observed monomodal distributions, but second modes in bimodal distributions could be formed by a second nucleation event, Ostwald ripening, or coalescence. The bimodality observed in some of our samples was likely produced by second nucleation event and/or coalescence (Sahagian 1985; Sahagian et al. 1989). We thus consider the most likely scenario for the majority of our samples to be a single or double nucleation/growth event followed by coalescence.

The paleoelevations obtained by our analysis are given in Table 2 and plotted in Figure 10. There is no discernible relationship between age and paleoelevation, suggesting that basalts were erupted throughout a range of elevations over the last 25 m.y. in all parts of the plateau. Initial paleoelevation of analyzed flows (Fig. 10) does not show any obvious relation with age, if paleoelevations for all localities are plotted together. Most lava flows (about 90%) were emplaced in the 800 to 2000 m range regardless of age. It is tempting to speculate about the preponderance of emplacement elevations near 1500 m. There may be a relation between the cause of uplift and magmatic source, modulated by lithospheric properties that control the elevation of emplaced lavas, but this is beyond the scope of the present paper.

Our analysis provides a general uplift history for the Colorado Plateau. Figure 11 includes two sets of data, isolating results from the downfaulted blocks of the Basin and Range transition zone (Marysvale). Aside from the transition zone samples, results indicate slow uplift of roughly 40 m/m.y. between 25 Ma and 5 Ma (only 800 m uplift during that time), and rapid uplift of 220 m/m.y. since 5 Ma (1100 m during that time). This suggests that the controversy between interpretations of ancient and recent uplift can be resolved by our results of a low rate of uplift early on, then a great increase in uplift in the last 5 million years. There is some suggestion in the results that the western and northern edge of the plateau may have begun a rapid uplift phase earlier than the eastern part, but this needs to be further explored by subsequent detailed sampling and analysis focused on relatively young flows (<9 Ma).

Table 2. Elevation calculations for Colorado Plateau basalt flows.

Site	Geological Unit	Age (Ma)	Present Elevation (m)	Calculated Paleoelevation and change (m)	Coordinates (W,N) Longitude	Latitude	Flow Thickness (m)	Location
3		10.2	3093	2270(+823)	107° 59.6'	39° 02.2'	5.74	Grand Mesa, CO
4		10.2	3262	1910(+1352)	108° 01.3'	39° 01.8'	3.07	Grand Mesa, CO
5		9.5	3201	1550(+1651)	108° 04.0'	39° 02.5'	3.18	Mesa Lakes, CO
6		9.5	3201	1710 (+1491)	108° 04.0'	39° 02.5'	3.15	Mesa Lakes, CO
8	Qtb	5	1993	1810(+183)	111° 59.6'	38° 08.5'	4.79	Antimony, UT
12	Tbas	10-15	2011	1020(+991)	112° 09.7'	38° 19.4'	3.28	Piute Reservoir, UT
14	Tpml	23	1827	670(+1157)	112° 11.4'	38° 14.2'	2.65	Junction, UT
15	Tbas	N_1	1978	980(+998)	112° 09.1'	38° 24.4'	2.57	Marysvale, UT
17	Tmc	22.8	3324	2010 (+1314)	112° 23.5'	38° 12.3'	2.49	LaBaron Lake, UT
18		1.9	1752	830(+992)	110° 09.5'	34° 06.4'	5.27	Corduroy Creek, AZ
19	Trc2	1.98	1933	1590(+343)	109° 23.1'	34° 20.9'	2.95	Lyman Lake, AZ
20	Trc2	1.98	1933	1130(+803)	109° 23.1'	34° 21.0'	3.73	Lyman Lake, AZ
21		4.0-4.7	2209	1390(+819)	105° 53.8'	36° 20.7'	5.09	Ojo Calliente, NM
22		3.4-4.1	2537	1410 (+1127)	105° 58.9'	36° 46.1'	2.44	Tres Piedras, NM
23	Thb	19.8	2522	710(+1812)	106° 12.4'	37° 11.2'	3.20	LaJara Creek, CO
24	Thb	17.6	2536	1140 (+1396)	106° 11.5'	37° 12.8'	2.14	LaJara Creek, CO
25	Thb	17.7	2485	950 (+1535)	106° 10.8'	37° 13.3'	2.77	LaJara Creek, CO
26	Thb	N_1 (20)	3322	1150 (+2172)	106° 24.7'	37° 21.1'	1.74	Devils Hole, CO
27	Thb	N_1 (20)	3320	1710 (+1610)	106° 24.7'	37° 21.1'	1.26	Devils Hole, CO
29	Thb	N_1 (20)	3330	1200 (+2130)	106° 24.7'	37° 21.1'	3.76	Devils Hole, CO
31		9.7	2929	1693 (+1236)	107° 02.8'	39° 25.9'	2.74	Basalt Mountain, CO
32		10.2	3170	1389 (+1781)	108° 03.7'	39° 03.1'	2.90	Grand Mesa, CO
33		10.2	3214	1646 (+1568)	108° 04.0'	39° 02.9'	2.16	Grand Mesa, CO
35		0.85	1055	1210 (-155)	113° 16.8'	37° 16.0'	3.05	Mogollon Rim, UT
36		0.85	1052	717 (+335)	113° 16.8'	37° 16.0'	2.57	Mogollon Rim, UT
38	Tby		1524	1346 (+178)	112° 24.3'	35° 04.9'	2.44	Hell Canyon, AZ
41	QTsf	1.6-2.0	1837	1043 (+794)	109° 57.8'	34° 21.1'	0.84	Silver Creek, AZ
47	Tnb	4.5	1996	933 (+1063)	105° 42.2'	36° 32.0'	1.91	Arroyo Hondo, NM
50	Ts	4.3	2280	1216 (+1054)	105° 45.4'	37° 04.8'	4.11	Rio Grande, CO

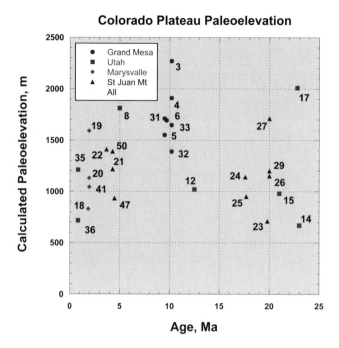

Figure 10. Paleoelevation of sampling localities throughout Colorado Plateau (Sahagian et al. 2002a).

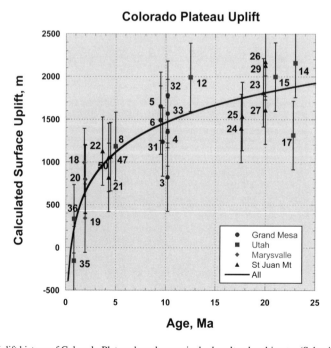

Figure 11. Uplift history of Colorado Plateau based on vesicular basalt paleoaltimeter (Sahagian et al. 2002a).

The Colorado Plateau has been tectonically active along its margins, and there is clear evidence of faulting and changes of local elevation and relief. This is particularly true in the western margin, in Utah, where the Colorado Plateau grades into the Basin and Range Province. Our technique provides only paleoelevation (Fig. 10). To infer epeirogeny or tectonics from this, a reference elevation must be subtracted. For most of the plateau, subtracting present elevation provides uplift and timing based on lava age. However, in the transition zone, more recent downfaulting makes present elevation an inappropriate reference for plateau uplift calculations. Rather, it is possible to invert the analysis. If we assume that the transition zone experienced the same uplift history as the rest of the plateau prior to Basin and Range downfaulting (Wolfe et al. 1997), and then we can use the difference between the results from downfaulted blocks and those of the rest of the plateau as a measure of magnitude of fault throw. On the basis of Figure 11, the Marysvale blocks have been subsequently downfaulted by about 0.6 - 1.0 km as their points are higher than the other points of the similar age in Figure 11, but not on Figure 10.

CONCLUSIONS

Analysis of vesicular basalts provides a robust measure of paleoatmospheric pressure and thus paleoelevation, providing a new and powerful tool for the existing arsenal of paleoelevation proxies to be used for paleogeographic, paleoclimatic, and paleohydrologic studies. This new tool is not affected by climate change or other confounding environmental factors, although it has its own unique set of limitations based on the preservation of vesicular lavas in complete cross-section and the need for sampling only flows with simple emplacement history.

An initial application of the technique to the Colorado Plateau indicates that although uplift began at least 25 m.y. before present with about 800 m displacement before 5 Ma, the main phase of rapid uplift occurred in the last 5 m.y., with an additional 1100 m since then. This suggests an early uplift rate of 40 m/m.y. and a recent rate of 220 m/m.y. With the available data, it is not possible to distinguish between a smooth curve (as drawn in Fig. 11) and a sudden increase in uplift rate between 9 and 5 Ma. However, the uplift curve for the Colorado Plateau (Fig. 11) resolves the controversy between the different interpretations of various other proxies. The "early" and "recent" uplift camps are reconciled by this analysis in that the stratigraphic and geomorphologic evidence for early activity is accommodated, while the evidence for rapid recent uplift is also consistent with our results.

In general, all available proxies and lines of evidence should be used to determine paleoelevation. While vesicular lavas may provide a single, powerful tool, as many approaches as possible should be employed to address specific cases of paleoelevation and epeirogeny.

ACKNOWLEDGMENTS

This work was supported by grants from the National Science Foundation (EAR-9614747; EAR-9909293; EAR-0509856). The HRXCT facility at the University of Texas at Austin is an NSF shared multi-user facility supported in part by grant EAR-0004082.

REFERENCES

Abbott LD, et al (1997) Measurement of tectonic surface uplift rate in a young collisional mountain belt. Nature 385:501-507
Abe Y, Matsui T (1986) Early evolution of the Earth: Accretion, atmosphere formation, and thermal history. J Geophys Res 91(B13):E291-E302
Allmendinger RW, Gubbels T (1996) Pure and simple shear plateau uplift, Altiplano-Puna, Argentina and Bolivia. Tectonophys 259(1-3):1-13

Amon M, Denson C (1986) A study of the dynamics of foam growth: Simplified analysis and experimental results of bulk density in structural foam molding. Polymer Eng Sci 26:255-267

Azbel IY, Tolstikhin IN (1990) Geodynamics, magmatism, and degassing of the Earth. Geochim Cosmochim Acta 54:139-154

Azbel IY, Tolstikhin IN (1993) Accretion and early degassing of the Earth: Constraints from PU-U-I-XE isotopic systematics. Meteoritics 28:609-621

Bondre NR (2003) Analysis of vesicular basalts and lava emplacement processes for application as a paleobarometer/paleoaltimeter: a discussion. J Geology 111: 499-502

Carlson WD (2006) Three-dimensional imaging of earth and planetary materials. Earth Planet Sci Lett 249(3-4):133–147

Chamberlain CP, Poage MA (2000) Reconstructing the paleotopography of mountain belts from the isotopic composition of authigenic minerals. Geology 28:115-118

Dickinson WR, Snyder WS (1979) Geometry of subducted slabs related to San Andreas transform. J Geology 87:609-627

Dullien F, Dhawan GK (1973) Photomicrographic size distribution determination of non-spherical objects. Powder Technology 7:305-313

Garzione CN, Dettman DL, Quade J, DeCelles PG, Butler RF (2000a) High times on the Tibetan Plateau: paleoelevation of the Thakkhola Graben. Geology 28:339-342

Garzione CN, Quade J, DeCelles PG, English NB (2000b) Predicting paleoelevation of Tibet and the Himalaya from 18O vs. altitude gradients in meteoric water across the Nepal Himalaya. Earth Planet Sci Lett 183:215-229

Garzione CN, Molnar P, Libarkin JC, MacFadden BJ (2006) Rapid late Miocene rise of the Bolivian Altiplano: Evidence for removal of mantle lithosphere. Earth Planet Sci Lett 241(3-4):543–556

Ghosh P, Garzione CN, Eiler JM (2006) Rapid uplift of the Altiplano revealed through ^{13}C-^{18}O bonds in paleosol varbonates. Science 311(5760):511-515

Gregory KM (1994) New prospects in old bubbles. Nature 372:407-408

Higgins MD (1994) Numerical modeling of crystal shapes in thin sections: Estimation of crystal habit and true size. Am Mineral 79:113-119

Ketcham RA, Carlson WD (2001) Acquisition, optimization and interpretation of X-ray computed tomographic imagery: applications to the geosciences. Comp Geosci 27:381-400

Lastowka LA, Sheehan AF, Schneider JM (2001) Seismic evidence for partial lithospheric delamination model of Colorado Plateau uplift. Geophys Res Lett 28(N7):1319-1322

Mangan M (1990) Crystal size distribution systematics and the determination of magma storage times: The 1959 eruption of Kilauea volcano, Hawaii. J Volc Geotherm Res 44:295-302

Marsh BD (1988) Crystal size distributions (CSD) in rocks and the kinetics and dynamics of crystallization: 1. Theory. Contrib Mineral Petrol 99:277-291

McQuarrie N, Chase CG (2000) Raising the Colorado Plateau. Geology 28(1):91-94

Parsons T, Thompson G, Sleep N (1994) Mantle plume influence on the Neogene uplift and extension of the United States Western Cordillera. Geology 22:83-86

Peterson T (1996) A refined technique for measuring crystal size distributions in thin section. Contrib Mineral Petrol 124:395-405

Proussevitch AA, Sahagian DL (2001) Recognition and separation of discrete objects within complex 3D voxelized structures. Comp Geosci 27:441-454

Proussevitch AA, Sahagian DL, Carlson WD (2007) Statistical analysis of bubble and crystal size distributions: Application to Colorado Plateau Basalts. J Volc Geotherm Res 164:112-126

Rowley DB, Currie BS (2006) Palaeo-altimetry of the late Eocene to Miocene Lunpola basin, central Tibet. Nature 439:677-681

Ruddiman W (1998) Early uplift in Tibet? Nature 394:723-725

Russ JC (1986) Practical Stereology. Plenum Press, New York

Sahagian DL (1985) Bubble migration and coalescence during the solidification of basaltic lava flows. J Geology 93:205-211

Sahagian DL, Anderson AT, Ward B (1989) Bubble coalescence in basalt flows: Comparison of a numerical model with natural examples. J Volc Geotherm Res 52:49-56

Sahagian DL, Maus JE 1994 Basalt vesicularity as a measure of atmospheric pressure and paleoelevation. Nature 372:449-451

Sahagian DL, Proussevitch AA (1998) 3D Particle Size Distributions from 2D Observations: Stereology for Natural Applications. J Volc Geotherm Res 84:173-196

Sahagian DL, Proussevitch AA, Carlson WD (2002a) Timing of Colorado Plateau uplift: Initial constraints from vesicular basalt-derived paleoelevations. Geology 30(9): 807-810

Sahagian DL, Proussevitch AA, Carlson WD (2002b) Analysis of vesicular basalts and lava emplacement processes for application as a paleobarometer/paleoaltimeter. J Geology 110:671-685

Sahagian DL, Proussevitch AA, Carlson WD (2003) Analysis of vesicular basalts and lava emplacement processes for application as a paleobarometer/paleoaltimeter: A reply. J Geology 111:502-504

Sahagian DL (2005) Paleoelevation measurement: Combining proxies and approaches. EOS Trans AGU 86(48):500

Saltikov SA (1967) The determination of the size distribution of particles in an opaque material from a measurement of the size distribution of their sections. *In:* Stereology. Elias H (ed) Springer-Verlag , New York, p 163-173

Shin H, Lindquist WB, Sahagian DL, Song SR (2005) Analysis of the vesicular structure of basalts. Comp Geosci 31(4):473-487

Song SR, Jones K, Lindquist WB, Dowd BA, Sahagian DL (2001) Synchrotron X Ray Computed Microtomography: Studies on vesiculated basaltic rocks. J Volc Geotherm Res 63:252-263

Spencer JE (1996) Uplift of the Colorado Plateau due to Lithosphere attenuation during Laramide low-angle subduction. J Geophys Res 101:13595-13609

Tajika E, Matsui T (1993) Degassing history and carbon cycle of the earth- from an impact-induced steam atmosphere to the present atmosphere. Lithos 30:267-280

Toramaru A (1990) Measurement of bubble size distributions in vesiculated rocks with implications for quantitative estimation of eruption processes. J Volc Geotherm Res 43:71-90

Williams DR, Pan V (1992) Internally heated mantle convection and the thermal and degassing history of the earth. J Geophys Res 97:8937-8950

Wolfe JA, Schorn HE, Forest CE (1997) Paleobotanical Evidence for High Altitudes in Nevada during the Miocene. Science 276:1672-1675

Reviews in Mineralogy & Geochemistry
Vol. 66, pp. 215-241, 2007
Copyright © Mineralogical Society of America

9

Stomatal Frequency Change Over Altitudinal Gradients: Prospects for Paleoaltimetry

Lenny L.R. Kouwenberg

Department of Geology
Field Museum of Natural History
Chicago, Illinois 60605, U.S.A.
lkouwenberg@fieldmuseum.org

Wolfram M. Kürschner

Laboratory of Palaeobotany & Palynology,
Utrecht University
3584 CD Utrecht, The Netherlands
w.m.kurschner@bio.uu.nl

Jennifer C. McElwain

UCD School of Biology and Environmental Science
University College Dublin
Belfield, Dublin 4, Ireland
Jennifer.McElwain@ucd.ie

ABSTRACT

Recently, a novel paleoaltimetry method was presented using leaf stomatal frequency response to the decline in CO_2 partial pressure with altitude, and tested on California black oak (*Quercus kelloggii*) (McElwain 2004). Here, we present new data detailing the influence of other climatic variables on leaf stomatal frequency change with altitude in the context of more fully characterizing how stomatal frequencies can be used to infer paleoelevations. A clear increase in stomatal density and stomatal index is observed with increasing elevation for *Q. kelloggii* (black oak) leaves, and *Nothofagus solandri* var. *cliffortioides* (mountain beech) growing over an altitudinal transect on the slope of Mt. Ruapehu (New Zealand). Modern leaves growing in full direct sunlight versus shaded diffuse light for both species show substantial differences in stomatal density and index, however, growth chamber experiments that vary light intensity have revealed that the magnitude of natural increase in radiation with altitude is likely insufficient to explain the overall increase in stomatal frequency (density and index) with elevation. Furthermore, temperature does not have a significant influence on black oak stomatal frequency in growth chamber experiments. Rather changes in stomatal density and index with altitude appear to reflect an adaptation to counteract the limited photosynthetic potential due to the CO_2 partial pressure decrease, further limited by shorter growing seasons and/or increased UV radiation. Our review of the uncertainties associated with the stomatal frequency paleoaltimeter from the literature, together with results from the new plant growth experiments indicate that if sea-level paleoatmospheric CO_2 concentration can be well-constrained, the stomatal frequency method has the potential for very low error margins.

1529-6466/07/0066-0009$05.00   DOI: 10.2138/rmg.2007.66.9

INTRODUCTION

The development of quantitative paleoaltimetry techniques over the last decade offers exciting opportunities to determine the timing and magnitude of topographic development. Quantitative elevation estimates allow testing of models of geodynamic processes, atmospheric circulation patterns and climate, and geochemical cycling (topography may increase weathering rates that affect the carbon cycle). Current paleoaltimetry proxy techniques involve hydrogen and oxygen isotopes (Mulch and Chamberlain 2007), preserved air bubbles in basalts (Sahagian and Proussevitch 2007) and reconstruction of temperature or enthalpy lapse rates with altitude using paleobotanical methods (Wolfe et al. 1997; Forest et al. 1999). Although great strides have been made in the improvement of these methods, problems remain due to a lack of knowledge about regional climate/precipitation patterns, difficulty in finding suitable material, and relatively large uncertainties. Recently, a novel paleobotanical method has been proposed, based on reconstructing the predictable decrease in CO_2 partial pressure with altitude using stomatal density (SD) analysis (McElwain 2004). This method has the potential for high accuracy, as the altitude of modern black oak trees in California from a range of 700-2100 m were predicted with an average error of prediction of ~300 m. Moreover, the stomatal density method can be used in concert with existing paleobotanical methods on a single flora to obtain a suite of independent elevation estimates for a single locality, thereby greatly increasing the confidence in the obtained estimates.

The positive relation between stomatal density and altitude in California black oak, the species used as a paleoaltimeter, has been reported for many other species (Körner and Cochrane 1985; Körner et al. 1986; Woodward 1986; Hovenden and Brodribb 2000; Woodward et al. 2002). However, this relationship is not present in all species and/or localities, which may limit the global applicability of the method (Körner et al. 1986; Hultine and Marshall 2000; Greenwood et al. 2003; Qiang et al. 2003). Furthermore, questions have been raised about the physiological basis of the response of stomatal density to a decrease in CO_2 partial pressure with altitude, considering that other environmental changes with altitude may alleviate the physiological carbon limitation imposed by lower CO_2 partial pressure (Johnson et al. 2004). By assessing the potential influence of different environmental factors changing with altitude on the stomata-altitude relationship, we develop a framework to help explain and predict the conditions under which the stomatal proxy-method can be confidently applied.

To unravel the effect of local climate conditions on the selective pressure driving the stomatal adjustment, we first focus on leaf morphological and stomatal frequency data of two species with contrasting leaf morphology and geographical distribution over altitudinal transects: California black oak (*Quercus kelloggii* Newberry), the species on which this method was originally based (McElwain 2004), and the mountain beech (*N. solandri* var. *cliffortioides* (Hook. f.) Poole) from New Zealand.

Second, we describe the influence of CO_2, irradiation and temperature, the most consistently changing climatic factors with altitude, on stomatal frequency in *N. solandri* var. *cliffortioides* (*N. solandri* var. *cliffortioides* will be referred to as *N. solandri* in the remainder of the text for brevity) and *Q. kelloggii* using material from (1) recent and historical field collections and (2) experiments where *Q. kelloggii* was grown in rigorously controlled environmental growth chambers under different light and temperature settings. These new data are used to consider (1) how environmental factors may change with altitude, (2) whether these factors are known to directly influence stomatal frequency, (3) how climatic factors singularly, and in combination, may affect leaf-air gas-exchange and drive stomatal adjustments, and (4) whether we can infer which leaf types and/or geographical regions may show these adjustments and thus show potential for paleoaltimetry.

Finally we will focus on aspects of the application of both stomatal density (SD: number of stomata per square millimeter) and stomatal index (SI: ratio of number of stomata to total number of epidermal cells plus stomata expressed as a percentage) for paleoelevation reconstruction. We discuss the key advantages, limitations and calculated uncertainties associated with stomatal-based paleoelevation proxies and make recommendations on the selection of fossil plant taxa, fossil localities and time periods, which will minimize paleoelevation errors using the stomatal-proxy approach.

In the following section, we present new data from California and New Zealand (Figs. 1 and 2) that help resolve contributions of several factors to SD and SI, including elevation, light levels, and temperature. We have chosen to present new results in this review for two reasons. First, we believe they more fully characterize complications that can arise in applying the method. Second, they illustrate the kinds of procedures and data needed to infer elevation. Note that materials and methods are presented in Appendix 1.

NEW RESULTS

Leaf morphology data

Leaf area shows a poorly fitted decreasing trend with altitude for *Quercus kelloggii* ($R^2 = 0.272$), and no significant trend in *N. solandri* (Fig. 3A; 4A). Epidermal cell density increases for both species (Fig. 3B; 4B), although the linear relations are not very consistent. *N. solandri* shows a significant and strong linear increase in stomatal density with altitude (Fig. 3C). Stomatal density in *Quercus kelloggii* also increases with altitude (Fig. 4C). SD in *N. solandri* leaves collected on an elevational transect on Mt. Ruapehu increases linearly from 460 to 600 stomata per mm between 600 and 1470 m (3.5% per 100 m altitude gain; Fig. 3C). In *Q. kelloggii* leaves, SD increases from 250 to 470 stomata per mm² between 1000 and 2400 m (6.3% per 100 m altitude gain; Fig. 4C). Stomatal index in both species also increases with elevation, but the response is more pronounced at higher elevations (Fig. 3D, 4D). In *N. solandri* the SI below 900 m does not change much, and below ~1500 m in *Q. kelloggii*. Non-linear regressions were fitted, as this type of non-linear curve with response limits has been observed before in stomatal index adjustment to increasing CO_2 levels (Kürschner et al. 1997), and the fit improved for both taxa. The response curves for stomatal frequency in *Q. kelloggii* in this data set from 2003 are highly similar to the altitudinal response of *Q. kelloggii* leaves collected in 1934/1935 from the same populations, with a linear response of SD above 1000 m ($R^2 = 0.74$; $p < 0.001$), and a strong SI response starting at ~1500 m ($R^2 = 0.59$; $p < 0.001$) (McElwain 2004). *Q. kelloggii* and *N. solandri* are distributed in different hemispheres and have different morphologies (*Q. kelloggii* has much larger leaves than *N. solandri*), yet still show remarkable similarities in direction and rate of their stomatal response to altitude.

Stomatal density and epidermal cell density, but not stomatal index, of *N. solandri* is found to be significantly higher in sun leaves than in shade leaves when data from three localities on the South Island are combined ($p < 0.05$; Table 1; Fig. 5). For the individual localities, however, stomatal density and index of sun and shade leaves are significantly different. For *Q. kelloggii*, sun leaves collected from a range of modern oak trees in California also show significantly higher stomatal density and index on sun leaves than shade leaves ($p < 0.01$; Table 2; Fig. 6). Under experimental settings, however, the leaves grown under higher light have a slightly increased stomatal density and index, but no significant differences are present in a nested mixed-model ANOVA (Table 3; Fig. 7). Linear regressions of stomatal density and index changes averaged per tree to light intensity are significant, but poorly fitted ($R^2 < 0.2$).

Q. kelloggii tree seedlings grown under experimentally controlled daytime temperatures of 22 °C (low temperature treatment) or 27 °C (high temperature treatment) do not show a significant difference in either stomatal density or stomatal index (Table 4; Fig. 8).

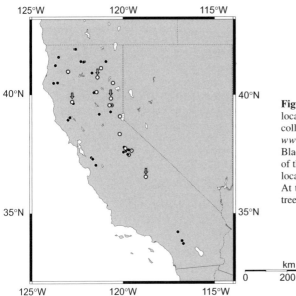

Figure 1. Map of California indicating the localities where *Q. kelloggii* leaves were collected in 2003. (map created on *http:// www.aquarius.ifm-geomar.de/omc/*). Black circles indicate sample locations of the complete data set, white circles the locations of the trees used in Figure 4. At the localities indicated by arrows two trees were sampled, instead of one.

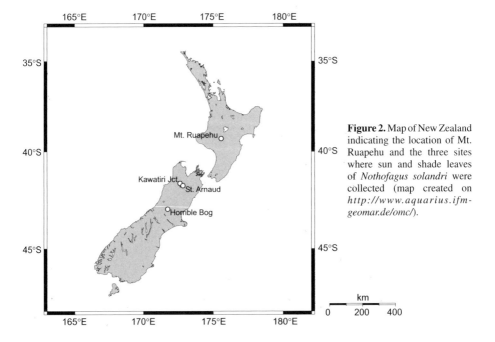

Figure 2. Map of New Zealand indicating the location of Mt. Ruapehu and the three sites where sun and shade leaves of *Nothofagus solandri* were collected (map created on *http://www.aquarius.ifm-geomar.de/omc/*).

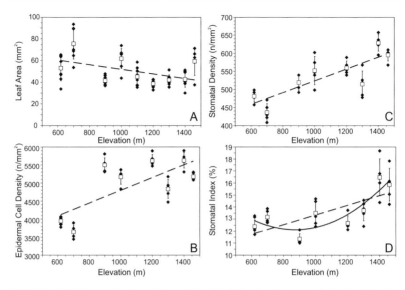

Figure 3. Relationship between leaf area (A), epidermal cell density (B), stomatal density (C) and stomatal index (D) versus altitude for *Nothofagus solandri* leaves growing on the slope of Mt. Ruapehu, New Zealand (collected in 1999). Black diamonds indicate the mean of ten counting fields on each leaf, white squares are the averages of five to eight leaves per elevation, with error bars of ±1 S.E.M. Nested mixed-model ANOVA with a general linear model indicates significant differences for all factors (p = 0.000). Averages per elevation were used for regression analysis: **A.** $y = -0.0212x + 73.1$; $R^2 = 0.276$; p = 0.147. **B.** $y = 1.70x + 3122$; $R^2 = 0.505$; p = 0.048. **C.** $y = 0.164x + 360$; $R^2 = 0.709$; p = 0.009. **D.** linear (dashed): $y = 0.004x + 9.33$; $R^2 = 0.540$; p = 0.038; non-linear (solid): $y = 0.00001x^2 - 0.0206x + 21.132$; $R^2 = 0.770$.

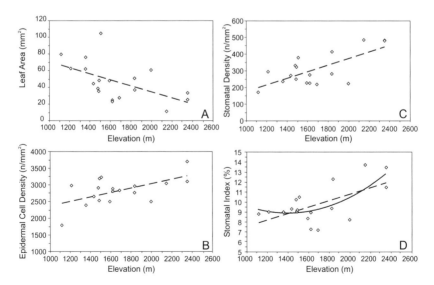

Figure 4. Relationship between leaf area (A), epidermal cell density (B), stomatal density (C) and stomatal index (D) versus altitude for *Quercus kelloggii* shade leaves (collected in 2003). Each point represents the mean of five random counts per leaf. Regressions: **A.** $y = -0.033x + 101.22$; $R^2 = 0.272$; p = 0.027. **B.** $y = 0.6608x + 1715$; $R^2 = 0.328$; p = 0.013. **C.** linear: $y = 0.1971x + 20.17$; $R^2 = 0.499$; p = 0.001. **D** linear (dashed): $y = 0.0032x + 4.3124$; $R^2 = 0.369$; p = 0.001; non-linear (solid): $y = 0.000004x^2 - 0.0123x + 17.55$; $R^2 = 0.466$.

Table 1. Stomatal density (SD), epidermal cell density (ED) and stomatal index (SI) of sun and shade leaves of *Nothofagus solandri var. cliffortioides*. Sun and shade leaves were collected at three localities (Fig. 2): Horrible Bog (HOR), Kawatiri Junction (KJ) and St. Arnaud (SA). Values are means of five leaves per light level (seven counts per leaf). The complete data set (total) was analyzed with a nested mixed-model ANOVA based on a general linear model, for comparisons within the individual localities a fully nested ANOVA was used.

locality	light	SD (n/mm²)	p-value	ED (n/mm²)	p-value	SI (%)	p-value
HOR	shade	376 ± 36	0.008	4727 ± 325	0.210	7.83 ± 0.61	0.029
HOR	sun	441 ± 21		4994 ± 293		9.22 ± 0.45	
KJ	shade	309 ± 28	0.000	4763 ± 244	0.034	6.95 ± 0.42	0.012
KJ	sun	409 ± 25		5305 ± 406		8.12 ± 0.75	
SA	shade	338 ± 13	0.000	4668 ± 199	0.002	8.07 ± 0.47	0.001
SA	sun	425 ± 16		5119 ± 112		8.61 ± 0.31	
total			0.017		0.011		0.108

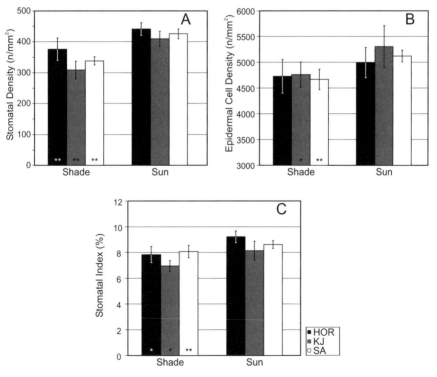

Figure 5. Stomatal density (SD; A), epidermal cell density (ED; B) and stomatal index (SI; C) of sun and shade leaves from modern *Nothofagus solandri* trees at three localities on the South Island of New Zealand (Fig. 2: Horrible Bog (HOR), Kawatiri Junction (KJ) and St. Arnaud (SA). Nested mixed-model ANOVA of the entire data set indicate significant differences between sun and shade leaves for SD (p = 0.017) and ED (p = 0.011) but not SI (p = 0.108) Asterisks indicate significant (*; p < 0.05) and highly significant (**; p < 0.01) differences from nested mixed-model ANOVA in that character between the means of sun and shade leaves per location (Table 1). Five sun and five shade leaves per tree were measured (seven random counts per leaf). Error bars represent ± 1 S.E.M.

Table 2. Stomatal density (SD), Epidermal cell density (ED) and stomatal index (SI) of modern *Quercus kelloggii* leaves, assigned to light regime during growth by degree of undulation, and p-values from a pairwise comparison using a nested mixed-model ANOVA based on a general linear model.

Light regime	No. of leaves	SD (n/mm²)	ED (n/mm²)	SI (%)
sun	16	414 ± 58	3059 ± 490	12.0 ± 1.4
neutral	23	352 ± 73	2963 ± 487	10.6 ± 1.0
shade	21	301 ± 94	2773 ± 426	9.7 ± 1.6
sun vs. shade (p-value)		0.001	0.099	0.000
sun vs. neutral (p-value)		0.023	0.592	0.005
neutral vs. shade (p-value)		0.063	0.222	0.042

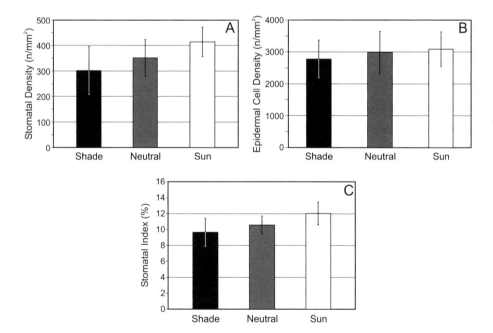

Figure 6. Stomatal density (SD; A), epidermal cell density (ED; B) and stomatal index (SI; C) of 61 modern *Quercus kelloggii* leaves collected in California in 2003 (Fig. 1). Leaves were assigned to sun, shade or neutral type by degree of undulation of the epidermal cell walls (e.g., Kürschner 1997). Nested mixed-model ANOVA showed significant differences for SD (p = 0.001) and SI (p = 0.000), but not ED (p = 0.217). Paired comparison of the means of stomatal density and index in a nested mixed-model ANOVA showed significant differences for SD and SI between sun and shade leaves (p < 0.01; Table 2). Error bars represent ± 1 S.E.M.

Table 3. Average stomatal density (SD), epidermal cell density (ED) and stomatal index (SI) of *Quercus kelloggii* leaves grown experimentally under low and high light conditions. The data set was analyzed with a nested mixed-model ANOVA based on a general linear model.

Light intensity (μmol$\cdot$m$^{-2}\cdot$s^{-1})	No. of leaves	SD (n/mm^2)	ED (n/mm^2)	SI (%)
49	8	178 ± 36	1706 ± 168	9.4 ± 1.2
85	8	226 ± 33	1997 ± 88	10.2 ± 1.3
156	11	206 ± 34	1733 ± 207	10.6 ± 1.6
204	10	237 ± 19	1949 ± 162	10.8 ± 0.6
p-value		0.202	0.032	0.654

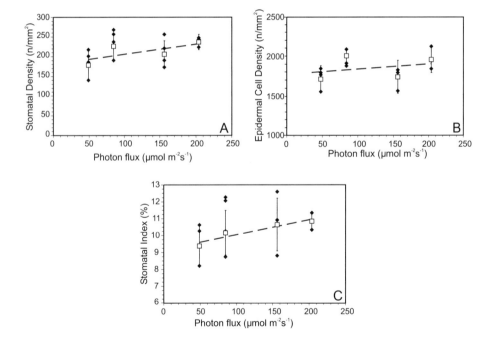

Figure 7. Stomatal density (SD; A), epidermal cell density (ED; B) and stomatal index (SI; C) of *Quercus kelloggii* leaves in relation to light intensity (photon flux) in four growth chambers. Black diamonds represent averages per tree (based on seven random counts per leaf on two to four leaves per tree), white squares averages per chamber. Nested mixed-model ANOVA based on a general linear model indicates no significant differences in SD (p = 0.202) or SI (p = 0.654), but in ED (p= 0.032). Linear regression of the means per tree is significant for SD (p = 0.011) and SI (p = 0.015), but R^2 values are very low (SD: R^2 = 0.173; SI: R^2 = 0.156). Linear regressions for the means per chamber: **A.** y = 0.2574x + 179.98, R^2 = 0.489; p = 0.301); **B.** y = 0.6316x + 1768, R^2 = 0.088; p = 0.703; **C.** y = 0.0089x + 9.1652, R^2 = 0.896; p = 0.054).

Table 4. Average stomatal density (SD), epidermal cell density (ED) and stomatal index (SI) of *Quercus kelloggii* leaves grown experimentally under low and high temperature regimes. The data set was analyzed with a nested mixed-model ANOVA based on a general linear model.

Temperature	No. of leaves	SD (n/mm²)	ED (n/mm²)	SI (%)
Low	11	345 ± 48	2395 ± 248	12.6 ± 1.0
High	9	290 ± 76	2224 ± 322	11.4 ± 1.7
p-value		0.323	0.442	0.313

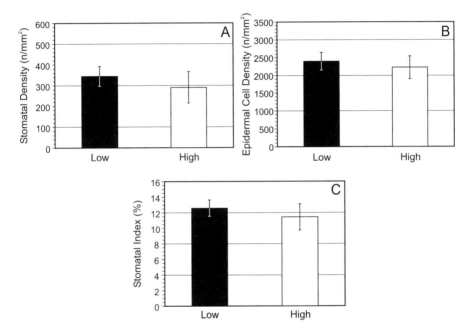

Figure 8. Stomatal density (SD; A), epidermal cell density (ED; B) and stomatal index (SI; C) of *Quercus kelloggii* leaves grown in growth chambers under low temperature (day: 20°C, night: 15°C) and high temperature regimes (day: 27°C, night: 22°C). Measurements for low temperature are based on eleven leaves (seven random counts each) from six trees and for the high temperature on nine leaves from five trees. Nested mixed-model ANOVA indicates no significant differences between low and high temperature treatments for SD (0.323), ED (0.442) or SI (0.313). Error bars represent ± 1 S.E.M.

CLIMATIC PARAMETERS, GAS EXCHANGE AND STOMATAL RESPONSE

Many abiotic factors change with elevation, such as temperature, humidity, irradiance (long-wave and UV-B) and wind speed, but atmospheric pressure decreases independent of micro-climatic conditions and can thus be the most accurately predicted without any substantial regional differences. The potential of each of these factors for explaining the observed changes in stomatal density and index will be discussed by comparing our experimental results and data from field-collected leaves with documented effects of climatic factors on stomatal frequency from the literature. Since stomatal frequency is closely linked to gas-exchange and water-use efficiency through its influence on stomatal conductance, the potential selective pressure of the combination of abiotic changes with altitude will then be discussed in terms of photosynthetic uptake and evaporative demand.

Changes in abiotic factors with altitude and direct effects on stomatal density and initiation

CO$_2$ partial pressure. Atmospheric pressure (P_{air}) decreases with altitude in a predictable manner (Jones 1992):

$$P_{air,z} = 101325^{\left[-\frac{MW_{air}\cdot g\cdot z}{RT}\right]} \tag{1}$$

where MW_{air} is the molecular weight of air (28.964×10^{-3} kg/mol), z is altitude in m, g is acceleration due to gravity in m/s, R is the gas constant (8.3144 J/mol) and T is mean July temperature in Kelvin.

The CO$_2$ partial pressure depends on the altitude as follows (McElwain 2004):

$$P_{CO_2,z} = \left[\frac{P_{air,z}}{101325}\right]P_{CO_2,sea\text{-}level} \tag{2}$$

where $P_{CO_2,sea\text{-}level}$ is the CO$_2$ partial pressure at sea-level and $P_{CO_2,z}$ is CO$_2$ partial pressure at unknown altitude z.

Many studies have shown genetically controlled adjustments in stomatal density and/or index for different plant taxa to changing CO$_2$ mixing ratios, although the degree and sign of response varies greatly between species (Peñuelas and Matamala 1990; Malone et al. 1993; Greenwood et al. 2003; Kouwenberg et al. 2003; Reid et al. 2003; Marchi et al. 2004; Wagner et al. 2005). The stomatal response to CO$_2$ concentration (at *constant* atmospheric pressure) and the developmental pathway involved has been reviewed extensively in recent literature (Gray et al. 2000; Brownlee 2001; Royer 2001; Lake et al. 2002; Woodward et al. 2002; Roth-Nebelsick 2005; Coupe et al. 2006) and a detailed discussion will thus not be included here. Stomatal density and index of fossil leaves have been used as a proxy to reconstruct paleoatmospheric CO$_2$ levels on timescales ranging from the Paleozoic to the last millennium (Van der Burgh et al. 1993; McElwain and Chaloner 1995; Rundgren and Beerling 1999; Wagner et al. 1999, 2004; Royer et al. 2001; McElwain et al. 2002, 2005; Roth-Nebelsick et al. 2004; Haworth et al. 2005; Kouwenberg et al. 2005; Rundgren et al. 2005; Van Hoof et al. 2006). Fewer studies have been carried out on SD and SI responses to CO$_2$ partial pressure decreases with elevation (i.e., *decreasing* atmospheric pressure), which is the focus of this paper because of the obvious application as a paleoaltimeter.

In a highly innovative experiment, Woodward and Bazzaz (1988) demonstrated that plants that were grown under lower air pressure (CO$_2$ partial pressure was changed, but CO$_2$ mixing ratios remained constant), showed significantly increased stomatal densities to retain comparable photosynthetic rates, under lower CO$_2$ availability. Since changes in CO$_2$ partial pressure have been shown experimentally to adjust stomatal frequencies, the decrease in CO$_2$ partial pressure can reasonably be invoked to have influenced the observed stomatal frequency increase with elevation for *N. solandri* and *Q. kelloggii*. The lack of stomatal index reduction at lower elevations in both species corresponds well with the documented response limits around 330-340 ppmV for other *Quercus* species (Kürschner et al. 1997). Considering the experimental evidence that CO$_2$ partial pressure under reduced air pressure can strongly affect stomatal density, and the simultaneous response of stomatal density and index for both species described here, decreasing CO$_2$ partial pressure is very likely to play a major role in the stomatal frequency increase with altitude.

Temperature. Temperature consistently decreases with altitude, although the specific temperature lapse rates are highly dependent on local geographical and climatic conditions. Humidity has a very prominent influence; theoretically, lapse rates can vary from ~10 °C/km in extremely dry conditions to 0 °C/km in extremely wet conditions, or can even be negative in special cases (Meyer 1992; Leuschner 2000). More realistic lapse rates are about 7.5 °C/km for dry and 5 °C/km for wetter mountain ranges (Meyer 1992; Leuschner 2000; Meyer 2007). In Tongariro National Park (New Zealand), where Mt. Ruapehu is located, the temperature lapse rate is close to the global mean of 6 °C/km (Druitt et al. 1990), and in northern California including the Sierra Nevada, where most *Q. kelloggii* grows, the temperature lapse rate is 5.9-6.0 °C/km (Meyer 1992).

Quercus kelloggii grown in a controlled environment under daytime temperatures of 20 °C and 27 °C did not show significant changes in either stomatal density or index (Table 4; Fig. 8). In temperature experiments on other species varying effects have been observed: no change in stomatal density is found (Apple et al. 2000; Luomala et al. 2005), or a negative correlation for certain age cohorts (Luomala et al. 2005), or a positive relationship for stomatal density (Ferris et al. 1996; Reddy et al. 1998) and stomatal index (Wagner 1998). The Tasmanian southern beech, *Nothofagus cunninghamii* shows an increase in stomatal density with altitude (Hovenden and Brodribb 2000), but no change in specific leaf area, stomatal density or stomatal index under an experimental controlled daytime temperature increase of 5 °C (Hovenden 2001). Since there is no experimental evidence for a consistent negative relationship between stomatal frequency and temperature, temperature decrease with altitude can not explain stomatal frequency increases.

Moreover, in natural habitats, any influence of temperature on stomatal initiation may be of little consequence, since most plants compensate for fluctuating temperatures by adjusting the timing of leaf development (Wagner 1998). At higher altitudes bud swelling and leaf expansion start later in the year (4.4 to 7 days/100 m for New Zealand beeches (Wardle 1984)), when temperatures are closer to temperatures at the start of the growing season at lower altitudes (Barrera et al. 2000 and references therein). Any potential direct effect of temperature on stomatal frequency is therefore likely negated by these phenological adjustments.

Precipitation. Water availability strongly affects stomatal density, but not stomatal index, because of its influence on epidermal cell expansion (Royer 2001). If lower precipitation and water availability with altitude strongly influenced stomatal density and epidermal cell density, then SD and cell density should be tightly correlated, and epidermal cell density and leaf size should show a large adjustment to altitude. Epidermal cell densities are indeed higher at higher altitude in both *N. solandri* and *Q. kelloggii*, and *Q. kelloggii* leaves are smaller, but the relations are weak or not consistently linear, and therefore do not reflect a predictable precipitation change with altitude. Moreover the increase in stomatal index indicates that stomatal initiation rates increased with altitude as well which can not be explained by water availability as only light and CO_2 are known to control stomatal initiation (Lake et al. 2002).

In general, cloud cover and rainfall do not show a general, consistent decrease with elevation. In many temperate mountain regions, as in California and New Zealand, cloud cover and precipitation actually increase with elevation (which should decrease stomatal densities), and in tropical systems rainfall increases up to 1000-1500 m and then decreases (Leuschner 2000). In the region of Mt. Ruapehu, rainfall increases strongly between elevations of ~600 and ~1120 m (Druitt et al. 1990). Precipitation data from 26 climate stations in Northern California (closest to the sample localities in Fig. 1) do not show a significant relation between precipitation and elevation in any direction ($R^2 = 0.07$; climate data from: *http://www.wrcc.dri.edu*). Thus, the increase in stomatal frequency in both *Q. kelloggii* and *N. solandri* to altitude is unlikely to be a direct result of changes in precipitation. If water availability were a primary factor affecting stomatal densities and caused larger variability due to microclimatic differences at sites, stomatal density and leaf size would be strongly correlated. The extremely weak direct correlation between stomatal density and leaf size for both species ($R^2 = 0.119$ for *Nothofagus* and 0.09 for *Quercus*) argues against water availability as an important influence on stomatal densities by inducing smaller epidermal cell size (and consequently smaller leaves) during drought (Fig. 9).

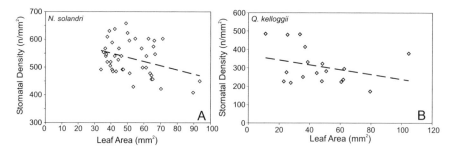

Figure 9. Relationship between stomatal density and leaf area for *Nothofagus solandri* leaves (A), grown at an altitudinal transect on the slope of Mt. Ruapehu, New Zealand, and *Quercus kelloggii* leaves (B) from California. **A.** Linear regression: $y = -1.5148x + 611.22$; $R^2 = 0.119$; $p = 0.023$ **B.** Linear regression: $y = -1.3121x + 369.36$; $R^2 = 0.090$; $p = 0.225$.

Light intensity. Regarding the influence of possible irradiation changes on stomatal frequency, a clearly higher SD and SI are observed in the field-collected sun leaves compared with shade leaves, for both *Q. kelloggii* (SD +38%; SI +24%) and *N. solandri* (SD +25%; SI +14%; averaged over the three sites). The positive effect of high light intensity on stomatal density and index of sun and shade leaves has been a long-standing observation in many other species (e.g., Kürschner 1997; Royer 2001; Lake et al. 2002). The light treatments in the *Q. kelloggii* light experiment covered a 300% increase in irradiation intensity from the equivalent of shade conditions to sunlight on an overcast day (because of their understorey habitat, black oak seedlings do not grow well under higher light intensities). These relatively low light levels are also reflected in the undulating epidermal cell walls observed in both treatments, lacking a clear "sun" type epidermal morphology. Under these light intensity increases, SD in *Q. kelloggii* increased by +16% and SI by +10%, but due to the large variability between trees in the chambers, this effect was not statistically significant (Table 3; Fig. 7). We can therefore conclude from these experiments that although irradiation clearly affects both stomatal density and index, the magnitude of change in light intensity required to elicit a significant stomatal frequency response in leaves far exceeds the typical irradiation changes observed along elevational gradients in either California or New Zealand.

In theory, solar radiation intensity gradually increases with altitude, but this trend can locally be very different due to changes in cloudiness with elevation. Long term instrumental measurements indicate that yearly global horizontal irradiation increases 10% with altitude between 200 and 1700 m (Delacasiniere et al. 1993), and measurements in the Swiss Alps show an altitude gradient over 370-3580 m for mean annual global radiation of 1.3 Wm^{-2}/100 m (Marty et al. 2002). However, in summer, this gradient is slightly negative above 2000 m due to increased cloudiness. Regional differences tend to override any expected irradiation increases with altitude, however. For example, the net radiation in the Austrian Alps stays constant with altitude (Leuschner 2000), because cloud cover gradually increases at higher elevations. In Hawaii, on the other hand, cloud cover increases up to 1000-1500 m, so radiation decreases over this range, and only at higher elevations does radiation begin to increase again. These regional trends illustrate that in mountain ranges proximal to oceans, such as in California and New Zealand, irradiance does not necessarily increase linearly with elevation, due to increased cloudiness and therefore can not explain the linear increase in SD and SI with altitude.

The increase in stomatal density and index with altitude could be a result of opening up of the higher elevation landscape, which increases the amount of incepted radiation, as has been suggested for *Nothofagus cunninghamii* forests in Tasmania (Hovenden and Vander Schoor 2006). Both *N. solandri and Q. kelloggii* show substantial differences in stomatal frequency between sun and shade leaves, and if the collected leaves become predominated by the sun morphotype, SD and SI increase with altitude (although the irradiation-related differences in stomatal density decrease for *Nothofagus cunninghamii* trees at higher elevations (Hovenden and Vander Schoor 2006). However, the sampling strategy for *N. solandri* and *Q. kelloggii* minimized local irradiation difference by sampling leaves from a similar position in the crown, and trees were selected that grew along an open road. In *Q. kelloggii* sun and shade leaves can be recognized and separated according to the degree of cell wall undulation, and the leaves analyzed in this study and depicted in Figure 4 are all shade leaves. Moreover, there is no relation between elevation and the degree of undulation of the oak leaves (One-way ANOVA on elevation vs. undulation class: p = 0.217), which unambiguously indicates that in this data set from California incepted light does not vary in a predictable manner with elevation.

Even though light intensity can influence stomatal density and stomata index, the weak experimental response of SD and SI to light intensity demonstrated for *Q. kelloggii*, together with the oceanic geographical settings of the study areas suggest that direct effects of irradiation on the overall altitudinal increase in stomatal frequency for both *Q. kelloggii* and *N. solandri* were probably not very pronounced. Microclimatic differences in incepted radiation per sampling site, however, may explain much of the variation found between sites.

UV-B. UV-B radiation increases with elevation by about 7-9% per km (Blumthaler et al. 1997; Alexandris et al. 1999; Zaratti et al. 2003). High UV-B levels are suggested to reduce the activity of photosynthetic enzymes and down-regulate photosynthetic genes (DeLucia et al. 1992; Jansen et al. 1998), but the effects vary for different species (Searles et al. 2001). Plants at high altitudes may suffer increased vulnerability to the enhanced UV-B levels as the short and cool growing seasons can delay the development of protective epidermis, cuticles and epicuticular wax (DeLucia et al. 1992; Turunen and Latola 2005). As a response to enhanced UV-B in experimental settings, most plant species exhibit a decrease in stomatal density (Dai et al. 1995; Visser et al. 1997; Keiller and Holmes 2001; Poulson et al. 2002; Gitz et al. 2005), but increases have also been observed (Stewart and Hoddinott 1993; Kostina et al. 2001). According to Nogues et al. (1998, 1999) the increased UV-B radiation at higher altitudes limits stomatal opening, and hence reduces stomatal conductance. An increase in SD with altitude could counterbalance this decreased stomatal conductance in order to retain optimized photosynthetic rates.

Effects of climatic changes with altitude on air-leaf gas exchange

CO_2 assimilation. The amount of CO_2 available for photosynthesis decreases with decreasing CO_2 partial pressure at higher elevations, but this effect is offset by the increase in diffusion speed at lower air pressure (Gale 1972, 1973). The lower temperature at higher altitudes, however, decreases diffusion speed, and therefore the temperature lapse rate of the particular mountain determines whether CO_2 availability decreases (dry-moist lapse rate) or stays relatively constant (very wet lapse rate) (Smith and Donahue 1991). The lower air pressure at altitude does not just decrease CO_2 partial pressure but also O_2 partial pressure, which results in lower photorespiration rates and more efficient photosynthesis. When all these effects are modeled, photosynthetic rates generally decrease with altitude, unless the temperature lapse rate is very low (which could occur in extremely wet mountain ranges), but the photosynthetic limitation is much less than expected based on just the partial pressure decrease (Terashima et al. 1995; Smith and Johnson 2007).

Growth experiments on three plant species under decreased air pressure, where CO_2 partial pressure was equivalent to high altitudes, but mixing ratio was unchanged, showed a significant increase in stomatal density, even though temperature was kept constant, creating a "wet" lapse rate situation, and CO_2 uptake was limited to a lesser degree than at high altitude under realistic lapse rates (Woodward and Bazzaz 1988). The increase in stomatal density in these species under low air pressure was as large as the experimental response to an equivalent

CO_2 decrease by adjusting the CO_2 mixing ratio (Woodward and Bazzaz 1988). These results strongly argue in favor of a central role of CO_2 partial pressure decreases with elevation on plant SD and SI. Under the relatively dry temperature lapse rates in California and New Zealand, therefore, the decrease in CO_2 partial pressure is expected to increase stomatal density and index, as observed in *Q. kelloggii* and *N. solandri* (Figs. 3, 4).

Transpiration. Two opposing effects work on transpiration at higher altitudes: the decreasing air pressure increases the diffusion speed of water (D_{wv}), but the lower temperature in turn decreases the diffusion speed. The importance of temperature lapse rates is illustrated by biophysical leaf models showing a significant increase in evaporation under wet lapse rates (3 °C), and a strong decrease in evaporation under dry lapse rates (8 °C) (Smith and Johnson 2007). However, an additional parameter stimulating transpiration is the increasing difference between leaf temperature (T_{leaf}) and air temperature (T_{air}). Therefore, factors affecting T_{leaf}, such as the radiation load, change D_{wv} too. The potentially higher radiation load at higher altitude will generally increase T_{leaf} (Smith and Johnson 2007) especially in large leaves, as they receive more irradiation than smaller leaves, to increase transpiration (Smith and Geller 1979). Perhaps the decrease in leaf size that is often seen at higher altitudes functions to keep the leaf cooler and cause less excess evaporation. Thus, transpiration effects of altitude depend greatly on local conditions, such as the temperature lapse rate (high/dry lapse rates cause lower evaporation) and radiation change (higher irradiation leads to increased evaporation), and on plant anatomical features such as leaf size (determining leaf temperature) and epidermal structure (affects the boundary layer thickness). For example, potential evapotranspiration for a 4 cm^2 leaf is modeled to be higher in equatorial mountains than in temperate areas, due to higher radiation loads and a larger leaf to air water vapor concentration gradient (Leuschner 2000). If radiation does not increase because of increased cloudiness, evaporation can easily decrease with altitude. High evaporation rates at high altitudes occur mainly in (1) dry tropical mountain chains because of high irradiation and low cloudiness and rainfall and (2) small oceanic mountains with strong upward convection of low air and thus low temperature lapse rates (Leuschner 2000).

Even if actual evaporation rates for the sites in this study are difficult to predict because of the dependence on the specific microclimatic conditions, we can infer general trends for the altitudinal transects in New Zealand and California. Both are located in temperate areas and have relatively dry temperature lapse rates (6 °C), and in California, cloudiness increases with altitude. Comparing these conditions to the modeled environments discussed above would suggest that evaporation is likely to decrease with altitude, or at least not increase significantly. The larger leaf size of the oak leaves may increase their evaporation rates relative to the smaller mountain beech leaves, but this remains speculative as no irradiation data available were available for either site.

Apart from air pressure, temperature and radiation, as discussed above, potential higher wind speed in the more open alpine environments can also influence evaporation. The boundary layer thickness of leaves is determined by the leaf dimensions and wind speed. Generally, wind speed increases with elevation (Leuschner 2000) and the greater wind speed will initially decrease the thickness of the boundary layer and enhance evapotranspiration (Gates 1976). However, the relation between transpiration rate and wind speed depends on the radiation absorption. In case of a high radiation absorption (and high leaf temperatures typical of larger leaves) or high leaf resistance, evaporation decreases with wind speed because of the convective cooling of the leaves (Baig and Tranquilini 1980). Under a high radiation load, transpiration can increase with wind speed in dry air, while decreasing in humid air (Gates 1968). Only when the low amounts of radiation are absorbed, for example by small leaves, resulting in relatively low leaf temperatures, does evaporation increase with wind speed due to the boundary layer thinning (Gates 1976). Leuschner (2000) modeled for a standard 4 cm^2 leaf that elevational increases in wind speed reduces potential transpiration by 11% in equatorial and 25% in middle-latitude mountain ranges. Thus, the increase in wind speed with altitude can offset the higher transpiration rates due to the larger leaf-air temperature difference created by the increase in radiation with elevation.

A decreased stomatal conductance, and thus transpiration under higher wind speeds has been reported in several experimental studies (Retuerto et al. 1996; Campbell-Clause 1998; Hoad et al. 1998), although no effect was detected in vegetation from high elevation in Puerto Rico (Cordero 1999), and grasses and shrubs showed enhanced transpiration despite decreased leaf temperatures (Yu et al. 1998). The varying effects of wind speed on transpiration in these studies could depend on the different leaf sizes studied. In summary therefore if wind speeds indeed increase with altitude in California and New Zealand, this could increase evaporation for the small *N. solandri* leaves, and decrease evaporation for the *Q. kelloggii* leaves, thus compensating for the higher radiation inception and leaf temperature of the larger leaves.

Gas exchange and stomatal density. Overall, the geographical setting of the studied areas and the subsequent relatively dry temperature lapse rates indicate that altitude affects gas-exchange in *Q. kelloggii* and *N. solandri* in similar fashion. The lower CO_2 partial pressure combined with the lower temperatures limit photosynthetic uptake at higher altitudes, although not as much as might be expected since the uptake is enhanced by the increased diffusion speed at higher altitude. Evaporation probably does not significantly increase, and may even be reduced, especially with increased cloudiness. Considering that experiments have shown that plants are as sensitive to CO_2 partial pressure as CO_2 mole fraction changes, the relatively small photosynthetic limitation due to reduced CO_2 availability at altitude is expected to increase stomatal densities and index in the two species studied, that are not hampered by excessive evaporation rates. The selective pressure for maintaining adequate carbon gain at high elevation due to decreased CO_2 partial pressure would obviously be higher in times of the geological past when sea-level CO_2 partial pressure was lower than current ambient levels.

However, even if photosynthetic capacity at higher altitudes might not be as severely hampered as the decrease in CO_2 partial pressure would imply, high altitude does restrict photosynthetic capacity by shortening the growing season due to low winter temperatures. Near treeline, in more open alpine landscapes, the thin snowpack is less effective at isolating the soil, and the resulting lower soil temperatures cause frost-drought to occur for a longer time (frost drought results in stomatal closure on trees living around treeline from November to April (Tranquilini 1976)). In extreme cases, such as the high elevation forests in the Basin and Range, Nevada, maximum photosynthetic rates are only attained for one month (Smith and Knapp 1990). Leaf gas exchange models indicate that stomatal aperture controls transpiration rates more effectively than photosynthetic rates (Pachepsky 1995), and therefore plants adapted to high elevation may increase stomatal density (and maximum stomatal conductance) to maximize photosynthetic carbon gain during the short period available (e.g., Woodward et al. 2002).

Higher stomatal densities allow the plants to maximize photosynthetic activity during the time when photosynthesis is not hampered by climatic conditions. Several aspects of higher altitude environments limit the time of photosynthetic activity, and the combination of these may explain the necessary compensation through increasing the maximum photosynthetic potential. Factors that change with increasing elevation affecting gas exchange are: (1) lower CO_2 partial pressure, especially under relatively high temperature lapse rates, (2) higher UV levels decreasing stomatal conductance, (3) higher irradiation that limits daytime photosynthetic hours because of extremely high transpiration rates in the middle of the day, and (4) the shorter growing seasons due to lower spring temperatures and later snowmelt than at lower elevations. The capability for higher maximum photosynthetic rates at altitude has been noted before, and is associated with higher nitrogen levels in leaves (Körner et al. 1986).

The two taxa studied here show a clear increase in stomatal density with altitude, as do many other species (Körner et al. 1986; Woodward 1986; Hovenden and Brodribb 2000), but this relation is not ubiquitous (Körner et al. 1986; Hultine and Marshall 2000; Greenwood et al. 2003; Qiang et al. 2003). It is obvious that the photosynthetic and gas exchange limitations vary for local conditions, such as temperature lapse rates, altitudinal irradiation profiles and leaf anatomy. In areas with low temperature lapse rates, for example, where the CO_2 exchange rate does not decrease with altitude and/or evaporation rates increase strongly, stomatal frequency changes with altitude may be much less pronounced, absent or even reversed in direction. These differences in selective pressure may explain the widely varying observed trends in stable carbon isotopes (reflecting water use efficiency), nitrogen content and stomatal parameters reported for different taxa over altitudinal ranges (Körner et al. 1986, 1991; Marshall and Zhang 1994; Sparks and Ehleringer 1997; Hultine and Marshall 2000; Kogami et al. 2001; Greenwood et al. 2003; Qiang et al. 2003).

APPLICATION OF STOMATAL FREQUENCY AS A PALEO-ALTIMETER: RECOMMENDATIONS AND LIMITATIONS

Stomatal frequency techniques

Stomatal density. The linear relation between stomatal density of *Q. kelloggii* and elevation above 1000 m (Fig. 4) suggests great potential for its use to reconstruct paleoelevations. As discussed above, however, it is unlikely that the stomatal density increase is the sole result of the decrease in CO_2 partial pressure. The stomatal increase may be an adjustment in gas exchange capability to counter the lower availability of CO_2 under shorter growing seasons, higher longwave radiation and higher UV radiation that would all decrease the maximum photosynthetic capacity. In this way, the observed stomatal density increase is definitely linked to the CO_2 partial pressure decrease, but can not be converted directly into a CO_2 partial pressure value using the relation in Figure 10 to calibrate leaves from different regions or sea-level CO_2 regimes. Indeed, when stomatal density data from Californian black oak leaves, collected between 1891 and recent are plotted against CO_2 partial pressure, an offset in the y intercepts is apparent between stomatal density–CO_2 partial pressure curves for each time period (Fig. 10). If CO_2 was the only factor directly determining the stomatal density decrease, the combined data from different sea-level CO_2 regimes and elevations would be expected to show a single linear response curve. Instead, all time periods, which only differ in sea-level CO_2 pressure, show the same SD-CO_2 slope, but with significant y offsets.

Therefore, to calibrate stomatal density to paleoelevation using the training set in Figure 4C, a correction factor is required. The elevation of eleven recent oak trees, geographically widely spread, could be predicted with an average error of estimation of ~300 m once $P_{CO_2,z}$ in Equation (2) was adjusted by adding the sea-level CO_2 concentration at the time of fossil growth minus the sea-level CO_2 concentration of the training set $[P_{CO_2, sea\text{-}level\ (fossil)} - P_{CO_2, sea\text{-}level\ (calibration)}]$ (McElwain 2004).

Stomatal index. The stomatal index represents the rate of stomatal initiation, and being independent of cell expansion, is less influenced by environmental factors that relate to gas exchange than stomatal density. In experiments, stomatal index is usually not influenced by factors other than CO_2 levels and, to a lesser extent, light intensity (Royer 2001; Lake et al. 2002). Thus, since stomatal index is directly related to CO_2 levels, prediction of paleoelevation using stomatal index should be feasible without extra corrections. The only other factor known to influence stomatal index apart from CO_2 is light intensity (Royer 2001). The additional error introduced by the potential influence of light intensity can be minimized when (1) a region is selected where light intensity does not increase much over altitude due to increasing cloud cover and/or fogginess, and (2) by distinguishing sun and shade leaf morphotypes which have a distinctive epidermal cell morphology, where anticlinal epidermal cell walls show more undulation (become less straight and more "jigsaw piece"-like), and

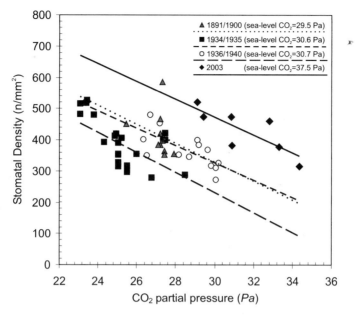

Figure 10. Historical and modern stomatal density (SD) data sets of *Quercus kelloggii* vs. CO_2 partial pressure from herbarium leaves (sun + shade) collected in (1) 1891-1900, (2) 1934-1935, (3) 1936-1940, and 2003, demonstrate similar SD response rates, but different intercepts. The 2003 data in this figure are not included in the present study [Used by permission of GSA, from McElwain (2004), *Geology*, Vol. 32, Fig. 3, p. 1019]).

can thus be separated for the calibration. The application of stomatal index instead of density has not been tested using *Q. kelloggii* leaves, because stomatal index in this species shows no response at altitudes under ~1500 m a.s.l., leaving a very short elevation range and data set for informative tests.

Constraints on the selection of method and material

Method. Whether to use stomatal density or stomatal index from leaf fossils for paleoelevation reconstruction depends on the availability of modern reference material and the quality of the fossil cuticle. Ideally both should be used on the same material to increase the confidence in the provided estimates. In cases, however, when the quality of the fossil cuticle does not allow epidermal cells to be recognized with the necessary accuracy, and stomatal index can not be determined, stomatal density measurements are the only available option. To use stomatal index, a training set of modern and herbarium material grown under different CO_2 pressure provides the necessary calibration. Because stomatal density is influenced by more factors than just the CO_2 partial pressure, the actual response of stomatal density of the chosen fossil taxon has to be confirmed by analyzing leaves from actual elevation transects. Stomatal density is physiologically more informative than stomatal index as SD is strongly related to maximum stomatal conductance.

Taxa. Although many species show the classic inverse relationship between stomatal frequency and CO_2, not all species do so. Therefore, the sensitivity of the species of fossil leaves needs to be tested when selecting fossil species to work with. Moreover, since plants have morphological and physiological constraints that do not allow adjustment of stomatal numbers to infinitely high or low CO_2, plant species have species-specific response limits to CO_2 (Kürschner et al. 1997). For instance, the nonlinearity of the stomatal index-elevation relation for both *Q. kelloggii* and *N. solandri* indicates that both species have already reached their response limit under modern ambient CO_2 levels. These species are therefore not suitable to reconstruct (1) low elevations under modern or higher CO_2 levels and (2) any elevations in time periods with much higher than present CO_2 levels, such as during the Mesozoic or Paleogene. To reconstruct paleoelevations during "greenhouse" ages, fossil taxa need to be used that are still sensitive at higher CO_2 levels, such as gymnosperms in general (Kouwenberg et al. 2003), or more specifically *Metasequoia glyptostroboides* or *Ginkgo biloba* (Royer et al. 2001). Contrary to CLAMP-based paleobotanical methods, the stomatal frequency method is not restricted to floras dominated by angiosperm taxa. A strong stomatal density response to maintain adequate carbon gain at high elevation due to decreased CO_2 partial pressure would be expected in species that do not suffer from high evaporative demands at higher altitudes, due to leaf anatomical adjustments such as increasing trichome density that can decrease transpiration by increasing boundary layer thickness.

Age range of applicability. Theoretically, this paleoaltimetric method could be used throughout the Late Cretaceous and Cenozoic, as long as the fossil species have extant counterparts for calibration purposes and cuticle is preserved. However, the accuracy and reliability of the method worsen further back in time. When selecting fossil floras for application of the stomatal frequency paleoelevation proxy, a number of potential setbacks need to be considered. First, the method requires calibration of stomatal frequency versus CO_2 pressure and/or altitude based on modern material of the same or a very closely related species. A few gymnosperm taxa, such as *Metasequoia* and *Ginkgo*, likely have evolved little since the Mesozoic. However, these are exceptions to the rule and for most species the Neogene may be the only period with a representative fossil record. Secondly, as mentioned earlier, finding suitable taxa, with modern equivalents for quantitative calibration, that also still respond to the (very) high CO_2 levels in greenhouse periods such as the Mesozoic and Paleogene might prove more of a challenge. Finally, sea-level CO_2 pressure, as the largest potential uncertainty associated with the method (see below), needs to be relatively well-constrained for the time period, requiring (1) good agreement on estimates of sea-level CO_2 pressure derived from other methods, or (2) in the ideal case, low-elevation floras containing the preferred taxon that are contemporaneous with the flora from unknown altitude. These requirements may be harder to fulfill further back in the Phanerozoic, as both stratigraphic control and certainty of paleoatmospheric CO_2 concentration decrease with age.

Sources of error and quantification of uncertainty

Error of prediction. McElwain (2004) used stomatal density counts to predict the known altitude of eleven modern *Q. kelloggii* trees across California from single leaves. This test revealed that, after using the correction for sea-level CO_2 differences between the calibration curve and the modern data, the average error in prediction was ~ 300 m. This prediction error is one of the smallest in any currently available paleoaltimetry method, but some other sources of uncertainty may increase this error.

Sun and shade leaves. Having a mixture of sun and shade leaves in the fossil assemblages could introduce a larger variability in the stomatal frequency data and subsequent elevation reconstruction, because they significantly differ in stomatal density and index (Royer 2001) and in sensitivity to CO_2 partial pressure (McElwain 2004). For *Quercus kelloggii*, the 38% difference in stomatal density between sun and shade leaves observed in this study would result in a 558 m difference in elevation, when converted to elevation estimates using Figure 4C. For *Nothofagus solandri*, converting stomatal frequency of either sun or shade leaves to estimate elevation, using Figure 3C and 3D, would result in a difference of 512 m for stomatal density and 250 m for stomatal index. This error may be smaller when using other species that show less difference in stomatal frequency between sun and shade morphotypes.

This added uncertainty can be greatly decreased, however, when sun and shade leaves can be distinguished by the undulation of the epidermal cell walls, as is the case in the oaks (Kürschner 1997). Often, sun-morphotypes are preferentially preserved in the fossil record (Kürschner 1997). All oak leaves in the light intensity experiment would be classified as shade leaves, and may provide us with an estimate of the effect of varying light intensity within the morphotypes. The observed change in SD of 16% equals an uncertainty in elevation estimates of 162 m. The exact light intensity related uncertainties vary for different species, but should be taken into account as an addition to the error margin.

Uncertainty in sea-level CO_2 estimates. The calculation of unknown paleo-elevations hinges on the difference in CO_2 partial pressure between sea-level and the site of unknown elevation (Eqn. 1). Therefore, not knowing exact sea-level CO_2 concentrations will introduce a significant additional error into the estimation. This uncertainty could be minimized by estimating sea-level CO_2 using stomatal frequency analysis on the same species from a contemporaneous low-elevation flora. However, if such fossil material is not available, CO_2 estimates based on other plant species or other proxies have to be used for calibration.

To explore the margin of error that uncertainty in sea-level CO_2 proxy data can introduce, we have calculated the offset in elevation estimates for a sea-level CO_2 regime of 365 ppmV with a hypothetical uncertainty of ±40 ppmV, and a sea-level CO_2 mixing ratio of 750 ppmV, in combination with uncertainties of ±40 ppmV and ±80 ppmV (Table 5; Fig. 11). These calculations indicate that under the low CO_2 regime, a 40 ppmV uncertainty results in an error ranging from ±900 m to ±1200 m, depending on the reconstructed elevation. The same uncertainty (±40 ppmV) in the high CO_2 regime results an error range of ±450 m to ±700 m, which would indicate that the stomatal method would be more precise for higher CO_2 regimes. However, usually error margins increase with the absolute value of the reconstructed CO_2 levels, and when the uncertainty in the high CO_2 regime is doubled to ±80 ppmV, the elevation error range increases to ±1000 m – ±1400 m.

This exercise clearly suggests that uncertainty in sea-level CO_2 estimates is potentially the largest source of error connected to the stomatal density paleoaltimeter, and when the uncertainty is very high, the utility of the method is severely compromised. Reliable and well-constrained CO_2 estimates or the availability of contemporaneous low-elevation leaf material can reduce or altogether eliminate this error source.

Table 5. Estimated elevation ranges using different uncertainties in sea-level CO_2 partial pressure for modern and high (Eocene) CO_2 levels.

Sea-level CO_2 pressure (Pa)		1000 m elevation	2000 m elevation	3000 m elevation	4000 m elevation
37.5	CO_2 decrease (Pa)	3.7	8	11.5	15
37.5	CO_2 decrease (%)	9.9	22.1	30.7	38.7
37.5	Elevation range (±40 ppmV) (m)	100-1950	1000-3150	1900-4150	2700-5300
77	CO_2 decrease (Pa)	8.3	16.5	23	29
77	CO_2 decrease (%)	10.8	21.4	29.9	37.7
77	Elevation range (±40 ppmV) (m)	550-1500	1520-2600	2360-3600	3300-4700
77	Elevation range (±80 ppmV) (m)	0-2000	930-3160	1770-4250	2600-5400

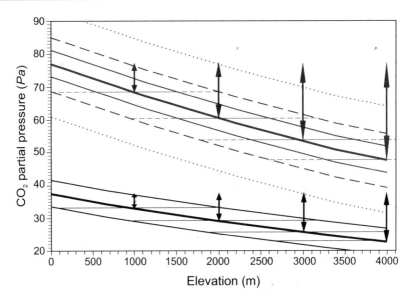

Figure 11. CO_2 partial pressure along an altitudinal range for different sea level CO_2 regimes. Thick black line was calculated using current global atmospheric CO_2 mixing ratios (365 ppmV), thick grey line represents global atmospheric CO_2 ratios comparable to estimated Eocene values (760 ppmV). Solid black and grey lines represent a sea-level CO_2 uncertainty of ±40 ppmV, grey dashed lines a sea-level uncertainty of ±80 ppmV. The arrows indicate the difference in CO_2 partial pressure between sea-level and the elevation. Horizontal solid lines indicates range in reconstructed elevation with the 40 ppmV uncertainty, horizontal dashed line the range for the 80 ppmV uncertainty. Dotted line indicates a ±160 ppmV uncertainty envelope, where the range in elevation estimates exceeds ±2000 m.

CONCLUSIONS

- The increase in stomatal density (SD) and stomatal index (SI) of *Quercus kelloggii* and *Nothofagus solandri* with elevation is most likely to be an adaptation to counteract the limited photosynthetic potential due to the CO_2 partial pressure decrease, further limited by shorter growing seasons and/or increased UV radiation. Growth experiments on *Q. kelloggii* show that temperature and light intensity differences comparable to gradients along elevational transects do not significantly affect stomatal density and index, and thus can not explain the increase in stomatal frequency with altitude. However, local differences in light inception, as reflected in typical sun and shade leaf morphology, may be an important source of variability in stomatal frequency.

- Leaf gas exchange rates are highly dependent on local climatic factors influencing CO_2 diffusion and evaporation rates, especially temperature lapse rates. The dependency of gas-exchange parameters on local climatic factors and leaf anatomy may account for the wide variability in leaf stomatal responses and stable isotope composition over elevation transects found in different species and different regions.

- For plants at high elevation, where evaporative demand is not excessively high (such as under high temperature lapse rates, increases in cloudiness and for small leaves) a strong SD response would be expected based on current ecophysiological understanding of plant growth along elevation gradients. Under low temperature lapse rates, strong irradiation increases and for large leaves, evaporative demand is high and selection pressure for an SD response to declining CO_2 partial pressure would not be expected.

- Stomatal density and index show great potential for paleoelevation reconstructions with low error margins, if additional error sources such as the presence of sun and shade morphotypes and especially uncertainty in sea-level CO_2 concentrations can be well constrained. Unlike other paleobotanical methods, stomatal frequency analysis is not restricted to angiosperm dominated floras, and has no requirements for a minimum amount of taxa present. The method will be most reliable when applied to fossil taxa that are closely related to extant species, and suitable taxa are most likely to be found for periods when CO_2 concentrations were not much higher than ambient (380 ppmV).

ACKNOWLEDGMENTS

We thank Madeleine McLeester and Shannon Loomis for the stomatal counts on the 2003 *Q. kelloggii* data set, and Rebekah Hines for assistance with the growth chamber experiments and in the field. Dana Royer and Peter Wilf graciously provided the leaf area data for *Q. kelloggii*. Personnel of the Klamath, Shasta and Stanislaus National Parks are thanked for permission to collect. Field assistance in New Zealand was lent by Rike Wagner, Ton van Druten, and David Feek, and equipment and advice was provided by John Flynn of the Massey University, Palmerston North. Ian Raine (GNS, New Zealand) collected the *Nothofagus* leaves from the South Island. Vincent Vos is much thanked for work on the Mt. Ruapehu altitudinal transect. D. Beerling (University of Sheffield) is thanked for assistance with the development of Equation 2. The comments of Ian Miller and two anonymous reviewers are very much appreciated and greatly improved the manuscript. The research presented here was financed by the Netherlands Organization for Scientific Research (NWO no. 750.198.07), the John Caldwell Meeker Foundation and the National Science Foundation (grant EAR-0207440).

REFERENCES

Alexandris D, Varotsos C, Kondratyev KY, Chronopoulos G (1999) On the altitude dependence of solar effective UV. Phys Chem Earth 24:515-517

Apple ME, Olszyk DM, Lewis J, Southworth D, Tingey DT (2000) Morphology and stomatal function of Douglas fir needles exposed to climate change: elevated CO_2 and temperature. Int J Plant Sci 161:127-132

Barrera MD, Frangi JL, Richter LL, Perdomo MH, Pinedo LB (2000) Structural and functional changes in *Nothofagus pumilio* forests along and altitudinal gradient in Tierra del Fuego, Argentina. J Veg Sci 11:179-288

Blumthaler M, Ambach W, Ellinger R (1997) Increase in solar UV radiation with altitude. J Photochem Photobiol B 39:130-134

Brownlee C (2001) The long and the short of stomatal density signals. Trends Plant Sci 6:441-442

Campbell-Clause JM (1998) Stomatal response of grapevines to wind. Aust J Exp Agr 38:77-82

Cordero RA (1999) Ecophysiology of *Cecropia schreberiana* saplings in two wind regimes in an elfin cloud forest: growth, gas exchange, architecture and stem biomechanics. Tree Physiol 19:153-163

Coupe SA, Palmer BG, Lake JA, Overy SA, Oxborough K, Woodward FI, Gray JE, Quick WP (2006) Systematic signalling of environmental cues in *Arabidopsis* leaves. J Exp Bot 57:329-341

Dai QJ, Peng SB, Chavez AQ, Vergara BS (1995) Effects of UVB radiation on stomatal density and opening in rice (*Oryza sativa* L). Ann Bot 76:65-70

Delacasiniere A, Grenier JC, Cabot T, Werneckfaga M (1993) Altitude Effect on the Clearness Index in the French Alps. Sol Energy 51:93-100

DeLucia EH, Day TA, Vogelmann TC (1992) Ultraviolet-B and visible light penetration into needles of two species of subalpine conifers during foliar development. Plant Cell Environ 15:921-929

Druitt DG, Enright NJ, Ogden J (1990) Altitudinal zonation in the mountain forests of Mt. Hauhungatahi, North Island, New Zealand. J Biogeogr 17:205-220

Ferris R, Nijs I, Behaeghe T, Impens I (1996) Elevated CO_2 and temperature have different effects on leaf anatomy of perennial ryegrass in spring and summer. Ann Bot 78:489-497

Forest CE, Wolfe JA, Molnar P, Emanuel KA (1999) Paleoaltimetry incorporating atmospheric physics and botanical estimated of paleoclimate. Geol Soc Am Bull 111:497-511

Gale J (1972) Availability of carbon dioxide for photosynthesis at high altitudes: theoretical considerations. Ecology 53:494-497

Gale J (1973) Experimental evidence for the effect of barometric pressure on photosynthesis and transpiration. Plant response to climatic factors. Proc of the Uppsala symposium, UNESCO, Ecology and Conservation 5:289-293

Gates DM (1968) Transpiration and leaf temperature. Ann Rev Plant Physiol 19:211-238

Gates DM (1976) Energy exchange and transpiration. *In:* Water and Plant Life. Lange OL, Kappan L, Schulze E-D (eds) Springer Verlag, Berlin, p 135-147

Gitz DC, Liu-Gitz L, Britz SJ, Sullivan JH (2005) Ultraviolet-B effects on stomatal density, water-use efficiency, and stable carbon isotope discrimination in four glasshouse-grown soybean (*Glycine max*) cultivars. Environ Exp Bot 53:343-355

Glover BJ, Martin C (2000) Specification of epidermal cell morphology. *In:* Plant Trichomes. Hallahan DL, Gray JC (eds) Academic Press, London, p 193-217

Gray JE, Holroyd GH, van der Lee FM, Bahrami AR, Sijmons PC, Woodward FI, Schuch W, Hetherington AM (2000) The HIC signalling pathway links CO_2 perception to stomatal development. Nature 408:713-716

Greenwood DR, Scarr MJ, Christophel DC (2003) Leaf stomatal frequency in the Australian tropical rainforest tree *Neolitsea dealbata* (Lauraceae) as a proxy measure of atmospheric $pCO(2)$. Palaeogeogr Palaeoclimatol Palaeoecol 196:375-393

Haworth M, Hesselbo SP, McElwain JC, Robinson SA, Brunt JW (2005) Mid-Cretaceous pCO_2 based on stomata of the extinct conifer *Pseudofrenelopsis* (Cheirolepidiaceae). Geology 33:749-752

Hoad SP, Marzoli A, Grace J, Jeffree CE (1998) Response of leaf surfaces and gas exchange to wind stress and acid mist in birch (*Betula pubescens*). Trees 13:1-12

Hovenden MJ (2001) The influence of temperature and genotype on the growth and stomatal morphology of southern beech *Nothofagus cunninghamii* (Nothofagaceae). Aust J Bot 49:427-434

Hovenden MJ, Brodribb T (2000) Altitude of origin influences stomatal conductance and therefore maximum assimilation rate in Southern Beech, *Nothofagus cunninghamii*. Aust J Plant Physiol 27:451-456

Hovenden MJ, Vander Schoor JK (2005) The response of leaf morphology to irradiance depends on altitude of origin in *Nothofagus cunninghamii*. New Phytol 169:291-297

Hultine KR, Marshall JD (2000) Altitude trends in conifer leaf morphology and stable carbon isotope composition. Oecologia 123:32-40

Jansen MAK, Gaba V, Greenberg BM (1998) Higher plants and UV-B radiation: balancing damage, repair and acclimation. Trends Plant Sci 3:131-135

Johnson DM, Smith WK, Silman MR (2004) Comment on "Climate independent paleoaltimetry using stomatal density in fossil leaves as a proxy for CO_2 partial pressure". Geology:e82-e83

Jones HG (1992) Plants and Microclimate. Cambridge University Press

Keiller DR, Holmes MG (2001) Effects of long-term exposure to elevated UV-B radiation on the photosynthetic performance of five broad-leaved tree species. Photosynth Res 67:229-240

Kogami H, Hanba YT, Kibe T, Terashima I, Masuzawa T (2001) CO_2 transfer conductance, leaf structure and carbon isotope composition of *Polygonum cuspidatum* leaves from low and high altitudes. Plant Cell Environ 24:529-538

Körner C, Bannister P, Mark AF (1986) Altitudinal variation in stomatal conductance, nitrogen content and leaf anatomy in different plant life forms in New Zealand. Oecologia 69:577-588

Körner C, Cochrane PM (1985) Stomatal responses and water elations of *Eucalyptus pauciflora* in summer along an elevational gradient. Oecologia 66:443-455

Körner C, Farquhar GD, Wong SC (1991) Carbon isotope discrimination by plants follows latitudinal and altitudinal trends. Oecologia 88:30-40

Kostina E, Wulff A, Julkunen-Tiitto R (2001) Growth, structure, stomatal responses and secondary metabolites of birch seedlings (*Betula pendula*) under elevated UV-B radiation in the field. Trees 15:483-491

Kouwenberg LLR, Hines RR, McElwain JC (2007) A new transfer technique to extract and process thin and fragmented fossil cuticle using polyester overlays. Rev Palaeobot Palyno 145:243-248

Kouwenberg LLR, McElwain JC, Kürschner WM, Wagner F, Beerling DJ, Mayle FE, Visscher H (2003) Stomatal frequency adjustment of four conifer species to historical changes in atmospheric CO_2. Am J Bot 90:610-619

Kouwenberg LLR, Wagner F, Kürschner WM, Visscher H (2005) Atmospheric CO_2 fluctuations during the last millennium reconstructed by stomatal frequency analysis of *Tsuga heterophylla* needles. Geology 33:33-36

Kürschner WM (1997) The anatomical diversity of recent and fossil leaves of the durmast oak (*Quercus petraea* Lieblein/*Quercus pseudocastanea* Goeppert): implications for their use as biosensors of paleoatmospheric CO_2 levels. Rev Palaeobot Palyno 96:1-30

Kürschner WM, Wagner F, Visscher EH, Visscher H (1997) Predicting the response of leaf stomatal frequency to a future CO_2 enriched atmosphere: constraints from historical observations. Geol Rundschau 86:512-517

Lake JA, Woodward FI, Quick WP (2002) Long-distance CO_2 signalling in plants. J Exp Bot 53:183-193

Leuschner C (2000) Are high elevations in tropical mountains arid environments for plants? Ecology 81:1425-1436

Luomala EM, Laitinen K, Sutinen S, Kellomaki S, Vapaavuori E (2005) Stomatal density, anatomy and nutrient concentrations of Scots pine needles are affected by elevated CO_2 and temperature. Plant Cell Environ 28:733-749

Malone SR, Mayeux HS, Johnson HB, Polley HW (1993) Stomatal Density and Aperture Length in Four Plant Species Grown Across a Subambient CO_2 Gradient. Am J Bot 80:1413-1418

Marchi S, Tognetti R, Vaccari FP, Lanini M, Kaligaric M, Miglietta F, Raschi A (2004) Physiological and morphological responses of grassland species to elevated atmospheric CO_2 concentrations in FACE-systems and natural CO_2 springs. Funct Plant Biol 31:181-194

Marty C, Philipona R, Frohlich C, Ohmura A (2002) Altitude dependence of surface radiation fluxes and cloud forcing in the alps: results from the alpine surface radiation budget network. Theor Appl Climatol 72:137-155

Marshall JD, Zhang J (1994) Carbon isotope discrimination and water-use efficiency in native plants of the North-Central Rockies. Ecology 75:1887-1895

McElwain JC (2004) Climate-independent paleoaltimetry using stomatal density in fossil leaves as a proxy for CO_2 partial pressure. Geology 32:1017-1020

McElwain JC, Chaloner WG (1995) Stomatal density and index of fossil plants track atmospheric carbon dioxide in the Palaeozoic. Ann Bot 76:389-395

McElwain JC, Mayle FE, Beerling DJ (2002) Stomatal evidence for a decline in the atmospheric CO_2 concentration during the Younger Dryas stadial: a comparison with Antarctic ice core records. J Quat Sci 17:21-29

McElwain JC, Mitchell FJG, Jones MB (1995) Relationship of stomatal density and index of *Salix cinerea* to atmospheric carbon dioxide concentrations in the Holocene. The Holocene 5:216-219

McElwain JC, Wade-Murphy J, Hesselbo SP (2005) Changes in carbon dioxide during an oceanic anoxic event linked to intrusion into Gondwana coals. Nature 435:479-482

Meyer HW (1992) Lapse rates and other variables applied to estimating paleoaltitudes from fossil floras. Palaeogeogr Palaeoclimatol Palaeoecol 99:71-99

Meyer HW (2007) A review of paleotemperature–lapse rate methods for estimating paleoelevation from fossil floras. Rev Mineral Geochem 66:155-171

Mulch A, Chamberlain CP (2007) Stable isotope paleoaltimetry in orogenic belts – the silicate record in surface and crustal geological archives. Rev Mineral Geochem 66:89-118

Nogues S, Allen DJ, Morison JIL, Baker NR (1998) Ultraviolet-B radiation effects on water relations, leaf development, and photosynthesis in droughted pea plants. Plant Physiol 117:173-181

Nogues S, Allen DJ, Morison JIL, Baker NR (1999) Characterization of stomatal closure caused by ultraviolet-B radiation. Plant Physiol 121:489-496

Pachepsky LB, Haskett JD, Acock B (1995) A two-dimensional model of leaf gas exchange with special reference to leaf anatomy. J Biogeogr 22:209-214

Peñuelas J, Matamala R (1990) Changes in N and S leaf content, stomatal density and specific leaf area of 14 plant species during the last three centuries of CO_2 increase. J Exp Bot 41:1119-1124

Poole I, Kürschner WM (1999) Stomatal density and index: the practice. *In:* Fossil Plant and Spores: Modern Techniques. Jones TP, Rowe NP (eds) The Geological Society, London, p 257-260

Poulson ME, Donahue RA, Konvalinka J, Boeger MRT (2002) Enhanced tolerance of photosynthesis to high-light and drought stress in *Pseudotsuga menziesii* seedlings grown in ultraviolet-B radiation. Tree Physiol 22:829-838

Sahagian D, Proussevitch A (2007) Paleoelevation measurement on the basis of vesicular basalts. Rev Mineral Geochem 66:195-213

Qiang W, Wang X, Chen T, Feng H, An L, He Y, Wang G.(2003) Variations of stomatal density and carbon isotope values of *Picea crassifolia* at different altitudes in the Qilian Mountains. Trees 17:258-262

Reddy KR, Robana RR, Hodges HF, Liu XJ, McKinion JM (1998) Interactions of CO_2 enrichment and temperature on cotton growth and leaf characteristics. Environ Exp Bot 39:117-129

Reid CD, Maherali H, Johnson HB, Smith SD, Wullschleger SD, Jackson RB (2003) On the relationship between stomatal characters and atmospheric CO_2. Geophys Res Lett 30:10.1029/2003GL017775

Retuerto R, Rochefort L, Woodward FI (1996) The influence of plant density on the responses of *Sinapsis alba* to CO_2 and windspeed. Oecologia 108:241-251

Roth-Nebelsick A (2005) Reconstructing atmospheric carbon dioxide with stomata: possibilities and limitations of a botanical pCO(2)-sensor. Trees 19:251-265

Roth-Nebelsick A, Utescher T, Mosbrugger V, Diester-Haass L, Walther H (2004) Changes in atmospheric CO_2 concentrations and climate from the Late Eocene to Early Miocene: palaeobotanical reconstruction based on fossil floras from Saxony, Germany. Palaeogeogr Palaeoclimatol Palaeoecol 205:43-67

Royer DL (2001) Stomatal density and stomatal index as indicators of paleoatmospheric CO_2 concentration. Rev Palaeobot Palyno 114:1-28

Royer DL, Wing SL, Beerling DJ, Jolley DW, Koch PL, Hickey LJ, Berner RA (2001) Paleobotanical evidence for near present-day levels of atmospheric CO_2 during part of the Tertiary. Science 292:2310-2313

Rundgren M, Beerling D (1999) A Holocene CO_2 record from the stomatal index of subfossil *Salix herbacea* L. leaves from northern Sweden. The Holocene 9:509-513

Rundgren M., Björck S, Hammarlund D (2005) Last interglacial atmospheric CO_2 changes from stomatal index data and their relation to climate variations. Glob Planet Change 49:47-62

Searles PS, Flint SD, Caldwell MM (2001) A meta-analysis of plants field studies simulating stratospheric ozone depletion. Oecologia 127:1-10

Smith WK, Donahue RA (1991) Simulated Influence of Altitude on Photosynthetic CO_2 Uptake Potential in Plants. Plant Cell Environ 14:133-136

Smith WK, Geller GN (1979) Plant transpiration at high elevations: theory, field measurements, and comparisons with desert plants. Oecologia 41:109-122

Smith WK, Johnson DM (2007) Biophysical effects of altitude on plant gas exchange. *In:* Perspectives in Biophysical Ecophysiology. De la Barrera E, Smith WK (eds) University of California Press

Smith WK, Knapp AK (1990) Ecophysiology of high elevation forests. *In:* Plant Biology of the Basin and Range. Osmond CB, Pitelka LF, Hidy GM (eds) Springer-Verlag, Berlin, p 87-144

Sparks JP, Ehleringer JR (1997) Leaf carbon isotope discrimination and nitrogen content for riparian trees along elevational transects. Oecologia 109:362-367

Stewart JD, Hoddinott J (1993) Photosynthetic acclimation to elevated atmospheric CO_2 and UV radiation in *Pinus banksiana*. Physiol Plantarum 88:493-500

Terashima I, Masuzawa T, Ohba H, Yokoi Y (1995) Is photosynthesis suppressed at higher elevations due to low CO_2 pressure? Ecology 76:2663-2668

Tranquilini W (1976) Water relations and alpine timberline. *In:* Water and Plant Life. Lange OL, Kappan L, Schulze E-D (eds) Springer Verlag, Berlin, p 473-491

Turunen M, Latola K (2005) UV-B radiation and acclimation in timberline plants. Environ Pollut 137:390-403

Van der Burgh J, Visscher H, Dilcher DL, Kürschner WM (1993) Paleoatmospheric signatures in Neogene fossil leaves. Science 260:1788-1790

Van Hoof TB, Bunnik FPM, Waucomont JGM, Kürschner WM, Visscher H (2006) Forest re-growth in medieval farmland after the Black Death pandemic - Implications for atmospheric CO_2 levels. Palaeogeogr Palaeoclimatol Palaeoecol 237:396-411

Visser AJ, Tosserams M, Groen MW, Kalis G, Kwant R, Magendans GWH, Rozema J (1997) The combined effects of CO_2 concentration and enhanced UV-B radiation on faba bean. 3. Leaf optical properties, pigments, stomatal index and epidermal cell density. Plant Ecol 128:209-222

Wagner F (1998) The influence of environment on the stomatal frequency in *Betula*. PhD Dissertation, Utrecht University, Utrecht

Wagner F, Bohnke SJP, Dilcher DL, Kürschner WM, van Geel B, Visscher H (1999) Century-scale shifts in early Holocene atmospheric CO_2 concentration. Science 284:1971-1973

Wagner F, Dilcher DL, Visscher H (2005) Stomatal frequency responses in hardwood-swamp vegetation from Florida during a 60-year continuous CO_2 increase. Am J Bot 92:690-695

Wagner F, Kouwenberg LLR, van Hoof TB, Visscher H (2004) Reproducibility of Holocene atmospheric CO_2 records based on stomatal frequency. Quat Sci Rev 23:1947-1954

Wardle JA (1984) The New Zealand Beeches. Ecology, Utilisation and Management. New Zealand Forest Service, Christchurch

Wolfe JA, Schorn HE, Forest CE, Molnar P (1997) Paleobotanical evidence for high altitudes in Nevada during the Miocene. Science 276:1672-1675

Woodward FI (1986) Ecophysiological studies on the shrub *Vaccinium myrtilis* L. taken from a wide altitudinal range. Oecologia 70:580-586

Woodward FI, Bazzaz FA (1988) The response of stomatal density to CO_2 partial pressure. J Exp Bot 39:1771-1781

Woodward FI, Lake JA, Quick WP (2002) Stomatal development and CO_2: ecological consequences. New Phytol 153:477-484

Yu YJ, Xin YY, Liu JQ, Yu ZH (1998) Effects of wind and wind-sand current on the physiological status of different sand-fixing plants. Acta Bot Sin 40:962-968

Zaratti F, Forno RN, Garcia Fuentes J, Andrade MF (2003) Erythemally weighted UV variations at two high-altitude locations. J Geophys Res 108:4263

APPENDIX 1
MATERIAL AND METHODS

Collected material

Leaves from 61 georeferenced *Q. kelloggii* trees were collected from multiple populations over an altitudinal range of 1100-2400 m across California in the summer of 2003 (Fig. 1). The leaf samples were divided into "sun," "neutral" or "shade" types based on the degree of undulation of the epidermal anticlinal cell walls, which is indicative of light levels during growth (Kürschner 1997).

One branch with fully expanded leaves was collected in September 1999 from nine *N. solandri* var. *cliffortioides* trees growing between 600 and 1460 m (treeline) on Mt. Ruapehu (North Island, New Zealand; 39°18'S 175°35'E; Fig. 2). All samples were taken at ~1.5 m height from the outer north side of the canopy to minimize non-altitude related variation in irradiation levels. Five to eight leaves per tree were processed for SD and SI investigation using standard protocols (McElwain et al. 1995; Poole and Kürschner 1999) and analyzed. *N. solandri* var. *cliffortioides* will be referred to as *N. solandri* in the remainder of the text for brevity.

Branches with sun and shade leaves from three *N. solandri* trees were collected in October 1999 from localities on the South Island, New Zealand: Horrible Bog (S 43°01'12.8" E 171°43'44.6"; 650 m a.s.l.), Kawatiri Junction (S 41°041' E 172°37'; 380 m) and St. Arnaud ((S 41°048' E 172°50'; 700 m) (Fig. 2). Five leaves from each branch were processed for cuticular analysis.

Light experiment

A light experiment was conducted on four-year-old *Q. kelloggii* seedlings in four CONVIRON E8 growth chambers from March to July 2006. Plants were grown in six liter containers filled with a mixture of peat moss, perlite and sand. Nutrients (Miracle Grow) were added twice a year, but not during the experiment. Average light intensity in the four chambers over the five month growth period was 49 (chamber B), 85 (chamber C), 156 (chamber A) and 204 $\mu mol \cdot m^{-2} \cdot s^{-1}$ (chamber D). Temperature in all four chambers was set to mimic a typical 24-hr cycle in the northern Californian habitat during the growing season (22 °C during the day and 17 °C at night with a 16 hr photoperiod). Humidity levels in all four chambers were maintained between 40 and 80%. Atmospheric CO_2 concentration in the chambers was ~400 ppmV. Plants were watered twice a week (0.25 l per seedling) using an automatic irrigation system. All four chambers contained 15 replicate plants from two different source populations in California.

Temperature experiment

A temperature experiment was conducted from February to May 2004 on two-year-old seedlings of *Quercus kelloggii* (the same plants used in the light experiment, but randomized between experiments). Plants in two Conviron E8 chambers were manually watered twice a week, and light intensity was set at a ~200 $\mu mol \cdot m^{-2} \cdot s^{-1}$ to mimic the natural understorey habitat of *Q. kelloggii* seedlings. In the low temperature treatment, temperature followed a 24-hr cycle, consisting of a 16 hour day of 22 °C and a 6 hour night of 15 °C, with a stepped temperature increment between day and night to simulate dawn and dusk transitions. In the high temperature treatment the same photoperiod settings in the high temperature setting were applied with daytime and nighttime temperature of 27 °C and 20 °C respectively. Growth room relative humidity and seedling nutrient and watering regime were the same for both chambers as in the light experiment.

Processing and analytical methods

Small leaf disks (~1 cm²) were cut out of the central part of the *Q. kelloggii* leaves using a hole punch. Edges were cut off of the *N. solandri* leaves (complete leaves were small enough to be processed whole), and, if necessary, the dense layer of trichomes on the abaxial surface was "shaved" by gentle scraping with a scalpel. All leaf samples were then treated with either 4% sodium hypochloride (*N. solandri*) or a 50/50 mixture of concentrated glacial acetic acid and 30% hydrogen peroxide (*Q. kelloggii*) at ~50 °C for one to two days to remove mesophyll tissue and separate the cuticle. The abaxial cuticle was stained with saffranin, mounted in glycerin jelly or water on a slide and computer-aided analysis was performed using transmitted light on a Leica DMLB epifluorescence microscope (*Q. kelloggii*) and a Leica Quantimet 500C/500+ Image Analysis system (*N. solandri*).

Stomatal density, epidermal cell density, and stomatal index were measured in five to seven digitally captured counting fields (0.068 mm^2) on each *Q. kelloggii* leaf and seven (sun and shade leaves) to ten (transect) counting fields (0.014 mm^2) on each *N. solandri* leaf. ImageJ freeware software was used to digitally stack cuticle images in the z plane and to assist in counting epidermal and stomatal cells.

Leaf area of the *N. solandri* leaves from Mt. Ruapehu was measured using a Wild Leitz model area meter. Three measurements were averaged for each leaf, and seven leaves from each elevation were analyzed. Leaf area of *Q. kelloggii* was measured digitally using SigmaScan software from scanned images (800 dpi) of leaves (including petiole).

The stomatal index reported for *N. solandri* leaves in this paper has been adjusted to solely reflect stomatal initiation rates without the influence of trichome density (TD). *N. solandri* leaves have widely varying numbers of trichomes, and since trichome density affects the number and distribution of epidermal pavement cells (Glover and Martin 2000), this also influences stomatal index values. In a data set of 480 counts on 88 *N. solandri* leaves, correlations between epidermal cell density and trichome numbers were significant and positive within groups of leaves with the same stomatal densities. This shows that the trichome-epidermal cell number correlation is not simply an effect of the degree of cell and leaf expansion. Using the regressed trichome-epidermal cell density correlation from the total data set, "trichome-free" epidermal cell numbers were calculated ($ED_{adjusted} = ED - 1.18 \times TD$) to express stomatal initiation rates, and the Stomatal Index for *N. solandri* in this paper is thus a "trichome-free" SI based on corrected epidermal cell numbers.

Results were analyzed by nested mixed-model ANOVA's using general linear procedures, in the MINITAB 15 statistical program. Nested mixed-model ANOVA was used when multiple leaves per tree and multiple trees per treatment were available. Additional analyses were linear and quadratic regressions (performed in MINITAB 15 and Excel), and when significant differences occurred, means were compared using Student's t-test or nested mixed-model ANOVA.

Temperature and precipitation data for selected climate stations in northern California was obtained from the Western Regional Climate Center (*http://www.wrcc.dri.edu/*).

APPENDIX 2
APPLICATION MANUAL

1. Identify cuticle-bearing fossil taxa in flora.

2. Select nearest living relative taxon.

3. Collect a modern training set from a reasonably similar geographical area (i.e. expecting comparable temperature lapse rates) to check if the taxon is indeed adjusting stomatal frequency to elevation and establish a calibration curve.

4. Collect modern reference material from different light intensity regimes (sun and shade leaves) and check for the possibility to recognize and separate these (for instance through cell wall undulation index (Kürschner 1997). Often, sun-morphotypes are preferentially preserved in the fossil record. If sun and shade leaves can be separated, develop morphotype-specific calibration curves to minimize potential error.

5. Analyze stomatal density and/or stomatal index on fossil material. (abaxial and/or adaxial). For processing methods, see Kouwenberg et al. 2007.

6. Convert measured SD and/or SI to paleo-CO_2 levels, and calculate paleoelevation using either (preferably) leaf material from the same taxon from a contemporaneous low elevation flora, or sea-level CO_2 reconstructions from other sources. Use CO_2 correction factor for stomatal density if necessary.

7. Calculate the applicable and taxon-specific error margins:

    a. error of prediction (uncertainty from curve fit in training set)

    b. additional error from potential mixing up of sun and shade leaves

    c. error introduced through uncertainty in sea-level CO_2 estimate (in absence of contemporaneous low elevation leaf material)

Reviews in Mineralogy & Geochemistry
Vol. 66, pp. 243-267, 2007
Copyright © Mineralogical Society of America

10

Thermochronologic Approaches to Paleotopography

Peter W. Reiners

Department of Geosciences
University of Arizona
Tucson, Arizona, 85721, U.S.A.
reiners@u.arizona.edu

ABSTRACT

Although thermochronology cannot directly constrain paleoelevation, it can provide estimates of the form, location, and scale of paleotopographic relief, i.e., paleotopography, and its change through time. Unique thermochronologic perspectives on paleotopography come from 1) spatial patterns of surface and subsurface cooling ages or cooling histories that reflect either the influence of topography on subsurface isotherm warping, or spatially focused erosion (including incision), and 2) the age-elevation relationship in paleolandscapes that may be preserved in detrital cooling age distributions. This chapter reviews the fundamental theory and results of these approaches and several example applications. Case studies show examples of both decreasing and increasing topographic relief through time, at the orogen scale and across short ridge-valley wavelengths, and significant modification of local topographic features in glaciated and fluvial settings. In some cases, thermochronologic evidence for fluvial incision at short wavelengths has also been argued to be the result of surface uplift at very long wavelengths. Although not yet used for such purposes, detrital approaches also have the potential to reconstruct paleotopographic relief and paleohypsometry in paleocatchments. In all cases, paleotopographic interpretations from thermochronology require important assumptions or case-by-case support from other lines of evidence. Central issues pertinent to thermochronologic interpretations of paleotopography are the nature of the shallow crustal thermal field through which the samples cooled (including the influence of fluid flow) and the role of rock uplift gradients in modifying simple relationships between erosion and topography.

INTRODUCTION

Continental topography links to some of the most fundamental questions about earth processes and history. Some modern topographic features, such as the Altiplano, Himalaya, or Tarim basin are among the most striking physical features on the planet and seem to suggest that profound insight into planetary dynamics resides in understanding their history through time and the processes that created them. Topography is also central to interactions between a wide range of phenomena, including mantle circulation (e.g., Cazenave et al. 1989; Forte et al. 1993; Ducea and Saleeby 1996), ocean chemical evolution (e.g., Raymo et al. 1988), biota and biotic evolution (e.g., Huey and Ward 2005; Dietrich and Perron 2006), and weather and climate on both local and global scales (e.g., Ruddiman and Kutzbach 1989; Molnar and England 1990; Montgomery et al. 2001; Roe et al. 2006; Smith 2006). The ability to reconstruct paleotopography would lead to improved understanding of all these phenomena and their interactions. As work in the last decade or so suggests, however, reconstructing paleotopography is one of the most difficult challenges facing Earth science. Nevertheless, tantalizing advances in the last few years, as discussed in this volume, have encouraged significant progress and innovation, making this one of the zeitgeists, if not holy grails, of modern Earth science.

1529-6466/07/0066-0010$05.00 · DOI: 10.2138/rmg.2007.66.10

Most of the progress in reconstructing paleotopography or paleoelevation thus far, and all of the chapters in this volume, focus on features either above or within about 3 meters of the Earth's surface. Low-temperature thermochronology, in contrast, contains information about the movement of rocks through the shallow part of the earth's crust, within a few kilometers of its interface with air or water. As this chapter shows, this information can be used to interpret variations in the form, location, and scale of relief at the surface of the earth's crust in the past.

For the purposes of this chapter, I take these characteristics—the form, location, and scale of topographic relief in the past—as the principal attributes of paleotopography. This is more information than simply paleorelief, which does not specify form, location, or scale of elevation differences in a landscape. But, at least as I define it here, paleotopography does not include paleoelevation—the absolute height (relative to sea level or some other supposedly relatively fixed reference) of a point on the earth's surface in the past. This is because thermochronology is concerned with movement of rocks relative to the earth's surface, which is a moving boundary relative to sea level, etc. Although currently paleotopography alone is enough of a challenge for any technique, at least with useful precision and accuracy, supporting information and inferences can be combined with thermochronology to indirectly constrain paleoelevation. Even for paleotopographic reconstructions, however, important assumptions must be made and caveats recognized for thermochronologic approaches. As always, the most rigorous, compelling, and ultimately useful paleotopographic reconstructions will come from combining approaches, such as the ones in this volume. Much fun will also be had (and hopefully insights achieved), by attempting to reconcile conflicting estimates from different approaches.

Thermochronology

Thermochronology begins with the measurement of ratios of isotopes, elements, or nuclear tracks that compose a parent-daughter radioisotopic decay system. These ratios reflect the thermal history of the sample in which they were measured. In commonly used low-temperature thermochronologic systems this is because retention of the radiogenic (daughter) product, which "keeps time," varies inversely with temperature. A simple example is a bulk crystal parent-daughter ratio, which relates to age through a decay equation. This age can be interpreted as the time elapsed since the crystal cooled below a critical temperature at which daughter retention switched from nearly nil, due to diffusive or annealing loss, to essentially complete, due to lack of thermal energy to drive diffusion or annealing. This hypothetical temperature can be thought of as the closure temperature (Dodson 1973), though in reality both the concept of closure and the mechanics of the increasing retentivity with decreasing temperature are more complicated. In slightly more sophisticated applications, thermochronology involves measurement of a series of parent-daughter ratios and their inferred or observed distribution within a sample. This distribution reflects something more complex and potentially powerful about the thermal history of the sample, and can be modeled as a finite time-temperature path, rather than a specific point in a monotonic cooling history. Examples of the former approach are bulk crystal dating by fission-track, K/Ar or $^{40}Ar/^{39}Ar$, or (U-Th)/He dating, whereas examples of the latter approach are track length distributions, multi-step or multi-spot *in situ* $^{40}Ar/^{39}Ar$ dating, or $^4He/^3He$ diffusion experiments combined with (U-Th)/He dating.

Thermal history constraints from thermochronology are usually interpreted as monotonic cooling histories, though more complex *t-T* paths, involving partial resetting can also be considered (e.g., Quidelleur et al. 1997; Lovera et al. 2002). Most importantly, the cooling history is typically reformulated into an exhumation history using a thermal model of the crust to transform $T(t)$ to $Z(t)$, where Z is depth. The focus of most thermochronologic studies is the determination of the timing and rate of exhumation, both across landscapes and through time. This exhumation may be the result of either erosional denudation or tectonic exhumation (in normal fault footwalls), or rarely by ductile thinning. In this chapter we are concerned primarily with thermal histories of rocks that are due to erosional exhumation. For more information on low-temperature thermochronology, including techniques and interpretations relevant to this chapter, the reader is directed to Reiners and Ehlers (2005) and references therein.

Erosion and paleotopography

Because thermochronology is very often used to infer spatial or temporal patterns of erosion rates (e.g., Reiners and Brandon 2006), a reasonable question is whether there are general relationships between erosion rate and either elevation or relief. Many tectonically active, high mountain ranges erode rapidly, exhuming rocks with extremely high cooling rates and young ages (e.g., Catlos et al. 2001; Lavé and Avouac 2001; Dadson et al. 2003; Kohn et al 2004; Thiede et al. 2004; Blythe et al. 2007). But other, often aerially extensive, regions with high elevation, such as the Tibetan plateau and Altiplano-Puna, erode extremely slowly (e.g., Kirby et al. 2002; Clark et al. 2005). Many parts of these plateaux are actually subsiding relative to adjacent highlands, resulting in burial by sediments and heating rather than cooling of subjacent rocks. Likewise, large variations in erosion rates of relatively low-elevation areas also exist (e.g., von Blackenburg 2006, and references therein).

If erosion does not simply relate to surface elevation, does it relate to topographic relief? (Here we follow the usage of Montgomery and Brandon (2002) and others in defining relief as an elevation difference over a specified lengthscale; e.g., mean local relief is often measured over a radial distance of 10 km). In general higher slopes lead to higher transport capacity and at least the potential for higher erosion rates. In tectonically active or geomorphically transient landscapes, there are correlations between local slope (or mean local relief) and erosion rate at relatively low slope (Riebe et al. 2000; Montgomery and Brandon 2002; von Blackenburg 2006). But in tectonically active regions with high relief (~1-2 km), erosion rates vary tremendously (e.g., 0.2 to 10 mm/yr), representing variations in rock uplift rate, climate, and rock strength (Montgomery and Brandon 2002). In tectonically inactive landscapes, over very large scales (such as orogen to orogen) relief and erosion rates are linearly correlated (Ahnert 1970; Montgomery and Brandon 2002), but at low mean local relief (<~0.5 km) there is much scatter in this correlation and large differences from region to region (von Blackenburg 2006). In the dataset used by Montgomery and Brandon (2002), for example, inactive orogens with mean local relief of about 0.3-0.4 km have erosion rates ranging from about 0.02 to 0.2 mm/yr. At regional scales, tectonically inactive landscapes may show no relationship between local slope and erosion, because landscapes approach steady-state in which erosion rates are constant across a range of slopes (Riebe et al. 2000).

The Sierra Nevada provide an example of the perils of relating erosion rates to relief. They are thought to have persisted since 60-80 Ma with high (1.5 km) relief over long wavelengths (~50 km), but erosion rates over these timescales are estimated to have been only 0.03-0.08 mm/yr (House et al. 1997, 2001; Clark et al. 2005). Incision rates may have been higher (~0.2 mm/yr) in the Pliocene (Stock et al. 2004), possibly aided by convective instability at depth (e.g., Ducea and Saleeby 1996), but this is still slow relative to erosion rates in many other tectonically active and inactive regions, and similar to some with much lower relief (e.g., Heffern et al. 2007).

Thus for practical use in reconstructing paleotopography, estimated erosion rates in the past do not provide very good constraints on either mean local relief or surface elevation. This is both because of potential uncertainties in whether a landscape was tectonically active, and because of wide ranges in erosion rate at a given index of local relief.

Cooling-age or cooling-history variations and paleotopography

Thermochronologic constraints on paleotopography generally come not from erosion rates, but from variations in cooling-ages or cooling-histories across a landscape, or proxies for them, in detrital cases. The simplest and most commonly noted type of spatial variation in cooling ages is the increase in age with elevation often seen over relatively short horizontal distances (i.e., the age-elevation relationship, or AER). An AER is expected to have a positive slope when erosion rates and rock uplift rates beneath it are uniform, and the horizontal distance between the samples is small or comparable to: 1) the vertical distance over which the samples were collected (as in high amplitude, low wavelength topography), and 2) the closure depth of the thermochronometer. In such cases, the closure isotherm at depth is sufficiently flat relative to the topography above it, so the slope of the AER is the erosion rate.

For elucidating paleotopography, however, it is more useful to consider cooling age variations over horizontal distances that are comparable to or larger than closure depths. Over these lengthscales, the AER is a convolution of both erosion rate and topography, and their changes with time. This is because subsurface isotherms, including closure isotherms of low-temperature thermochronometers, are bent by topography. The thermochronologic legacy of paleotopography may be present not only in spatial variations of bulk cooling ages, but also in variations of bedrock cooling histories, for example from $^{4}He/^{3}He$, track length models, mineral pairs, or crystal-size-age correlations. Finally, in some cases, cooling ages of detrital minerals from paleolandscapes can also be used to estimate paleotopographic relief.

As discussed earlier, paleotopography is not the same thing as paleoelevation. It is worrisome, for example, that the bathymetric relief of, say, the modern Monterey Canyon off California, could be interpreted as high subaerial paleoelevation and paleorelief by geologists of the future, were it moved relative to sea level and its sediments and sedimentary rocks stripped away. However, in many cases, it may be either reasonable to assume subaerial exposure of the landscape, or to reference a particular point in the paleolandscape to sea level, providing a minimum surface elevation for parts of it. Thus, as usual, paleotopographic relief estimates from thermochronology are probably best interpreted with the aid of additional geologic information.

Having outlined the rules of the thermochronologic paleotopography game, most of the rest of this chapter is organized around two main types of approaches: bedrock and detrital. The bedrock approaches include studies of regional variations in surficial cooling ages and/or histories, and studies from subsurface tunnel samples. The detrital approach involves the (potential) use of cooling age distributions in detritus from ancient landscapes.

BEDROCK-BASED APPROACHES

Interpretation of paleotopography from spatial patterns of cooling ages at or near (in the case of tunnels) the surface, requires a general understanding of the interaction of topography with the shallow crustal thermal field. This section begins with an outline of this, followed by several case studies and their interpretational foundations for understanding bedrock cooling-age and cooling-history variation in terms of paleotopography, including two examples from tunnel studies. Finally, I highlight important assumptions and caveats in all of these approaches.

Topographic bending of isotherms

Many studies have modeled topographic bending of subsurface isotherms and either discussed its importance for interpreting spatial patterns of cooling ages across landscapes or applied it in this way (Turcotte and Schubert 1982, 2002; Stüwe et al. 1994; Mancktelow and Grasemann 1997; House et al. 1998, 2001; Stüwe and Hintermüller 2000; Ehlers et al. 2001; Braun 2002a,b; Ehlers and Farley 2003; Braun 2005; Ehlers 2005). I do not present the mathematics of the thermal models here, but note that Mancktelow and Grasemann (1997) provide a comprehensive treatment of the problem. Their work, and many others, present equations for temperature as a function of depth in a layer of prescribed thickness (with fixed upper and lower temperature boundary conditions), heat capacity, density, internal heat production, and surface lapse rate, as well as horizontal distance beneath steady, periodic topography of a specified wavelength and amplitude. Some of the models also account for steady advection of material and heat towards the surface via erosion. More sophisticated, usually numerical, models are required to account for transient erosion and three-dimensional topography (e.g., Braun 2005).

A simplified sketch of the steady-state, two-dimensional thermal field beneath static topography illustrates critical points about topographic effects on cooling ages at the surface (Fig. 1). Low temperature isotherms at shallow depths follow topography closely; higher temperature isotherms at greater depths are much less strongly warped by topography. This results in varying depths of the closure isotherm of low-temperature thermochronometers beneath topography, both affecting interpretations of erosion rates, and also posing the potential to extract paleotopographic information from cooling-age patterns. The magnitude of isotherm warping can be easily understood using the admittance ratio α, (Braun 2002a), which is the amplitude of an isotherm beneath surface topography of wavelength λ_1, relative to the amplitude of the topography h_0. The admittance ratio is only weakly sensitive to topographic amplitude, and instead depends mostly on the wavelength of the topography and the temperature of the isotherm. This is summarized in Figure 2, which shows α as a function of isotherm temperature and topographic wavelength for a particular set of model parameters. Note that α never approaches unity in this example, even for extremely large topographic wavelengths, because the surface lapse rate causes lower temperatures at higher elevations.

In interpreting cooling ages across a landscape, the most relevant isotherm is that of the thermochronometer's closure temperature. Figure 2 shows that the 60 °C isotherm, a typical T_c for the apatite (U-Th)/He system in a slowly eroding landscape, will be deflected to about 45%

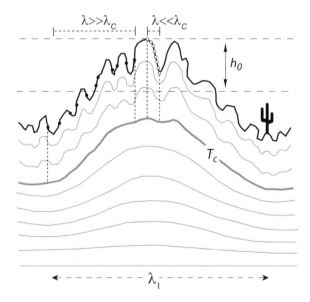

Figure 1. Cross-sectional cartoon of isotherm distribution under two-dimensional topography. Relatively low-temperature isotherms that are close to the surface follow topography more closely than higher temperature isotherms at greater depth. In this example there are two dominant wavelengths of topography, λ_1, the long wavelength, and a superimposed one, about twelve times shorter. T_c is the closure isotherm, h_0 is the amplitude of the long-wavelength topography, λ_c is the critical wavelength for this hypothetical thermochronometer, and λ is the wavelength over which samples are collected. Samples collected over the long wavelengths ($\lambda \gg \lambda_c$) show little or no age-elevation relationship, because the closure isotherm follows the long wavelength topography closely. Samples collected over the short wavelength topography ($\lambda \ll \lambda_c$) show a positive age-elevation relationship because the closure isotherm is nearly flat over this wavelength. See Braun (2002a, 2005) for details.

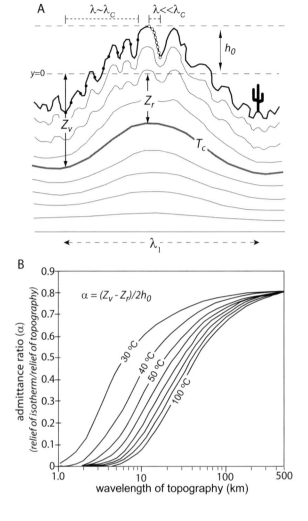

Figure 2. A. Warping of isotherms beneath topography with different wavelengths, showing the depth difference for the closure isotherm (T_c) beneath the mean surface elevation under the valley (Z_v) and ridgetop (Z_r) composing the dominant long wavelength topography. Note that the closure isotherm shown here is deeper and less bent by topography than that shown in Figure 1 (here, λ_1 of the long-wavelength topography is close to λ_c, whereas in Figure 1, λ_1 was significantly larger than λ_c. B. The difference between Z_v and Z_r, normalized to the amplitude of the topography at this wavelength, is the admittance ratio, α—the amplitude of the closure isotherm relative to that of topography. Longer wavelength topography and higher temperature isotherms have higher α. See text for details. After Reiners et al. (2003a), but the 30 and 40 °C isotherm admittance ratios have been added, because new ^{4}He/^{3}He methods are sensitive to temperatures this low.

of the topographic amplitude, for a topographic wavelength of 20 km. In contrast, the 100 °C isotherm, close to the T_c for the apatite fission-track system in a slowly eroding setting, will only be deflected by about 25%, and the 30 °C isotherm, to which the ^{4}He/^{3}He method is sensitive in some cases, is deflected by about 70%, close to the maximum allowed by the surface lapse rate (~80%). This shows that, in general, the lowest temperature thermochronometers will be the most sensitive paleotopography. Figure 2 also shows the importance of topographic wavelength. Whereas 20-km wavelength topography will typically bend the bulk grain apatite He closure isotherm by about 45%, 2-3 km wavelength topography will have essentially no effect, and topography with wavelengths >50 km will bend the isotherm to nearly the maximum amount allowed by the lapse rate and topographic amplitude.

The sensitivity of a thermochronometer to topography of a given wavelength can be represented by Braun's (2002a; 2005) critical topographic wavelength, which is the ratio of the thermochronometer's closure temperature to the mean geothermal gradient beneath the landscape ($\lambda_c = T_c \times dZ/dT$). λ_c for the apatite He system is approximately 2-3 km. As shown in Figure 1, if samples are collected at a range of elevations across topography with a wavelength much less than this ($\lambda \ll \lambda_c$), there is essentially no bending of the closure isotherm at depth, and the age-elevation relationship (AER) among the samples reflects the erosion rate (elevation difference divided by age difference). In contrast, if samples are collected at a range of elevations across topography with a wavelength much greater than the critical wavelength ($\lambda \gg \lambda_c$), they will show no age difference (or a difference that is only proportional to the lapse rate), because the closure isotherm follows the topography closely. These statements hold true only if there has been no change in topography between the time the samples cooled through their closure isotherms and emerged at the surface.

The House et al. (1998, 2001) Sierra Nevada studies

One of the earliest, if not the first, application of these principles to detecting paleoto-pography was the work of House et al. (1998, 2001) in the Sierra Nevada of California. The essence of the strategy is outlined in Figure 3. House et al. measured apatite (U-Th)/He ages in samples from two parallel transects along orogenic strike of the northwest-southeast trending part of the central-southern Sierra Nevada range (Fig. 4). Both transects crossed several major, orogen-normal canyons, including those of the San Joaquin and Kings rivers, that create long-wavelength topography at a scale of about 50 km. House et al. aimed to collect samples at simi-lar elevations across this topography by exploiting superimposed short-wavelength relief (and collecting out of the trace of the main transect to some degree). Apatite He age variations among these samples were assumed to arise from two sources: 1) elevation differences across short topographic wavelengths, caused by a regional age-elevation relationship reflecting the mean erosion rate, and 2) position relative to long wavelength topography, reflecting varying depths of the apatite He closure isotherm beneath the long-wavelength ridge-canyon topography. By considering age-elevation relationships over short wavelengths (House et al. 1997), House et al. corrected each age to a constant elevation in the transect to remove the effects of varying sample elevations. Corrections were also made for large-scale tilting of the range (House et al. 2001).

The pattern of ages in House et al.'s western transect shows a distinct anti-correlation with surface elevation across the major long-wavelength topography, especially in the southern por-tion of the range (Fig. 4). This was attributed to depth variation of the closure isotherm of the apatite He system at the time these samples were exhumed through it, roughly 60-80 Ma. The

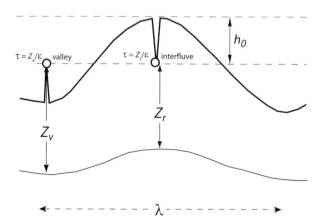

Figure 3. The essence of the House et al. (1998; 2001) approach to detecting paleotopographic relief in the Sierra Nevada. Very short wavelength topography superimposed on long wavelength (λ) topography with amplitude h_0 is used to sample at similar elevations in major valleys and interfluves in an orogen-parallel transect. If the regional erosion rate ε is uniform, ages of valley samples will be older than those of interfluve samples, in proportion to the difference in closure isotherm depths beneath each location (Z_v-Z_r).

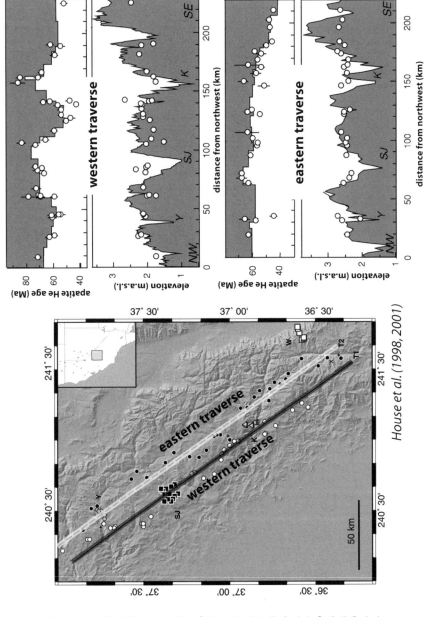

Figure 4. Sample locations and results of the House et al. (1998, 2001) studies, after House et al. (2001). Two parallel sample transects traversed major orogen-perpendicular river valleys (map view at left; cross section views at right). Along the western transect, elevation- and tilt-corrected apatite He ages are systematically older in the San Joaquin and Kings river valleys, and younger in the interfluves. These results were used to suggest the presence of valleys even larger than the modern ones, but in the same locations, at the mean age of the samples, 60-80 Ma. Ages along the eastern transect, in contrast, showed no systematic change relative to topography. Together with results from the western transect, House et al. (2001) interpreted this to mean decreasing relief to the northeast in the Late Cretaceous.

House et al. (1998, 2001)

simplest explanation for this is that the modern valleys are in approximately the same positions as they were at 60-80 Ma. The calculated relief in these paleovalleys necessary to cause these age differences at the erosion rates indicated by the short-wavelength age-elevation relationships (0.04-0.05 mm/yr) was estimated at 1.5 ± 0.5 km, higher than modern relief. Given other assumptions about the distribution of long and short-wavelength relief in the Sierra and by analogy with orogenic plateaux in other locations, House et al. inferred a late Cretaceous mean elevation for the Sierra Nevada of about 3 km. The eastern of the two transects did not show the same distribution of elevation-corrected ages with respect to long-wavelength topography (Fig. 4). From this House et al. inferred that it lay in a less deeply dissected portion of the orogenic plateau.

Aside from illustrating a thermochronologic approach to paleotopographic reconstruction, the House et al. results have spurred vigorous debate about the paleoelevation of the Sierra Nevada. House et al. noted that paleobotanical data were consistent with high elevations in the Early Cenozoic (Forest et al. 1995; Gregory-Wodzicki 1997; Wolfe et al. 1997, 1998). Stable isotope compositions of soils east of the Sierra also suggest a major rain shadow there since at least the Middle Miocene (Poage and Chamberlain 2002), and hydrogen isotope records in Eocene sediments on the west flank of the range also indicate topography and surface elevations much like today's (Mulch et al. 2006). Nevertheless, other researchers point to various geologic lines of evidence for middle or late Cenozoic uplift (Huber 1981; Unruh 1991; Wakabayashi and Sawyer 2001; Stock et al. 2004). An additional question is how erosion rates as low as 0.04-0.05 mm/yr could have been maintained since the late Cretaceous if relief and elevation were as high or higher than today's.

Spectral approaches

As mentioned above, the expectation that AERs collected over short wavelength topography reflect the regional erosion rate, whereas AERs collected over long wavelength topography yield no age variations (Fig. 1), only holds true for steady topography. Deviations of observed spatial cooling age patterns from this expectation provide a measure of relief change in the landscape (Braun 2002a,b, 2005). For example, decreasing relief (i.e., resulting from locally higher erosion rate at high elevations), yields younger ages at high elevation (Fig. 5). At the shortest possible topographic wavelength, AERs still record the mean regional erosion rate, but at wavelengths approaching the critical wavelength, AERs will have slopes steeper than the actual erosion rate, and at long wavelengths, AERs will have negative slopes. Conversely, increasing relief (i.e., locally higher erosion rates at low elevations), yields younger ages at low elevation. At the shortest possible topographic wavelength, AERs still record the mean regional erosion rate, but at wavelengths approaching critical, AERs will have shallower slopes than the actual erosion rate, and at long wavelengths AERs will have positive slopes (Braun 2002b).

Spatial variations in erosion rate that result in topographic relief changes will therefore be preserved in AERs. Specifically, AERs over different topographic wavelengths contain information about relief change at various wavelengths, as well as the mean erosion rate over the whole landscape. These principles have been formalized in the context of landscape evolution by Braun (2002a,b, 2005) as the "spectral method," in which the relationship between age and elevation, called the gain function G, is defined over a continuous three-dimensional spectrum of topographic wavelengths ω. As described above, the AER at the shortest wavelength reflects the mean erosion rate. As formalized by Braun, the gain function at the shortest wavelengths (G_S) is the reciprocal of erosion rate. If relief has not changed, the gain function at long wavelengths (relative to the critical wavelength) (G_L) is zero, as there is no relationship between age and elevation. But as described above and shown in Figure 5, if relief has decreased, ages at high elevation will be younger than those at low elevation, so G_L is negative. If relief has increased, ages at high elevation will be older than those at low elevation, so G_L is positive. Braun (2002a, 2005) showed that relief change is $\beta = 1/(1 - G_L/G_S)$.

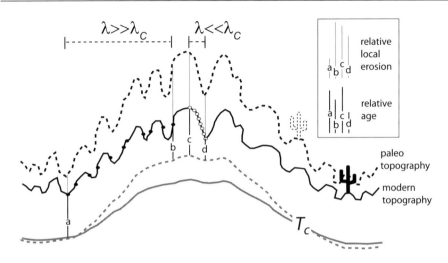

Figure 5. An example of the effect of decreasing topographic relief (i.e., locally higher erosion rate at high elevations) on cooling ages at the surface. Dashed line under dashed cactus is paleosurface, and solid line under solid cactus is post-relief-change surface. Grey dashed line is paleo-closure isotherm; solid grey line is post-relief-change closure isotherm. Vertical grey lines above circles (which are sample locations) represent relative amounts of local erosion during relief-reduction period. Vertical black lines under sample circles represent relative vertical distances traveled by, and ages of, each sample. Over long topographic wavelengths ($\lambda \gg \lambda_c$), ages of samples at different elevations are no longer equal (see Fig. 1), and higher elevation samples are younger than lower elevation ones (compare the lengths of lines a and b, which represent the distance traveled by samples under low and high elevation portions of the landscape, respectively). Over short ($\lambda \ll \lambda_c$) wavelengths, the age-elevation relationship is steeper than predicted by the broad scale erosion rate.

Figure 6 shows the spectral method (from Braun 2002a) applied to both synthetic data associated with real topography from the southern Alps of New Zealand, and real data with real topography from House et al.'s (1998) Sierra Nevada study. In the synthetic data example, the inverse of the prescribed erosion rate is recovered for the gain function at the shortest wavelengths. At long wavelengths, the gain function is either positive or negative for a factor of two increased or decreased relief, respectively. Although the geographic sample distribution in the House et al. (1998) study is not well suited to the spectral method, several first order results emerge from this analysis. First, the ~0.04-mm/yr erosion rate found by House et al. in the Sierra Nevada is consistent with the inverse of the gain function at short wavelengths (Fig. 6), though there is a large degree of scatter due to few data collected over short wavelengths. Second the gain function at long wavelengths is a large negative number, implying decreased relief. Combined with the short wavelength data, this analysis suggests an approximate relief decrease of about a factor of two since the mean age of the samples (~60-80 Ma). This is consistent with House et al.'s (1998, 2001) interpretations of Late Cretaceous relief greater than modern.

The efficiency and accuracy of the spectral method requires sampling a sufficient number and density of samples over appropriate lengthscales and elevation ranges. Braun (2002a) estimated that the optimum sampling interval for this method is about five times smaller than the critical topographic wavelength (ω_c) of the thermochronometer being used, and the optimum lengthscale of the sampling area is about five times larger than ω_c.

In general, the spectral approach is most sensitive to the most recent relief changes and those occurring over the longest topographic wavelength. Important assumptions are also required. Among these are a steady and uniform (except as perturbed by topography) geothermal gradient,

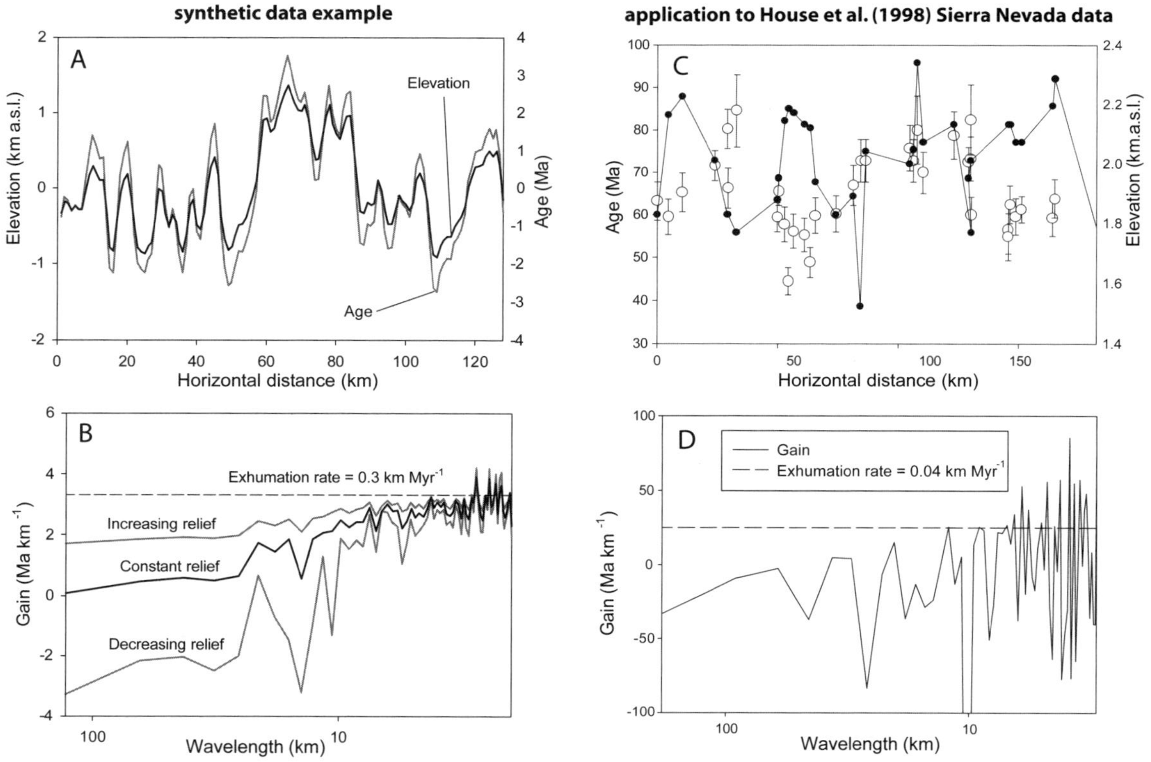

Figure 6. Application of the Braun (2002a,b, 2005) spectral method to quantifying topographic relief change, after Braun et al. (2002a). **A.** Elevation profile from a 1-km digital elevation model of the Southern Alps of New Zealand, along with synthetic (predicted) apatite (U-Th)/He ages from the same region. Mean values of 1.58 km and 10.78 Ma have been subtracted from the elevation and age profiles, respectively. **B.** The gain function between age and elevation calculated by assuming different scenarios for relief evolution in panel A. The exhumation rate of 0.3 km/m.y. is represented by the inverse of the gain function at the shortest wavelengths. At longer wavelengths, the gain function is negative for decreasing relief and positive for increasing relief. **C** and **D.** Application of the approach outlined in panels A and B to House et al.'s (1998) Sierra Nevada data. In D, the inverse of the gain function at short wavelengths is close to the 0.04 km/m.y. erosion rate inferred from short-wavelength AERs in the region (House et al. 1997, 1998, 2001), but at longer wavelengths, the gain function is negative, implying decreasing relief since the mean age of the samples (~60-80 Ma). See text and Braun et al. (2002a; 2005) for more details.

stationary landforms that may grow or shrink over time, and no horizontal advection of rock (which would result in a phase shift of the gain function) (Braun 2002a). Most importantly, because this approach implicitly associates age variations over long wavelengths with transient topography, it requires uniform rock uplift rates throughout the domain of analysis. If active structures or isostatic rebound partition uplift over the landscape domain, age variations over long wavelengths may reflect erosion rate variations not associated with relief change. Young cooling ages in the core of a mountain range, relative to the flanks, for example, may not necessarily be caused by decreasing relief of the range. Instead it may reflect high rates of rock uplift there due to focused crustal strain or isostatic rebound of a deep crustal root.

Braun and Robert (2005) included the effects of isostasy and flexural rebound in a fully 3D model of the combined paleorelief, erosion, and isostatic rebound history of the post-orogenic Dabie Shan of eastern China, using data from multiple thermochronometers (Reiners et al. 2003). AERs in this orogen show positive slopes over short wavelengths, and negative slopes over longer wavelengths (older ages at lower elevations on the range flanks). A simple spectral approach might interpret this as decreased relief of the range since the mean age of the samples. But incorporation of isostatic rebound shows that there is a trade-off in acceptable solutions to the observations between the long term relief change and elastic thickness of the rebounding crustal section—a greater reduction in relief has similar effects as a thinner elastic lithosphere. When combined with other model parameters in a fully coupled 3D model, however, this ambiguity is reduced, and a best-fit solution to a six-parameter search yielded an elastic plate thickness of 9 km, a post-orogenic erosional decay phase of ~70 m.y. with an e-folding time scale of ~250 m.y., a basal temperature of ~650 °C, relief loss of ~2.5, and a mean exhumation rate of ~0.02 km/m.y (Braun 2005; Braun and Robert 2005).

Other bedrock-based approaches to understanding topographic change

Several studies have used variations in 1) regional cooling ages, 2) intrasample crystal ages, or 3) intracrystalline ages to interpret topographic change through time, in particular incision or lateral shifts in high topography. Examples of several of these are described here.

Glacial modification of topography in the Coast Mountains, British Columbia. Few geomorphic processes are capable of modifying topography as quickly as glaciation. Three examples from the same region in the Coast Mountains of British Columbia provide examples of the dramatic changes late Neogene glaciation has had in modifying the landscape.

Ehlers et al. (2006) measured apatite (U-Th)/He ages in two ~60-km horizontal transects at roughly constant elevation across rugged (~3 km relief) glaciated topography in this region. The apatite He ages ranged from ~1.5 to 8 Ma, but the youngest ages were in a region about 15-20 km to the southwest of the area with the highest modern mean elevation and highest peaks. Using three-dimensional thermokinematic modeling, this was interpreted as a result of a ~16-km northeastward shift in the position of the highest topography within the last 1.5-4.0 m.y., most likely due to intense glacial erosion towards the southwest.

The most compelling example of thermochronologic analysis of paleotopography thus far comes from Shuster et al.'s (2006) application of the ^{4}He/^{3}He method to a glacial valley in the same region studied by Ehlers et al. (2006). This method relies on proton irradiation of apatite crystals to create a uniform ^{3}He distribution in each grain. Samples are then degassed by step-heating, and the intracrystalline ^{4}He distribution and diffusion kinetics are revealed by the evolution of the ^{4}He/^{3}He and fractional He yield through the degassing (Shuster et al. 2003; Shuster and Farley 2003, 2005). Accounting for complications from alpha ejection phenomena, samples that exhibit slowly changing ^{4}He/^{3}He during step heating indicate a high intracrystalline ^{4}He concentration gradient and protracted cooling through or residence in, the He partial retention zone. In contrast, little change in the ^{4}He/^{3}He through the step heating experiment represents rapid cooling (Shuster and Farley 2005).

An illustration of the power of this technique comes from samples in the 2-km deep, glacially carved, Kliniklini Valley in the B.C. Coast Mountains. $^4He/^3He$ step heating results from apatites in the bottom of the valley indicate very rapid cooling from about 80 °C to surface temperatures at 1.8 ± 0.2 Ma (Fig. 7). Samples from higher elevations up the side of the valley also require rapid cooling at this time, but the magnitude of this cooling decreases systematically with elevation, with the sample at the top of the valley instead requiring only slow cooling between 6 and 1.8 Ma, and residence at the surficial temperatures by 1.8 Ma with no cooling thereafter. The simplest explanation for these collective results is that essentially the entire 2-km-deep valley was incised at 1.8 ± 0.2 Ma at a rate of about 1.8 mm/yr, about six times faster than rates prior to that time. These results demonstrate the power of glaciers to create high topographic relief in very short periods of time. It is also noteworthy that in this case the onset of localized incision was interpreted as a result of climate change, rather than long wavelength surface uplift, as argued for the fluvial incision cases described later.

In another study from the same area, Densmore et al. (2007) found that bulk-grain apatite (U-Th)/He ages in subvertical transects in the Kliniklini Valley showed a break in the slope of the AER at ~6 Ma for the apatite He system. This is important because a simple interpretation of these data assuming spatially uniform regional erosion rates would likely invoke an increase in erosion rates at ~6 Ma. In reality, this break in slope in the AER is due to incision, not erosion over a wide area, and the incision occurred much more recently than the age implied by the break in slope, as valley widening and deepening exposed samples in the apatite He partial retention zone on the sides of the valley (Densmore et al. 2007).

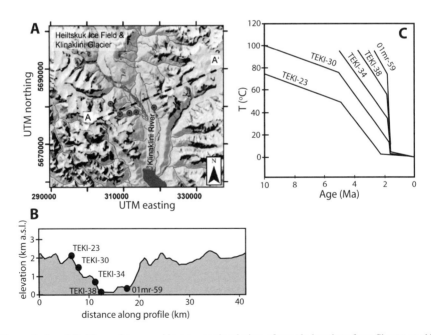

Figure 7. A and **B**. Map and topographic cross-sectional view of sample locations from Shuster et al.'s (2005) study of incision of the Kliniklini valley, Coast Mountains, British Columbia. **C**. Model thermal histories for each sample, derived from $^4He/^3He$ evolution of step-heating experiments on proton-irradiated samples, and bulk grain (U-Th)/He dates. Samples from the valley bottom require rapid cooling, from 80 °C to surface temperatures, at 1.8 ± 0.2 Ma, and samples from higher elevations require thermal histories with progressively smaller extents of cooling (beginning at 1.8 Ma) with elevation. The highest sample (TEKI-23) was at surface temperature before the 1.8 Ma cooling event experienced by the other samples. Collectively, these data are interpreted to be the result of ~2 km incision at 1.8 Ma. After Shuster et al. (2005).

Taken together, these studies from the Coast Mountains leave little doubt that glacial erosion can rapidly reorganize topographic configurations, even in postorogenic landscapes, at a range of lengthscales, including incising 2-km deep valleys in less than 2 m.y., and causing ~10-20-km lateral shifts of the regionally highest parts of a mountain range.

Crystal-size-age correlations and incision of Southeast Greenland fjords. In ways analogous to track length modeling in the AFT system (e.g., Ketcham 2005), and both inter- and intracrystalline cooling ages in the (U-Th)/Pb (e.g., Mezger et al. 1989; Grove and Harrison 1999; Hawkins and Bowring 1999; Schmitz and Bowring 2003; Schoene and Bowring 2006) and $^{40}Ar/^{39}Ar$ systems (Wright et al. 1991; McDougall and Harrison 1999; Harrison et al. 2005), variations in fractional He retentivity among apatite crystals in the same rock has the potential to constrain the range of possible cooling paths of rocks through low temperatures sensitive to overlying topography. Intercrystalline He retentivity variations that are understood to the degree that they can be used in this way arise from varying extents of radiation damage (Shuster et al. 2006; Flowers et al. 2007) or varying crystal size (Reiners and Farley 2001). For example, apatite crystals from samples that cooled relatively quickly show no age-size correlations, whereas those from samples that cooled slowly and resided in a partial retention zone for long periods of time show age-size correlations over a wide range of ages.

An example of how this has been used to infer paleotopographic change comes from apatite He and fission-track data from the fjords of southeast Greenland (Hansen and Reiners 2006). With increasing elevation from 0.6 to 2.1 km above sea level, AFT ages increase from 64 to 836 Ma and average apatite He ages increase from about 55 to 230 Ma. High elevation samples with old ages show no correlation between crystal size and apatite He age, suggesting rapid cooling below their T_c in the earliest Mesozoic. But two samples from lower elevation in the fjords show a wide range of ages, from about 100-170 Ma in one sample, and 20-100 Ma in the lowest one, and both samples show correlations between crystal age and size. These correlations require protracted cooling or residence in the He partial retention zone until at least 20 Ma and possibly much more recently. A relatively simple explanation for these data is that rapid cooling in the earliest Mesozoic created a broad low-relief surface that was then incised by glaciers in the late Cenozoic, producing the fjords and the modern topography (Hansen and Reiners 2006).

Fluvial incision on orogenic plateaux flanks. Low-temperature thermochronologic studies of major river gorges in Eastern Tibet and the western Andes have also been used to interpret the timing and rates of rapid fluvial incision on the flanks of large orogenic plateaux, and by inference, the timing and rates of plateau uplift. Kirby et al. (2002) used multiple thermochronometers to model the cooling histories of samples near the edge and in the interior of the Tibetan plateau. They noted that samples near the southeast margin of the plateau experienced rapid cooling from roughly 200 °C to less than ~70 °C in the late Miocene (~5-12 Ma), whereas interior samples showed older ages for a given thermochronometer, and more gradual cooling histories. To the south of Kirby et al.'s (2002) study area, Clark et al. (2005) compared apatite He ages to their depths in fluvial valleys below the local plateau surface. The resulting composite AERs show a break in slope in the late Miocene (9-13 Ma) consistent with rapid incision in rivers throughout the region. Meanwhile, along the flanks of the planet's *second* largest subaerial orogenic plateau (the Altiplano-Puna) in the Peruvian Andes, Schildgen et al. (2007) interpreted apatite He ages collected along strike of a major river canyon as evidence for onset of ~2.4 km of rapid incision beginning at ~9 Ma.

All three of these studies used thermochronologic data to interpret the timing and rates of fluvial incision and development of major canyon topography (interestingly, at roughly the same time). However, they are also important because all three interpreted the incision as resulting from surface uplift of the plateaux surrounding the canyons in the late Miocene-early Pliocene. Thus these studies provide examples of paleotopographic inferences on two contrasting lengthscales: fluvial valleys and orogenic plateaux surrounding them. It should probably be remembered, however, that although each study marshals compelling regional and geologic considerations to support the contemporaneity of, and direct geomorphic linkage between, regional uplift and incision, the role of climate change in initiating incision in pre-existing high topography is difficult to rule out entirely (e.g., Molnar and England 1990).

Tunnel studies. Long (~10-20 km) tunnels in mountainous regions provide opportunities to measure cooling ages in subsurface horizontal transects beneath high-relief topography. Differences between observed cooling age patterns and those predicted from the modern topography can then be used to infer paleotopographic features of the region along the length of the tunnel.

Foeken et al. (2007) examined apatite FT and He ages through the Malta tunnel in the Austrian Alps. This 20-km tunnel was excavated under modern topography as much as ~1 km above it and through the core of the Hochalm-Ankogel Dome, a structural feature that, if projected into the air above the region, suggests the form of a dome with ~3–4-km amplitude and ~20-km wavelength (Fig. 8). Foeken et al. (2007) addressed whether this structural dome bore a topographic manifestation of similar magnitude in the past. Their apatite He ages, summarized in Figure 8 along with other regional features, show a concave-up pattern resembling a mirror image of the inferred structural dome. These data were used to suggest dome formation and topographic expression until about 7-10 Ma, when the modern topography was created. Interestingly, these data also suggest relatively minimal modification of the topography by Plio-Pleistocene glaciation.

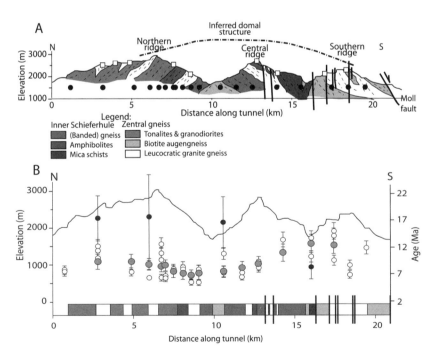

Figure 8. Cross-sectional view (**A**), and apatite FT and He ages (**B**) from the Malta tunnel, Swiss Alps, after Foeken et al. (2007). The dashed line in B represents the approximate topography that would be predicted from a structurally continuous dome through the region. In B, the black symbols are apatite FT ages, and the grey symbols are mean ages of multiple aliquots for each sample (open symbols). The apatite He ages show a broad concave-up pattern through the tunnel that bears little or no relationship to the modern topography, but is similar to a mirror image of the inferred structural dome. Modeling of these data indicated the larger-scale domal structure existed as late as 7-10 Ma (Foeken et al. 2007).

At least one other tunnel also shows patterns of cooling ages requiring recent large changes in overlying topography. The Simplon tunnel, in the Italian-Swiss Alps underlies topography as much as 2.5 km above it. Pignalosa et al. (unpublished data; FT ages from Univ. of Bologna; He ages from Univ. of Arizona) showed that surficial apatite He ages from above the tunnel and subsurface apatite FT ages from within the tunnel are roughly similar, and show much older ages under the deepest points in the tunnel, the opposite of what would be expected from steady topography (Fig. 9). In contrast, apatite He ages from the tunnel are oldest (~11 Ma) at the shallowest depth near the northern portal, decrease to ~3 Ma about 2-3 km from the portal, and are uniformly young (1-2 Ma) through the rest of the tunnel, even at the south portal, where the tunnel is only a few hundred meters below the surface. The only exception to this is ages as old as 3.3 Ma about 4.5 km from the south portal, precisely where channelized groundwater flow and anomalously low water temperatures were recorded (Fig. 9) (Maréchal and Perrochet 2001). Taken as a whole, these data suggest spatially focused erosion (on a wavelength of at least 2-3 km), near the southern portal of the tunnel at sometime between about 4-7 Ma and ~2 Ma, and increased topographic relief, possibly by Quaternary glacial erosion. The data also suggest either relatively little erosion over the last ~10 m.y. near the point of modern highest topography and the northern portal, or else focused fluid flow in these areas between about 2 and 10 Ma.

Caveats in interpreting spatial patterns of cooling ages and cooling histories

Rock uplift gradients. As the fluid-flow example from the Simplon tunnel shows, and as noted earlier, other phenomena besides topographic changes may cause variations in cooling ages and cooling histories in landscapes. Setting aside for a moment the question of the shallow crustal thermal field, gradients in rock uplift can also generate such spatial patterns. For example, systematic patterns of low-temperature thermochronometric cooling ages with and without clear relationships to modern topography have been found in many areas. Just a few of these are the Olympics (Brandon et al. 1998; Batt et al. 2001), the Himalayan front (Wobus et al. 2003, 2005; Hodges et al. 2004), the Cascades (Reiners et al. 2003b), the Apennines (Thomson et al. in prep), and the Dabie Shan (Reiners et al. 2003a). None of these studies invoked topographic change as an explanation for the spatial pattern of ages, although in some cases it cannot be ruled out. Instead, these studies interpreted age patterns as reflecting spatial variation in time-averaged erosion rates, without specifically implying accompanying topographic change. The reason for this is that spatial variations in erosion rates (as inferred from spatial variations in cooling ages or cooling histories) need not require topographic change, if rock uplift gradients are present. These gradients may exist either through isostatic rebound or strain partitioning (across folds, faults, plutons, or broad distributed uplift) that compensate for erosion to preserve topographic form, or act independently to modify it in ways unrelated to erosion.

Gradients in rock uplift as complications to paleotopographic interpretations can most likely be ignored if spatial patterns of cooling ages or cooling histories are observed over short lengthscales (10-20 km) in tectonically inactive settings. This is because over short lengthscales, flexure prevents gradients in isostatic rebound, and in tectonically inactive settings, faults, folds, and magmatism are not active, preventing gradients in localized strain, as well as the potential for focused erosion and coupled strain partitioning. In this respect, interpretations of topographic change across deep valleys (such as in the Coast Mountains, Greenland, and the Sierra Nevada) are probably reliable. Nonetheless, it should be recognized that topography often bears some relationship to spatial patterns of kinematics in the crust beneath it, and faults or other features can accommodate uplift or subsidence across modern topographic features, creating cooling age variations that may otherwise be interpreted signs of the topography's youth or antiquity.

The thermal field and fluid flow. Finally, as with most thermochronologic interpretations, paleotopographic interpretations based on spatial patterns of cooling ages requires important assumptions about the shallow crustal thermal field at the time of cooling through relevant isotherms. The expected relationships between topography on isotherms may be modified by

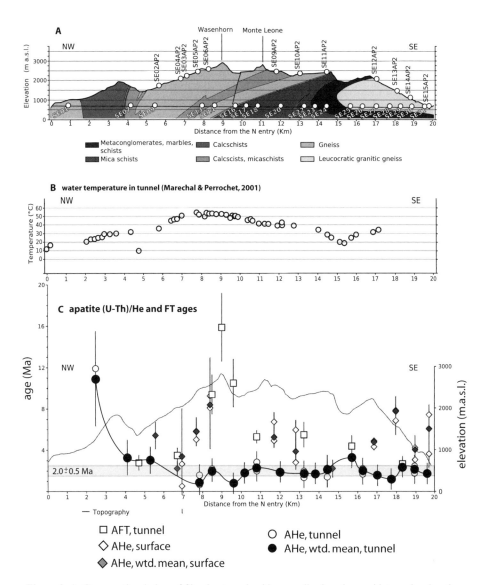

Figure 9. A. Cross sectional view of Simplon tunnel, with generalized geology, with tunnel and surface sample locations (Pignalosa unpublished data). **B**. Water temperatures in the tunnel (Maréchal and Perrochet 2001). **C**. Apatite (U-Th)/He (AHe) and fission-track (AFT) ages from surface and tunnel samples. Weighted mean ages of multiple replicates of apatite He ages are shown joined by solid black line. Apatite He ages from the surface and AFT ages from within the tunnel are roughly similar, and show unexpectedly old ages under the highest topography. Within the tunnel, all apatite He ages southeast of the first three samples are ~1-2.5 Ma, with the exception of one sample at distance 15 km, which corresponds to anomalously low groundwater temperatures in the tunnel.

a number of factors, including variations in thermal properties, radiogenic heat production, magmatism, and, perhaps most importantly, groundwater flow.

Although the effects of fluid flow on heat advection and the thermal field in the shallow crust have received considerable attention (e.g., Smith and Chapman 1983; Forster and Smith 1989; Manga 1998, 2001; Manga and Kirschner 2004), there are relatively few studies focusing on its potential importance in thermochronologic interpretations (e.g., Ehlers 2005; Dempster and Persano 2006; Whipp and Ehlers 2007). Hot springs may be dramatic evidence of this issue, but they may also be restricted in space and time, and may localize thermal effects rather than produce pervasive landscape-scale thermal field perturbations. In this respect, simple topographically driven groundwater flow and discharge to cold springs may have very important (and potentially insidious) effects on isotherm warping in areas of high topographic relief by advectively redistributing thermal power (e.g., Ehlers and Chapman 1999). The general effect of topographically driven fluid flow is to flatten isotherms, as heat is captured from surrounding rock along its flow path and redistributed at low elevations in valleys as springs. This acts in opposition to, canceling out, or overwhelming, topographic warping of isotherms (Forster and Smith 1989). Ehlers (2005) and Braun (pers comm) showed that for reasonable hydraulic conductivities, shallow fluid flow can warp closure isotherms for the apatite He system sufficiently so as to cancel out or invert the thermal field underneath topographic relief as high as 2 km. For studies seeking to use spatial patterns of cooling ages and cooling histories to understand the history of either erosion rates or paleotopography, this is a serious challenge. In general, one thing that is needed to improve understanding of the effects of fluid flow is studies from modern settings attempting to distinguish cooling-age spatial patterns from the expected effects of fluid flow, modern topography, and paleotopography on subsurface isotherms.

DETRITAL APPROACHES

In 1996, Stock and Montgomery noted that the relationship between elevation and thermochronometric cooling age in landscapes leads to the potential to estimate paleorelief of ancient landscapes by using measured ages of paleodetritus as tracers of surfaces at particular elevations in their catchment prior to erosion. In essence, the range of cooling ages (Δt) for a given thermochronometric system in detritus from a landscape could, in combination with a constraint on the paleoerosion rate in the landscape (dZ/dt), be used as an estimate of the minimum range of elevations ($dZ/dt \times \Delta t \sim \Delta Z$), or paleorelief, of the landscape. This approach relies on a number of assumptions such as horizontality of the contours of equal age in the landscape ("chrontours"), purely vertical exhumation, and capturing a range of ages in the detritus that is sufficiently representative of the full age (and elevation) range in the landscape.

The Stock and Montgomery (1996) approach formed the basis for several subsequent studies of *modern* detritus (Brewer et al. 2003; Hodges 2005; Ruhl and Hodges 2005; Stock et al. 2006; Huntington and Hodges 2006; Vermeesch 2007). All of these studies are based on the fact that, as long as there is a unique relationship between age and elevation within a catchment and exhumation was purely vertical, the probability distribution of cooling ages in detritus from it represents three features of the catchment: a) the age-elevation relationship within it, b) the probability distribution of elevation within it (the hypsometry), and c) the spatial distribution of erosion (or, at least, erosional yield of the thermochronometric phase). Given a probability distribution of cooling ages from a sufficiently large number of detrital grains, knowledge of any two of these features allows for solution of the third. For example, Stock et al. (2006) combined a cooling age distribution with knowledge of the age-elevation relationship from nearby bedrock subvertical transects and catchment hypsometry to estimate the relative spatial focusing of erosion within two small catchments on the east flank of the Sierra Nevada. Ruhl and Hodges (2005) used probability distributions of cooling ages combined with observed hypsometry and an assumption of spatially uniform erosion rates to derive an estimated erosion rate for the Nyadi catchment in the Nepal Himalaya of ~0.7 mm/yr from 11 to 2.5 Ma, and an increase in this rate after 2.5 Ma. Both Ruhl and Hodges (2005) and Brewer et al. (2003) noted that several Himalayan catchments had normalized cooling age probability distributions that were distinct from those of hypsometry in the same catchments. This requires either a variable slope of the AER (indicating a change in erosion rate at some time in the past) in the catchment, or spatially nonuniform erosion within it (or some other factor producing spatially variable yield of the thermochronometric phase, such as lithologic heterogeneity).

Detritus from a paleolandscape contains information on the paleorelief (in fact, the paleohypsometry), of its former catchment, through its probability distribution of cooling ages, provided the paleoerosion rate can be estimated, and some assumption about the spatial uniformity of erosion (and bedrock abundance of the thermochronometric mineral) can be made. The former could potentially be estimated from mineral-pair methods, such as using coupled apatite He and apatite FT ages from the same clast, or from AERs of bedrock samples that record the erosion rate through the time interval in question. In general, the latter constraint must be assumed, however.

Figure 10 shows an example of this approach, using modern fluvial sand from a catchment with known hypsometry and some constraints on the AER from bedrock data. The Icicle Creek catchment on the eastern flank of the Washington Cascades drains an area of about 540 km^2 and contains elevations from 360 to 2760 meters above sea level (2.4 km of modern relief). Most of the catchment is underlain by plutonic rocks of the Mt. Stuart batholith (presumably with a reasonably uniform modal abundance of apatite), and bedrock apatite He ages in the eastern part of the catchment range from about 18 to 45 Ma from about 500 to 2700 meters elevation, implying a mid Cenozoic erosion rate of about 0.08 mm/yr—approximately the same rate as determined from apatite FT-He age differences in single samples in the area (Reiners et al. 2002, 2003b). The hypsometry of the plutonic portion of the catchment and the cumulative probability distribution of 57 single-grain apatite He ages are shown in Figure 10. They are somewhat similar, and differences probably represent failure of the assumptions of spatially uniform erosion and lithology, or a nonunique AER in the catchment.

If these apatite He ages were from paleodetritus instead of modern river sand, they could be used to estimate paleorelief of its paleocatchment. First, the range of cooling ages is 35 m.y., or 29 m.y. if the four outlying grains with 9-11 Ma ages are excluded. Using the Stock and Montgomery approach, $\Delta Z = \Delta t \times dZ/dt$. An estimate for the last term of about 0.08 mm/ yr can be estimated from the age-elevation relationship or coupled AFT-apatite He ages of samples in the bedrock. This predicts a relief of approximately 2.4-2.8 km, which compares favorably with the observed relief of 2.4 km. Second, although it is not shown here, one could theoretically reconstruct the paleohypsometry of the catchment by scaling the cumulative probability distribution of cooling ages to the erosion rate estimated by bedrock or mineral pair constraints. Once again this assumes spatially uniform erosion and lithology, and a unique age-elevation relationship. The intent here is not to suggest that these assumptions are met, but simply to outline the potential use of paleodetritus in estimating paleorelief and paleohypsometry of paleocatchments.

Of course the paleodetrital approach to paleotopography carries with it many of the same assumptions and caveats as described above for bedrock approaches. A particular complication for these methods using apatite is the apparent difficulty of finding suitable grains in detritus that has experienced significant amounts of time near the surface and has been subject to corrosion, wildfire and/or diagenesis (e.g., Reiners et al., in press). Another, more general issue arising from attempts to estimate paleorelief/hypsometry from paleodetritus using

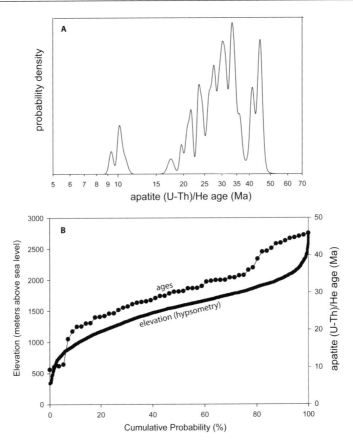

Figure 10. A. Probability density of 57 single-grain apatite (U-Th)/He ages from the Icicle Creek catchment, east-central Washington Cascades (data from Reiners et al., in press), with logarithmic x-axis, to prevent apparent overrepresentation of probability density at young ages. Subject to several assumptions and combined with a constraint on erosion rate, the simple range of ages seen here can be used to estimate relief in the catchment, as outlined by Stock and Montgomery (1996). **B**. Cumulative probability of elevation in the Icicle Creek catchment (black solid line), and the apatite He ages shown in A (filled circles). Again subject to assumptions and an additional constraint on erosion rate, the cumulative probability curve of detrital ages could be used as a proxy for hypsometry of a catchment (see text for details).

low-temperature thermochronometry comes from the uncertainty of the horizontal extent of paleocatchments. Chrontours (contours of equal age across a landscape) will only be sufficiently flat, resulting in a unique age-elevation relationship, if the wavelength of topographic relief in the catchment is sufficiently small. For example, a very large catchment may contain a wide range of elevations, but if this topographic relief is expressed over a horizontal distance much greater than that of the critical wavelength of the thermochronometer, there may be little or no variation in cooling ages, because the closure isotherm lies at the same depth beneath the surface anywhere in the catchment. In effect, the distribution of cooling ages in paleodetritus reflects best the topographic relief with wavelengths shorter than the critical wavelength. For example, apatite He cooling ages in paleodetritus most faithfully reflect the <2-3-km-scale roughness of the paleolandscape, whereas relief at longer wavelengths will have a subdued expression in the detrital record. Estimating topographic relief at different wavelengths would require paleodetrital records using multiple thermochronometers.

SUMMARY

Spatial patterns of ages corresponding to cooling through low temperatures characteristic of the shallow crust can be used to infer paleotopography in several ways. Thermochronologic approaches to paleotopography require important assumptions, not the least of which is that the thermal field in the shallow crust is affected mostly by topography and not by variations in advection by fluids, heat production, or other factors. Also important for most approaches involving spatial variations of surficial bedrock cooling ages is that rock uplift beneath a landscape is uniform, an assumption that is probably most likely to be met over short wavelengths in tectonically inactive settings.

Some examples of paleotopographic constraints from thermochronology include the following. The antiquity and amplitude of long-wavelength topography can be estimated by dating samples collected across it at similar elevations, by exploiting superimposed short wavelength topography, as in the House et al. (1998, 2001) studies. Changes in topographic relief at a variety of wavelengths over the timescale of the mean cooling ages can be estimated with spectral analysis of coupled AERs, as in the Braun (2002a,b, 2005) studies, and this can be combined with rock uplift variations as in the Braun and Robert (2005) study. Across relatively short wavelengths, variations in cooling ages between samples, between grains from the same sample, and within individual grains have also been used to elucidate recent incision or shifting of topographic highs. Recent examples include Ehlers et al. (2006); Shuster et al. (2005); Clark et al. (2005); Hansen and Reiners (2006); Schildgen et al. (2007). Similar approaches have also been taken with subsurface cooling ages from Alpine tunnels under high topography (e.g., Foeken et al. 2007; Pignalosa unpublished data), and these studies have interpreted major Late Cenozoic changes in overlying topography above the tunnels. Finally, although Stock and Montgomery (1996) proposed using ranges of detrital cooling ages to estimate paleotopographic relief of ancient landscapes, and variations on this approach have proliferated in modern settings, it has not yet been applied to real samples.

ACKNOWLEDGMENTS

I thank Mark Brandon for help with the hypsometry calculations and discussions about detrital approaches, and I thank Antonio Pignalosa and Max Zattin for allowing me to show unpublished fission-track and (U-Th)/He data from the Simplon tunnel. Todd Ehlers and David Shuster provided constructive reviews of this chapter, and Danny Stockli provided the initial opportunity for its writing. This work was supported by NSF EAR grant 0236965 from the Petrology & Geochemistry and Tectonics programs. This work benefitted from discussions with members of the Earth System Evolution Program of the Canadian Institute for Advanced Research.

REFERENCES

Ahnert F (1970) Functional relationships between denudation, relief, and uplift in large mid-latitude drainage basins. Am J Sci 268:243–263

Batt GE, Brandon MT, Farley KA, Roden-Tice M (2001) Tectonic synthesis of the Olympic Mountains segment of the Cascadia wedge, using two-dimensional thermal and kinematic modeling of thermochronological ages. J Geophys Res 106:26731-46

Blythe AE, Burbank DW, Carter A, Schmidt KL, Putkonen J (2007) Plio-Quaternary exhumation history of the central Nepalese Himalaya: 1. Apatite and zircon fission-track and apatite [U-Th]/He analyses. Tectonics 26(3): TC3002, doi:10.1029/2006TC001990

Brandon MT, Roden-Tice MK, Garver JI (1998) Late Cenozoic exhumation of the Cascadia accretionary wedge in the Olympic Mountains, Northwest Washington State. Geol Soc Am Bull 110:985-1009

Braun J (2002a) Quantifying the effect of recent relief changes on age-elevation relationships. Earth Planet Sci Lett 200:331–343

Braun J (2002b) Estimating exhumation rate and relief evolution by spectral analysis of age-elevation datasets. Terra Nova 14:10–214

Braun J (2005) Quantitative constraints on the rate of landform evolution derived from low-temperature thermochronology. Rev Mineral Geochem 58:351–374

Braun J, Robert X (2005) Constraints on the rate of post-orogenic erosional decay from low-temperature thermochronological data: application to the Dabie Shan, China. Earth Surf Proc Land 30:1203-1225

Brewer ID, Burbank DW, Hodges KV (2003) Modelling detrital cooling-age populations: insights from two Himalayan catchments. Basin Res 15:305-320

Catlos EJ, Harrison TM, Kohn MJ, Grove M, Ryerson FJ, Manning CE, Upreti BN (2001) Geochronologic and thermobarometric constraints on the evolution of the Main Central Thrust, central Nepal Himalaya. J Geophys Res 106:16777-16204

Cazenave A, Souriau A, Dominh K (1989) Global coupling of earth surface topography with hotspots, geoid, and mantle heterogeneities. Nature 340:54-57

Clark MK, House MA, Royden LH, Whipple KX, Burchfiel BC, Zhang X, Tang W (2005) Late Cenozoic uplift of southeastern Tibet. Geology 33:525-528

Clark MK, Maheo G, Saleeby J, Farley KA (2005) The non-equilibrium landscape of the southern Sierra Nevada, California. GSA Today 15:4-10

Dadson SJ, Hovius N, Chen HG, Dade WB, Hsieh ML, Willett SD, Hu JC, Horng MJ, Chen MC, Stark CP, Lague D, Lin JC (2003) Links between erosion, runoff variability and seismicity in the Taiwan orogen. Nature 426:648– 651

Dempster TJ, Persano C (2006) Low-temperature thermochronology: Resolving geotherm shapes or denudation histories. Geology 34:73-76

Densmore MS, Ehlers TA,Woodsworth GJ (2007) Effect of Alpine glaciation on thermochronometer age-elevation profiles. Geophys Res Lett 34:L02502. doi:10.1029/2006GL028371

Dietrich WE, Perron JT (2006) The search for a topographic signature of life. Nature 439:411-418

Dodson MH (1973) Closure temperature in cooling geochronological and petrological systems. Contrib Mineral Petrol 40:259-274

Ducea MN, Saleeby JB (1996) Buoyancy sources for a large unrooted mountain range, the Sierra Nevada, California: Evidence from xenolith thermobarometry. J Geophys Res 101:8229-8241

Ehlers TA (2005) Crustal thermal processes and the interpretation of thermochronometer data. Rev Mineral Geochem 58:315–350

Ehlers TA, Armstrong PA, Chapman DS (2001) Normal fault regimes and the interpretation of low-temperature thermochronometers. Phys Earth Planet Inter 126:179–194

Ehlers TA, Chapman D (1999) Normal fault thermal regimes; conductive and hydrothermal heat transfer surrounding the Wasatch Fault, Utah. Tectonophysics 312:217-234

Ehlers TA, Farley KA (2003) Apatite (U-Th)/He thermochronology: methods and applications to problems in tectonics and surface processes. Earth Planet Sci Lett 206:1–14

Ehlers TA, Farley KA, Rusmore ME, Woodsworth GJ (2006) Apatite (U-Th)/He signal of large-magnitude accelerated glacial erosion, southwest British Columbia. Geology 34:765-768

Flowers RM, Shuster DL, Wernicke BP, Farley KA (2007) Radiation damage control on apatite (U-Th)/He dates from the Grand Canyon region, Colorado Plateau. Geology 35:447-450

Foeken JPT, Persano C, Stuart FM, ter Voorde M (2007) Role of topography in isotherm perturbation: Apatite (U-Th)/He and fission track results from the Malta tunnel, Tauern Window, Austria. Tectonics 26(3): TC3006, doi:10.1029/2006TC002049

Forest CE, Molnar P, Emmanuel KA (1995) Paleoaltimetry from energy conservation principles. Nature 374:347-350

Forster C, Smith L (1988b) Groundwater flow systems in mountainous terrain. 2. Controlling factors. Water Resour Res 24(7):1011-1023

Forster C, Smith L (1989) The influence of groundwater flow on thermal regimes in mountainous terrain: a model study. J Geophys Res 94(B7):9439-9451

Forte AM, Peltier WR, Dziewonski AM, Woodward RL (1993) Dynamic surface topography - A new interpretation based upon mantle flow models derived from seismic tomography. Geophys Res Lett 20:225-228

Gregory-Wodzicki KM (1997) The late Eocene House Range flora, Sevier Desert, Utah: Paleoclimate and paleoelevation. Palaios 12:552–567

Grove M, Harrison TM (1999) Monazite Th/Pb age-depth profiling. Geology 27:487-490

Hansen K, Reiners PW (2006) Low temperature thermochronology of the southern East Greenland continental margin: Evidence from apatite (U-Th)/He and fission track analysis and implications for intermethod calibration. Lithos 92:117-136

Harrison TM, Grove M, Lovera OM, Zeitler PK (2005) Continuous thermal histories from inversion of thermal profiles. Rev Mineral Geochem 58:389-409

Hawkins DP, Bowring SA (1999) U–Pb monzaite, xenotime and titanite geochronological constraints on the prograde to post-peak metamorphic thermal history of Paleoproterozoic migmatites from the Grand Canyon, Arizona. Contrib Mineral Petrol 134:150-169

Heffern EL, Reiners PW, Naeser CA, Coates DA, (2007) Geochronology of clinker and implications for evolution of the Powder River Basin landscape, Wyoming and Montana, in Stracher, G., ed., Reviews in Engineering Geology: Geol Soc Am Spec Pap. (in press)

Hodges KV, Ruhl KW, Wobus CW, Pringle MS (2005) $^{40}Ar/^{39}Ar$ thermochronology of detrital minerals. Rev Mineral Geochem 58:239-257

Hodges KV, Wobus C, Ruhl K, Schildgen T, Whipple K (2004) Quaternary deformation, river steepening, and heavy precipitation at the front of the Higher Himalayan ranges. Earth Planet Sci Lett 220:379-89

House MA, Wernicke BP, Farley KA (1998) Dating topography of the Sierra Nevada, California, using apatite (U-Th)/He ages. Nature 396:66-69

House MA, Wernicke BP, Farley KA (2001) Paleogeomorphology of the Sierra Nevada, California, from (U-Th)/He ages in apatite. Am J Sci 301:77-102

House MA, Wernicke BP, Farley KA, Dumitru TA (1997) Cenozoic thermal evolution of the central Sierra Nevada, CA from (U-Th)/He thermochronometry. Earth Planet Sci Lett 151:167-179

Huber NK (1981) Amount and timing of late Cenozoic uplift and tilt of the central Sierra Nevada, California— Evidence from the upper San Joaquin River basin: U.S. Geological Survey Prof Pap 1197, 28 p.

Huey RB, Ward PD (2005) Hypoxia, Global warming, and terrestrial Late Permian extinctions. Science 308:398-401

Huntington KW, Hodges KV (2006) A comparative study of detrital mineral and bedrock age-elevation methods for estimating erosion rates. J Geophys Res 111:F03011, doi:10.1029/2005JF000454

Ketcham RA (2005) Forward and inverse modeling of low-temperature thermochronometry data. Rev Mineral Geochem 58:275-314

Kirby E, Reiners PW, Krol M, Hodges K, Whipple K, Farley K, Tang W, Chen Z (2002) Late Cenozoic uplift and landscape evolution along the eastern margin of the Tibetan Plateau: Inferences from $^{40}Ar/^{39}Ar$ and (U-Th)/He thermochronology. Tectonics doi:10.1029/2000TC001246

Kohn MJ, Wieland MS, Parkinson CD, Upreti BN (2004) Miocene faulting at plate tectonic velocity in the Himalaya of central Nepal. Earth Planet Sci Lett 228:299-310

Lavé J, Avouac JP (2001) Fluvial incision and tectonic uplift across the Himalayas of central Nepal. J Geophys Res 106:26561-26591

Lovera OM, Grove M, Harrison TM (2002) Systematic analysis of K-feldspar $^{40}Ar/^{39}Ar$ step heating results II: Relevance of laboratory argon diffusion properties to nature. Geochim Cosmochim Acta 66:1237-1255

Mancktelow NS, Grasemann B (1997) Time-dependent effects of heat advection and topography on cooling histories during erosion. Tectonophysics 270:167-95

Manga M (1998) Advective heat transport by low-temperature discharge in the Oregon Cascades. Geology 26:799-802

Manga M (2001) Using springs to study groundwater flow and active geologic processes. Ann Rev Earth Planet Sci 29:201-228

Manga M, Kirschner JW (2004) Interpreting the temperature of water at cold springs and the importance of gravitational potential energy. Water Resources Res 40:W05110, doi:10.1029/2003WR002905

Maréchal J-C, Perrochet P (2001) Theoretical relation between water flow rate in a vertical fracture and rock temperature in the surrounding massif. Earth Planet Sci Lett 194:213-219

McDougall I, Harrison TM (1999) Geochronology and Thermochronology by the $^{40}Ar/^{39}Ar$ Method, Second Edition. Oxford University Press, Oxford

Mezger K, Hanson GN, Bohlen SR (1989) High-precision U–Pb ages of metamorphic rutile: application to the cooling history of high-grade terranes. Earth Planet Sci Lett 96:106-118

Molnar P, England P (1990) Late Cenozoic uplift of mountain ranges and global climate change: chicken or egg? Nature 346:29-34

Montgomery DR, Balco G, Willett SD (2001) Climate, tectonics, and the morphology of the Andes. Geology 29:579-582

Montgomery DR, Brandon MT (2002) Topographic controls on erosion rates in tectonically active mountain ranges. Earth Planet Sci Lett 201:481-489

Mulch A, Graham SA, Chamberlain CP (2006) Hydrogen isotopes in Eocene river gravels and paleoelevation of the Sierra Nevada. Science 313:87-89

Poage MA, Chamberlain CP (2002) Stable isotopic evidence for a Pre-Middle Miocene rain shadow in the western Basin and Range: Implications for the paleotopography of the Sierra Nevada. Tectonics 21, doi:10.1029/2001TC001303

Quidelleur Z, Grove M, Lovera OM, Harrison TM, Yin A, Ryerson FJ (1997) The thermal evolution and slip history of the Renbu Zedong Thrust, southeastern Tibet. J Geophys Res 102:2659-2679.

Raymo ME, Ruddiman WF, Froelich PN (1988) Influence of late Cenozoic mountain building on ocean geochemical processes. Geology 16:649-653

Reiners PW, Ehlers TA (eds) (2005) Low-Temperature Thermochronology: Techniques, Interpretations, and Applications. Rev Mineral Geochem Volume 58. Mineralogical Society of America

Reiners PW, Ehlers TA, Garver JI, Mitchell SG, Montgomery DR, Vance JA, Nicolescu S (2002) Late Miocene exhumation and uplift of the Washington Cascades. Geology 30:767-770

Reiners PW, Ehlers TA, Mitchell SG, Montgomery DR (2003b) Coupled spatial variations in precipitation and long-term erosion rates across the Washington Cascades. Nature 426: 645-47,

Reiners PW, Farley KA (2001) Influence of crystal size on apatite (U-Th)/He thermochronology: An example from the Bighorn Mountains, Wyoming. Earth Planet Sci Lett 188:413-420

Reiners PW, Thomson SN, McPhillips D, Donelick, RA, Roering JJ (in press) Wildfire thermochronology and the fate and transport of apatite in hillslope and fluvial environments. J Geophys Res-Earth Surf (in press)

Reiners PW, Zhou Z, Ehlers TA, Xu C, Brandon MT, Donelick RA, Nicolescu S (2003a) Post-orogenic evolution of the Dabie Shan, eastern China, from (U-Th)/He and fission-track dating. Am J Sci 303:489-518

Riebe CS, Kirchner JW, Granger DE, Finkel RC (2000) Erosional equilibrium and disequilibrium in the Sierra Nevada, inferred from cosmogenic ^{26}Al and ^{10}Be in alluvial sediment. Geology 28:803-806

Roe GH, Stolar D, Willett SD (2006) The sensitivity of a critical wedge orogen to climatic and tectonic forcing. *In:* Tectonics, Climate, and Landscape Evolution. Geological Society of America Special Paper # 398. Willett SD, Hovius N, Brandon M, Fisher DM (eds), Geological Society of America, Boulder, p 227-239

Ruddiman WF, Kutzbach JE (1989) Forcing of late Cenozoic Northern Hemisphere climate by plateau uplift in southern Asia and the American west. J Geophys Res 94:18409-18427

Ruhl KW, Hodges KV (2005) The use of detrital mineral cooling ages to evaluate steady-state assumptions in active orogens: an example from the central Nepalese Himalaya. Tectonics 24:TC4015, doi:10.1029/2004TC001712

Schildgen TF, Hodges KV, Whipple KX, Reiners PW, Pringle MS (2007) Uplift of the Altiplano and Western Cordillera revealed through canyon incision history, Southern Peru. Geology 35: 523-537

Schmitz MD, Bowring SA (2003) Constraints on the thermal evolution of continental lithosphere from U–Pb accessory mineral thermochronometry of lower crustal xenoliths, southern Africa. Contrib Mineral Petrol 144:592-618

Schoene B, Bowring SA (2006) Determining accurate temperature–time paths from U–Pb thermochronology: An example from the Kaapvaal craton, southern Africa. Geochim Cosmochim Acta 71:165-185

Shuster DL, Farley KA (2003) ^{4}He/^{3}He thermochronometry. Earth Planet Sci Lett 217:1-17

Shuster DL, Farley KA (2005) ^{4}He/^{3}He thermochronometry. Rev Mineral Geochem 58:181-203

Shuster DL, Farley KA, Sisterson JM, Burnett DS (2003) Quantifying the diffusion kinetics and spatial distributions of radiogenic ^{4}He in minerals containing proton-induced ^{3}He. Earth Planet Sci Lett 217:19-32

Shuster DL, Flowers RM, Farley KA (2006) The influence of natural radiation damage on helium diffusion kinetics in apatite. Earth Planet Sci Lett 249:148-161

Smith L, Chapman D (1983) On the thermal effects of groundwater flow; 1. regional scale systems. J Geophys Res 88:593-608

Smith RB (2006) Progress on the theory of orographic precipitation. *In:* Tectonics, Climate, and Landscape Evolution. Geological Society of America Special Paper #398. Willett SD, Hovius N, Brandon M, Fisher DM (eds), Geological Society of America, Boulder, p 1-16

Stock GM, Anderson RS, Finkel RC (2004) Pace of landscape evolution in the Sierra Nevada, California, revealed by cosmogenic dating of cave sediments. Geology 32:193-196

Stock JD, Montgomery DR (1996) Estimating paleorelief from detrital mineral age ranges. Basin Res 8:317-327

Stüwe K, Hintermüller M (2000) Topography and isotherms revisited: The influence of laterally migrating drainage divides. Earth Planet Sci Lett 184:287-303

Stüwe K, White L, Brown R (1994) The influence of eroding topography on steady-state isotherms. Applications to fission track analysis. Earth Planet Sci Lett 124:63-74

Thiede R, Bookhagen B, Arrowsmith J., Sobel ER, Strecker MR (2004) Climatic control on rapid exhumation along the Southern Himalayan Front. Earth Planet Sci Lett 222:791-806

Thomson SN, Brandon MT, Zattin M, Reiners PW, Isaacson, P (in prep) Thermochronologic evidence of orogen-parallel differences in wedge kinematics during extending convergent orogenesis in the northern Apennines, Italy

Turcotte DL, Schubert G (1982) Geodynamics. 1st Edition. Cambridge Univ. Press

Turcotte DL, Schubert G (2002) Geodynamics. 2nd Edition. Cambridge Univ. Press

Unruh JR (1991) The uplift of the Sierra Nevada and implications for late Cenozoic epeirogeny in the western Cordillera. Geol Soc Am Bull 103:1395-1404

Vermeesch P (2007) Quantitative geomorphology of the White Mountains (California), using detrital apatite fission track thermochronology. J Geophys Res doi:10.1029/2006JF000671 (in press)

von Blanckenberg F (2006) The control mechanisms of erosion and weathering at basin scale from cosmogenic nuclides in river sediment. Earth Planet Sci Lett 242:224-239

Wakabayashi J, Sawyer TL (2001) Stream incision, tectonics, uplift, and evolution of topography of the Sierra Nevada, California. J Geol 109:539–562

Whipp DM Jr, Ehlers TA (2007) Influence of groundwater flow on thermochronometer-derived exhumation rates in the central Nepalese Himalaya. Geology (in press)

Wobus C, Heimsath A, Whipple K, Hodges K (2005) Out-of-sequence thrust faulting in the central Nepalese Himalaya. Nature 434:1008-11

Wobus CW, Hodges KV, Whipple KX (2003) Has focused denudation sustained active thrusting at the Himalayan topographic front?. Geology 31:861-64

Wolfe JA, Forest CE, Molnar P (1998) Paleobotanical evidence of Eocene and Oligocene paleoaltitudes in midlatitude western North America. Geol Soc Am Bull 110:64–678

Wolfe JA, Schorn HE, Forest CE, Molnar P (1997) Paleobotanical evidence for high altitudes in Nevada during the Miocene. Science 276:1672–1675

Wright N, Layer PW, York D (1991) New insights into thermal history from single grain ^{40}Ar/^{39}Ar analysis of biotite. Earth Planet Sci Lett 104:70–79

Reviews in Mineralogy & Geochemistry
Vol. 66, pp. 269-278, 2007
Copyright © Mineralogical Society of America

11

Terrestrial Cosmogenic Nuclides as Paleoaltimetric Proxies

Catherine A. Riihimaki

Department of Geology
Bryn Mawr College
101 N. Merion Ave.
Bryn Mawr, Pennsylvania, 19010, U.S.A.
criihima@brynmawr.edu

Julie C. Libarkin

Department of Geological Sciences &
Division of Science and Mathematics Education (DSME)
Michigan State University
206 Natural Science Building
East Lansing, Michigan, 48824-1115, U.S.A.
libarkin@msu.edu

ABSTRACT

The production rates of terrestrial *in situ* cosmogenic nuclides depend on the altitude and can therefore be used to constrain paleoaltitudes if the history of cosmogenic-nuclide production in a region can be constrained. This chapter discusses the theory behind cosmogenic-nuclide paleoaltimetry, sampling strategies, and practical limitations of the technique. Three exposure scenarios may allow for the calculation of past altitudes: 1) exposure for a finite time period at a single elevation and without erosion or burial, followed by immediate shielding from further production of cosmogenic nuclides; 2) steady uplift of a surface throughout nuclide production, without erosion or burial; and 3) exposure of a sample without erosion or burial for a sufficient duration that the concentration of a cosmogenic radionuclide has reached equilibrium. To constrain paleoelevation, all exposure scenarios require independent evidence of the depth-history of a sample during exposure to cosmic rays because production rates attenuate rapidly in rock. Depth profiles and measuring multiple nuclides allow for better constraints of parameters in paleoaltitude calculations.

INTRODUCTION

The production rate of cosmogenic nuclides in Earth materials is sensitive to elevation. If the production-rate history of a region can be constrained by the concentration of cosmogenic nuclides in bedrock or sediment deposits, the production rate can potentially be used to deduce paleoelevations. However, because production rates also scale with erosion, burial, exposure time, and cosmic ray flux, the accuracy and precision of this paleoaltimeter strongly depends on the ability to constrain other variables that affect production of cosmogenic nuclides over timescales relevant to paleoaltimetry research. This chapter summarizes the background theory of the cosmogenic nuclide paleoaltimeter, its major sources of uncertainty, and potential approaches to limit these uncertainties. Summaries of other geological applications of cosmogenic isotopes include Gosse and Phillips (2001) and Bierman (1994).

1529-6466/07/0066-0011$05.00

DOI: 10.2138/rmg.2007.66.11

Rock or mineral composition will directly influence which nuclides are produced and in what quantities (Table 1). Most nuclides are produced from common rock-forming elements like O, Si, or Ca, although some nuclides, like ^{129}I, result from neutron spallation of rare heavy elements. We focus on the small number of nuclides that show promise for paleoaltimetry research (italicized text in Table 1). Other nuclides are impractical for paleoaltimetry for a variety of reasons: 1) the isotope is too short-lived to be of use in paleoaltimetry (e.g., ^{14}C, ^{24}Na) because its half-life is much shorter than the million-year timescale over which significant changes in elevation can occur, 2) the low production rate produces nuclide concentrations that are just at or below analytical detection limits (e.g., ^{81}Kr), 3) the nuclide is difficult to detect above natural background levels because of significant natural occurrence (e.g., ^{7}Li), and 4) other isotopes of the same element interfere with analytical precision. For example, the latter two conditions explain the limited measurement of terrestrial cosmogenic ^{36}Ar and ^{38}Ar that has occurred to date (Renne et al. 2001).

Table 1. Table of terrestrial cosmogenic nuclides of interest to paleoaltimetry.

Nuclide	Targets in Rock	Production rate (atoms g^{-1}target a^{-1})*	Half-life (yr)
^{3}He	*O, Si, Ca, Mg, Fe*	*(olivine) = 115;*	*Stable*
^{21}Ne	*Mg, Al, Si*	*(quartz) = 20* *(olivine) = 169*	*Stable*
^{36}Ar	Cl, K, Ca, Fe	Not known Decay product from ^{36}Cl	Stable
^{38}Ar	Cl, K, Ca, Fe	Not known Decay product from ^{38}K	Stable
^{126}Xe	Te, Ba		Stable
^{10}Be	*O, Mg, Si, Fe*	*(quartz) = 5.6*	*1.52×10^6***
^{14}C	O, Mg, Si, Fe	(quartz) = 20	5730
^{26}Al	*Si, Al, Fe*	*(quartz) = 35*	*7.1×10^5*
^{36}Cl	*Cl, K, Ca, Fe*	*(Ca) = 67* *(K) = 74 to 235*	*3.01×10^5*
^{41}Ca	Ca, Ti, Fe		1.04×10^5
^{53}Mn	Fe		3.7×10^6
^{81}Kr	Rb, Sr, Zr		2.3×10^5
^{129}I	Te, Be, Rare Earths		1.56×10^7

Note: After Bierman (1994), Gosse and Phillips (2001), and Lal (1988). See Gosse and Phillips (2001) for details about radionuclide decay constants and production rate uncertainties. Isotopes in italics are those that have been used routinely in cosmogenic nuclide studies. Stable isotopes will resolve paleoaltimetry for Cenozoic and older rocks. *High-latitude, sea-level rates from empirical studies only. Production mostly via neutron spallation unless noted. **Recent debate on the half-life of ^{10}Be has focused on whether this number is off by >10%.

PRODUCTION OF COSMOGENIC NUCLIDES IN ROCK

Cosmogenic nuclides are produced in rock when secondary cosmic rays, the result of interactions between primary cosmic rays and atomic nuclei in the Earth's atmosphere, interact with rock at the Earth's surface. The primary mechanism for nuclide production at Earth's surface is neutron spallation, where a high-energy cosmic ray neutron bombards an atomic nucleus, causing the nucleus to disintegrate into smaller nuclei, creating isotopes of lower atomic number. The change in concentration N (atoms g^{-1}) of a particular nuclide through time t can be expressed as:

$$\frac{dN}{dt} = P - \lambda N \tag{1}$$

where P is the production rate (atoms g^{-1} yr^{-1}) at a particular time and λ is the decay constant (yr^{-1}) for the nuclide ($\lambda = 0$ for stable nuclides). The measured concentration is the integral of production and decay through time, which can be determined analytically for certain scenarios. For example, for a bedrock surface with steady erosion rate E (cm yr^{-1}) and no change in elevation,

$$N = \frac{P}{\lambda + \frac{E\rho_{rock}}{\Lambda}}\left(1 - \exp\left[-t\left(\lambda + \frac{E\rho_{rock}}{\Lambda}\right)\right]\right) \tag{2}$$

where ρ_{rock} is rock density (g cm^{-3}), and Λ is the attenuation pathlength (g cm^{-2}) for production in rock (Lal 1991; Bierman 1994). If the exposure time of sample can be constrained, the long-term production rate can be calculated and long-term changes in elevation can be deduced.

Production rates increase with elevation of a target because cosmic rays at higher elevations have traveled through less mass in the atmosphere. Several methods for scaling production with elevation have been developed (e.g., Lal and Peters 1967; Lal 1991; Dunai 2000; Stone 2000). In the simplest case, changes in production of a nuclide with elevation can be approximated by an exponential function that mimics the exponential pattern of secondary cosmic ray flux for both neutrons and muons (Gosse and Phillips 2001):

$$P(y) = P_{n_0}(0)\exp\left[\frac{(z(0) - z(y))}{L_n}\right] + P_{\mu_0}(0)\exp\left[\frac{(z(0) - z(y))}{L_\mu}\right] \tag{3}$$

where $P(y)$ is production rate at the surface at elevation y; $P_{n_0}(0)$ and $P_{\mu_0}(0)$ are the production rates (atoms g^{-1} yr^{-1}) for nucleons and muons at the Earth's surface at sea level; $z(y)$ and $z(0)$ are atmospheric depth (g cm^{-2}) at elevation y and at sea level; and L_n and L_μ are the cosmic ray attenuation pathlengths for nucleons and muons. Most researchers choose an intermediate value for L_n of ~165 g cm^{-2} (Reedy et al. 1994). L_μ can only be approximated as a single number because the attenuation pathlengths for different mechanisms of muogenic production differ (Heisinger et al. 2002a). L_μ for negative muon capture in the atmosphere has been estimated at ~1350 g cm^{-2} (Heisinger et al. 2002b) while L_μ for nuclide production by fast muons in the atmosphere has been estimated at ~4320 g cm^{-2} (Heisinger et al. 2002a).

Atmospheric density must be known to convert atmospheric depth to altitude. For mid-latitudes, elevation can be converted to atmospheric depth via the standard atmosphere (Lide 1999):

$$z(y) = z(0) \cdot \exp\left\{\left[\frac{-gM}{R\xi}\right]\left[\ln T_s - \ln(T_s - \xi y)\right]\right\} \tag{4}$$

where $z(y)$ is atmospheric depth at elevation y, $z(0)$ is the atmospheric depth at sea level (1033.2 g cm^{-2}), T_s is sea level temperature (288.15 K), ξ is the adiabatic lapse rate (0.0065 K m^{-1}), g

is acceleration due to gravity, M is molar weight of water and R is the gas constant ($gM/R =$ 0.03417 K m^{-1}). Based on 50-year historical barometric records, Stone (2000) has shown that differences in modern atmospheric density between the polar and equatorial regions influences production of cosmogenic nuclides. However, long-term changes in regional atmospheric pressure resulting from changes in climatic conditions, tectonic activity, or uplift of mountain belts disturbing global circulation patterns are difficult to assess.

Variations in production rates caused by factors other than elevation must be constrained for successful application of the cosmogenic-nuclide paleoaltimeter. Production rates are particularly sensitive to the depth of a sample below Earth's surface, because production rates decrease with increasing mass through which cosmic rays have traveled and the density of rock is orders of magnitude greater than the density of the atmosphere. Therefore, any local erosion and/or deposition at a sample site during the duration of exposure must be constrained. Production of cosmogenic nuclide with depth in rock can be approximated as an exponential, with different scaling factors for production by nucleons and muons:

$$P(x) = P_{n_0}(0)\exp\left[\frac{-\rho_{rock}x}{\Lambda_n}\right] + P_{\mu_0}(0)\exp\left[\frac{-\rho_{rock}x}{\Lambda_\mu}\right] \tag{5}$$

where $P(x)$ is the nuclide production rate at depth x below the surface, $P_{n_0}(0)$ and $P_{\mu_0}(0)$ are the spallogenic and muogenic production rates at the surface, ρ_{rock} is rock density, and Λ_n and Λ_μ are the attenuation pathlengths for nucleons and muons. As with attenuation in the atmosphere, the nucleon attenuation pathlength is ~165 g cm^{-2} (Reedy et al. 1994). Λ_μ can only be approximated by a single number because different processes of muogenic production have different attenuation pathlengths, with Λ_μ for negative muon capture of ~1500 g cm^{-2} (Bilokon et al. 1989; Gosse and Phillips 2001) and Λ_μ for production by fast muons of ~4320 g cm^{-2} (Heisinger et al. 2002a). Equation (5) may be refined by incorporating separate terms for each muogenic production mechanism. For example, for heavily eroded rock where material originating below 10 meters depth has been brought to the surface, the polynomial approximations of Stone et al. (1998) should be used for muon attenuation in place of the second half of (5).

COSMOGENIC-NUCLIDE PALEOALTIMETRY

The sensitivity of nuclide production rates to changes in elevation (Fig. 1) was recognized very early in the history of cosmognic-nuclide research (e.g., Lal and Peters 1967), and several researchers have suggested that cosmogenic nuclides would make useful paleoaltimeters (e.g., Brook et al. 1995; Gosse and Phillips 2001; Gosse and Stone 2001; Libarkin et al. 2002; Blard et al. 2005). For successful application of the cosmogenic-nuclide paleoaltimeter, the production rate history of a sample must be constrained, and thus the duration of exposure and the depth history of a sample must be known. Stable nuclides likely hold the most promise for paleoaltimetry, at least in the short term, because they are preserved in buried samples over millions of years, sufficiently long for significant changes in elevation for a region. We will limit our discussion to stable nuclides except for the special case in which the concentration of unstable nuclides on a surface with no erosion or deposition has reached saturation, in which additional production is balanced by radioactive decay. We recommend that interested readers access the reviews by Lal (1988, 1991), Bierman (1994), and Gosse and Phillips (2001) for more comprehensive discussions of cosmogenic-nuclide applications to geologic problems.

The concentration of a stable cosmogenic nuclide for a surface sample that experienced a period of continuous exposure over a known duration t, with no erosion or deposition or change in altitude during exposure, followed by subsequent burial and immediate shielding from further production, can be approximated as

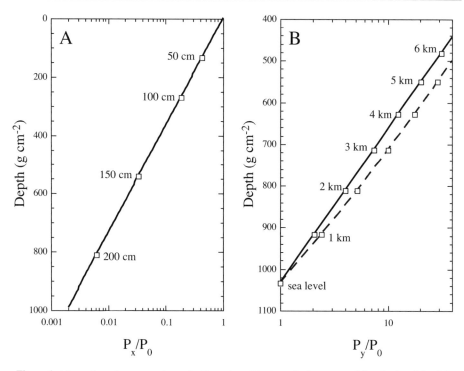

Figure 1. Attenuation of cosmogenic production rates with atmospheric pressure (elevation) and depth in rock. A) Log-linear plot of depth as a function of normalized production rate as per equation (5), assuming a rock density of 2.7 g cm^{-3}. Slope is rock attenuation coefficient, Λ. B) Log-linear plot of atmospheric depth as a function of normalized production rate as per equation (3) (solid line) and as per the scaling function presented in Stone (2000) for 40° latitude (dashed line). The offset between the two lines indicates the importance of the elevation scaling relationships when reconstructing high paleoaltitudes.

$$N = P_{n_0}(0) \cdot t \cdot \exp\left[\frac{\left(z(0) - z\left(y_{paleo}\right)\right)}{L_n}\right] + P_{\mu_0}(0) \cdot t \cdot \exp\left[\frac{\left(z(0) - z\left(y_{paleo}\right)\right)}{L_\mu}\right] \quad (6)$$

where $z(y_{paleo})$ is the paleo-atmospheric depth at the paleoaltitude y_{paleo}. If muogenic production is negligible,

$$z\left(y_{paleo}\right) = z(0) - L_n \ln\left[\frac{N}{P_{n_0}(0) \cdot t}\right] \quad (7)$$

Equation (7) can then be used to determine y_{paleo}. Otherwise, Equation (6) can be solved numerically for $z(y_{paleo})$ and y_{paleo}.

For production of radionuclides during the case of single-stage, continuous exposure without erosion, burial or uplift during nuclide production, Lal (1991) has shown that concentration of the radionuclide scales as:

$$N = \frac{P(0)}{\lambda} \exp\left[\frac{\left(z(0) - z\left(y_{paleo}\right)\right)}{L}\right] \left(1 - \exp[-\lambda t]\right) \quad (8)$$

For a sample that has been continuously exposed for sufficiently long that production is balanced by decay, N becomes insensitive to the magnitude of t, and therefore $z(y_{paleo})$ can be determined. Brook et al. (1995) used this methodology to show that high concentrations of [10]Be and [26]Al in Neogene glacial deposits of the Dry Valleys, Antarctica, are inconsistent with recent uplift of the region. However, they caution that the methodology can only be used to show the absence of uplift, particularly if erosion is poorly constrained.

The concentration of a stable cosmogenic nuclide in a sample that has experienced steady uplift during continuous exposure without erosion or burial requires integration over the range of elevations the sample experienced during exposure. For uplift rate $\dot{u}$, modern elevation y_{modern}, and total exposure time T,

$$N = P_{n_0}(0) \cdot \int_0^T \exp\left[\frac{\left(z(0) - z\left(y_{modern} - \dot{u}t\right)\right)}{L_n}\right] dt$$

$$+ P_{\mu_0}(0) \cdot \int_0^T \exp\left[\frac{\left(z(0) - z\left(y_{modern} - \dot{u}t\right)\right)}{L_\mu}\right] dt \qquad (9)$$

where $z(y_{modern} - \dot{u}t)$ is the elevation at time t. This equation can be evaluated numerically to determine $\dot{u}$.

In general, the initial elevation of the sample can only be determined if the production-rate history, a function of rock and surface uplift as well as erosion and burial history, can be deduced. Theoretically, Equations (1), (3) and (5) can be used to relate measured nuclide concentrations of a paleosurface to paleoelevation. The primary limitation of the cosmogenic-nuclide paleoaltimeter is the requirement that several parameters be constrained with great precision. Therefore, estimating elevation of paleosurfaces (e.g., Libarkin et al. 2002) requires special consideration when choosing sampling localities and collecting samples (Fig. 2). Suites of samples associated with a single paleosurface can provide constraints on erosion and burial histories, noncosmogenic production of nuclides, and inherited concentrations prior to *in situ* exposure. Measuring concentrations of multiple cosmogenic nuclides in each sample can also improve the interpretation because the concentration of each nuclide is the product of the same *in situ* exposure history. Libarkin et al. (2002) and Blard et al. (2005) describe specific sampling methods.

Sites that are most likely to provide ideal cases for cosmogenic-nuclide paleoaltimetry are in igneous provinces where stacked lava or pyroclastic flows resulted in instantaneously buried paleosurfaces and also provide for dating of exposure intervals. The duration of exposure of a given flow is the difference in ages between the studied layer and the overriding flow. Post-burial nuclide accumulation for cases where overlying flows are thin can be constrained by sampling overlying flows (Fig. 2), and the extent of erosion can be determined from flow features at the top of the deposit.

POTENTIAL SOURCES OF ERROR IN
COSMOGENIC-NUCLIDE PALEOALTIMETRY

Error in the cosmogenic-nuclide paleoaltimeter comes from uncertainty in atmospheric parameters that cause temporal variations in P and z, geologic parameters that cause changes in the depth of a sample, and laboratory parameters that cause measurement uncertainty. Section 6 of Gosse and Phillips (2001) provides a detailed discussion of sources of error for cosmogenic-radionuclide applications, and any attempt to use the cosmogenic-nuclide paleoaltimeter should address these as well as the factors influencing nuclide production

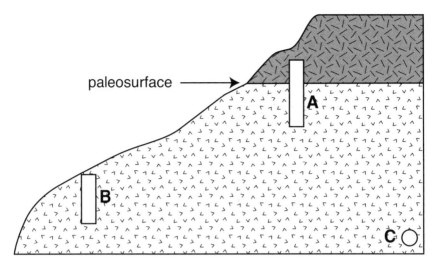

Figure 2. Schematic of sampling strategy for a buried paleosurface. A) Core into and below paleosurface to constrain paleo-production of cosmogenic nuclide. B) Core into and below modern surface to constrain production during recent exposure. C) Sample(s) shielded from both modern and ancient exposure to cosmic rays to constrain non-cosmogenic production of nuclides. Not to scale.

discussed here. We recommend a suite of Monte Carlo simulations to assess the sensitivity of any paleoaltimetry calculation to uncertainty in parameters of Equations (1), (3), (4) and (5). Analytical error propagation is not appropriate except in certain special cases in which paleoelevation can be calculated analytically. Each geologic setting will have unique sources of uncertainty depending upon erosive conditions, geomagnetic field records for the exposure interval, and noncosmogenic sources of the nuclide.

Secondary cosmic ray flux and cosmic ray composition at the Earth's surface are complex quantities to evaluate, and in practice assumptions about the constancy of cosmic rays over timescales relevant to paleoaltimetry research. Short time scale variations in production rates, such as might result from the 11-year cyclicity in the cosmic ray flux due to solar flares (Raisbeck et al. 1990), will average out of the data over million-year timescales. Likewise, assumptions about the constancy of atmospheric density must be made so that atmospheric depth can be converted to elevation.

Uncertainty in the depth history of a sample is a primary source of uncertainty for the cosmogenic-nuclide paleoaltimeter. Because of the >2000-fold difference in rock density versus atmospheric density, a 0.5-m uncertainty in depth is equivalent to >1-km uncertainty in altitude. Uncertainty in the depth of a sample during exposure is particularly problematic in regions where loess deposits may episodically bury a surface. For example, Hancock et al. (1999) find cosmogenic evidence of an ephemeral 0.5-1.5 m silt cap on currently uncapped, 600-ka terraces in the Wind River basin, Wyoming. The duration of time required to deposit a sedimentary layer may also result in a complex exposure history that can only be deduced with depth profiles and multiple nuclides (Riihimaki et al. 2006). However, Dunai et al. (2005) suggest that some deposits in the hyperarid Atacama Desert, Chile, have remained at the same depth without erosion or deposition for >20 Ma.

In ideal situations in which erosion rate is well constrained, cosmogenic-nuclide paleoaltimetry currently will be exact to no more than ~200 m at best, given current uncertainty in scaling functions and long-term atmospheric conditions, analytical limitations, as well as uncertainties in the depth-history of a sample and duration of exposure. For example, using Equation (7), paleoelevation is calculated by first converting a measured nuclide concentration to atmospheric depth, and then converting atmospheric depth to elevation. Uncertainty on atmospheric depth (from Taylor series expansion) will be (similar to Blard et al. 2005)

$$\sigma z(y) = \sqrt{\begin{array}{l} \left(\dfrac{\partial z(y)}{\partial z(0)}\sigma z(0)\right)^2 + \left(\dfrac{\partial z(y)}{\partial N}\sigma N\right)^2 + \left(\dfrac{\partial z(y)}{\partial P_{n_0}(0)}\sigma P_{n_0}(0)\right)^2 \\[4mm] + \left(\dfrac{\partial z(y)}{\partial L_n}\sigma L_n\right)^2 + \left(\dfrac{\partial z(y)}{\partial t}\sigma t\right)^2 \end{array}} \tag{10}$$

and on elevation will be:

$$\sigma y = \sigma z(y) \cdot \frac{RT_s}{gMz(y)}\left(\frac{z(y)}{z(0)}\right)^{\frac{R\xi}{gM}} \tag{11}$$

For a sample measured for ^{3}He with an exposure time t of 100 ± 1 kyr, L_n of 160 ± 10 g cm^{-2}, P_{n_0} of 103 ± 4 atoms g^{-1} yr^{-1}, $z(0)$ of 1033.2 g cm^{-2}, and expected ^{3}He concentrations with 10% uncertainty, the 1σ uncertainties of y range from 200 m at low paleoelevations to 500 m at high paleoelevations (Fig. 3). The elevation dependence of the error bars stems from the uncertainty in L_n. The 10% uncertainty in N accounts for 1000 m of uncertainty in y. We caution that this simple error analysis represents only a minimum assessment of uncertainty since it does not include any uncertainty in the sample's depth history.

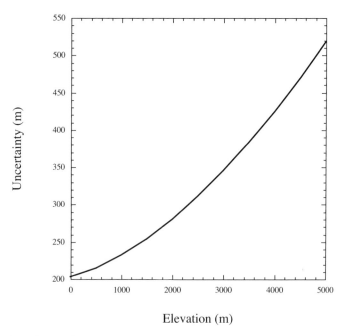

Figure 3. Estimated uncertainty on paleoaltimetry evaluated via Equations (10) and (11) for a particular set of parameters. See text for details. The plot demonstrates 1σ error bars >200 m and dependency of uncertainty on elevation.

CONCLUSIONS

The dependence of cosmogenic-nuclide production rates on elevation makes the cosmogenic-nuclide paleoaltimeter an intriguing tool for elevation studies, particularly for investigations into the development of high elevation plateaus. The relationship between elevation and climate change (Molnar and England 1990) presents concerns for paleoaltimeters that depend on temperature-elevation relationships such as botanical (e.g., Gregory and Chase 1992; Forest et al. 1999; Kouwenberg et al. 2007; Meyer 2007) and oxygen isotope (Garzione et al. 2000; Rowley et al. 2001; Rowley and Garzione 2007; Rowley 2007; Quade et al. 2007; Kohn and Dettman 2007; and Mulch and Chamberlain 2007) approaches. Cosmogenic-nuclide paleoaltimetry coupled with another pressure-related paleoaltimeter such as basalt vesicularity (e.g., Sahagian and Maus 1994; Sahagian and Proussevitch 2007) would be a powerful tool for comparison with more common proxies. The challenges for cosmogenic-nuclide paleoaltimetry cannot be understated, however. The application of this tool to pre-Holocene rocks will require careful sampling, analysis of multiple samples per altitude estimate, and acknowledgement of uncertainty in scaling functions.

Future studies should collect and process samples in a careful order to ensure project success. Initial determination of exposure time followed by mass spectrometric analysis of shielded and modern surface samples will ensure early on that conditions for a successful paleoaltimetry have been met. An ideal study using cosmogenic-nuclide paleoaltimetry should ensure that: 1) nuclide inventories are abundant enough for mass spectrometric analyses; 2) background levels of nucleogenic or alphagenic nuclides are low relative to cosmogenic; 3) modern nuclide build-up is low relative to ancient; 4) the exposure age of the target surface is measurable using ^{40}Ar/^{39}Ar or a similar technique; and 5) the erosion and burial history of the study locality is well understood or measurable.

ACKNOWLEDGMENTS

CAR was supported by NSF Grant EAR-0518754. This chapter would not have been possible without the seminal works on terrestrial cosmogenic nuclides of Lal, Peters, Gosse, Phillips, Stone, Granger, and many others. We are indebted to these researchers for their clarity of vision and efforts to synthesize a complex and diverse science. A brief conversation with Peter Molnar and Greg Houseman suggested the importance of log-linear plots in presenting cosmogenic nuclide ideas to the general geologic community. The manuscript was significantly improved by comments from Greg Balco and reviews by John Gosse, Pierre-Henri Blard, and Marc Caffee.

REFERENCES

Bierman PR (1994) Using *in situ* produced cosmogenic isotopes to estimate rates of landscape evolution: A review from the geomorphic perspective. J Geophys Res 99:13885-13896
Bilokon H, Castagnoli GC, Castellina S, Piazzoli BD, Mannocchi G, Meroni E, Picchi P, Vernetto S (1989) Flux of vertical negative muons stopping at depths 0.35-1000 hg/cm². J Geophys Res 94:12,145-12,152
Blard P, Lave J, Pik R, Quidelleur X, Bourles D, Kieffer G (2005) Fossil cosmogenic ³He record from K-Ar dated basaltic flows of Mount Etna volcano (Sicily, 38N): Evaluation of a new paleoaltimeter. Earth Planet Sci Lett 236:613-631
Brook EJ, Brown ET, Kurz MD, Ackert RP Jr, Raisbeck GM, Yiou F (1995) Constraints on age, erosion, and uplift of Neogene glacial deposits in the Transantarctic Mountains determined from *in situ* cosmogenic ¹⁰Be and 26Al. Geology 23:1063-1066
Dunai TJ (2000) Scaling factors for production rates of *in situ* produced cosmogenic nuclides: A critical reevaluation. Earth Planet Sci Lett 176:157-169
Dunai TJ, González López GA, Juez-Larré J (2005) Oligocene–Miocene age of aridity in the Atacama Desert revealed by exposure dating of erosion-sensitive landforms. Geology 33:321-324

Forest CE, Wolfe JA, Molnar P, Emanuel KA (1999) Paleoaltimetry incorporating atmospheric physics and botanical estimates of paleoclimate. Bull Geol Soc Am 111:497-511

Garzione CN, Quade J, DeCelles PG, English NB (2000) Predicting paleoelevation of Tibet and the Himalaya from $\delta^{18}O$ vs. altitude gradients in meteoric water across the Nepal Himalaya. Earth Planet Sci Lett 183:215-229

Gosse JC, Phillips FM (2001) Terrestrial *in situ* cosmogenic nuclides: theory and applications. Quat Sci Rev 20:1475-1560

Gosse JC, Stone JO (2001) Terrestrial cosmogenic nuclide methods passing milestones toward paleo-altimetry. EOS Trans Am Geophys Union 82:82, 86-89

Gregory KM, Chase CG (1992) Tectonic significance of paleobotanically estimated climate and altitude of the late Eocene erosion surface, Colorado. Geology 20:581-585

Hancock GS, Anderson RS, Chadwick OA, Finkel RC (1999) Dating fluvial terraces with ^{10}Be and ^{26}Al profiles: application to the Wind River, Wyoming. Geomorphology 27:41-60

Heisinger B, Lal D, Jull AJT, Kubik P, Ivy-Ochs S, Neumaier S, Knie K, Lazarev V, Nolte E (2002a) Production of selected cosmogenic radionuclides by muons: 1. Fast muons. Earth Planet Sci Lett 200:345-355

Heisinger B, Lal D, Jull AJT, Kubik P, Ivy-Ochs S, Neumaier S, Knie K, Lazarev V, Nolte E (2002b) Production of selected cosmogenic radionuclides by muons: 2. Capture of negative muons. Earth Planet Sci Lett 200:357-369

Kohn MJ, Dettman DL (2007) Paleoaltimetry from stable isotope compositions of fossils. Rev Mineral Geochem 66:119-154

Kouwenberg LLR, Kürschner WM, McElwain JC (2007) Stomatal frequency change over altitudinal gradients: prospects for paleoaltimetry. Rev Mineral Geochem 66:215-242

Lal D (1988) In situ-produced cosmogenic isotopes in terrestrial rocks. Ann Rev Earth Planet Sci 16:355-388

Lal D (1991) Cosmic ray labeling of erosion surfaces: *in situ* nuclide production rates and erosion models. Earth Planet Sci Lett 104:424-439

Lal D, Peters B (1967) Cosmic-ray produced radioactivity on the earth. *In:* Handbook of Physics 46. Springer-Verlag, Berlin, p 551-662

Libarkin JC, Quade J, Chase CG, Poths J, McIntosh W (2002) Measurement of ancient cosmogenic ^{21}Ne in quartz from the 28 Ma Fish Canyon Tuff, CO. Chem Geol 186:199-213

Lide DR (ed) (1999) CRC Handbook of Chemistry and Physics (79th edition). CRC Press, Boca Raton. p 14-16-14-22

Meyer HW (2007) A review of paleotemperature–lapse rate methods for estimating paleoelevation from fossil floras. Rev Mineral Geochem 66:155-171

Molnar P, England P (1990), Late Cenozoic uplift of mountain ranges and global climate change: chicken or egg? Nature 346:29-34

Mulch A, Chamberlain CP (2007) Stable isotope paleoaltimetry in orogenic belts – the silicate record in surface and crustal geological archives. Rev Mineral Geochem 66:89-118

Quade J, Garzione C, Eiler J (2007) Paleoelevation reconstruction using pedogenic carbonates. Rev Mineral Geochem 66:53-87

Raisbeck GM, Yiou F, Jouzel J, Petit JR (1990) ^{10}Be and d^2H in polar ice cores as a probe of the solar variability's influence on climate. Phil Trans Royal Soc London A 330:471-480

Reedy RC, Nishiizumi K, Lal D, Arnold JR, Englert PAJ, Klein J, Middleton R, Jull AJT, Donahue DJ (1994) Simulations of terrestrial in-situ cosmogenic nuclide production. Nucl Instrum Methods Phys Res B 92:297-300

Renne PR, Farley KA, Becker TA, Sharp WD (2001) Terrestrial cosmogenic argon. Earth Planet Sci Lett 188:435-440

Riihimaki CA, Anderson RS, Safran EB, Dethier DP, Finkel RC, Bierman PR (2006) Longevity and progressive abandonment of the Rocky Flats surface, Front Range, Colorado. Geomorphology 78:265-278

Rowley DB (2007) Stable isotope-based paleoaltimetry: theory and validation. Rev Mineral Geochem 66:23-52

Rowley DB, Garzione CN (2007) Stable isotope-based paleoaltimetry. Ann Rev Earth Planet Sci 35:463-508

Rowley DB, Pierrehumbert RT, Currie BS (2001) A new approach to stable isotope-based paleoaltimetry: Implications for paleoaltimetry and paleohypsometry of the high Himalaya since the Late Miocene. Earth Planet Sci Lett 18:253-268

Sahagian DL, Maus JE (1994) Basalt vesicularity as a measure of atmospheric pressure and paleoelevation. Nature 372:449-451

Sahagian D, Proussevitch A (2007) Paleoelevation measurement on the basis of vesicular basalts. Rev Mineral Geochem 66:195-213

Stone JO (2000) Air pressure and cosmogenic isotope production. J Geophys Res 105:23753-23759

Stone JOH, Evans JM, Fifield LK, Allan GL, Cresswell RG (1998) Cosmogenic chlorine-36 production in calcite by muons. Geochim Cosmochim Acta 63:433-454